UNIVE 10 0748372 3 NOTTINGHAM

WITHDRAWN

KU-080-489

FROM THE LIBRARY

The Neuron

THE NEURON

Cell and Molecular Biology

FOURTH EDITION

IRWIN B. LEVITAN

Thomas Jefferson University

LEONARD K. KACZMAREK

Yale University

MEDICAL LIBRARY
QUEENS MEDICAL CENTRE

OXFORD
UNIVERSITY PRESS

Oxford University Press is a department of the University of
Oxford. It furthers the University's objective of excellence in research,
scholarship, and education by publishing worldwide.

Oxford   New York
Auckland   Cape Town   Dar es Salaam   Hong Kong   Karachi
Kuala Lumpur   Madrid   Melbourne   Mexico City   Nairobi
New Delhi   Shanghai   Taipei   Toronto

With offices in
Argentina   Austria   Brazil   Chile   Czech Republic   France   Greece
Guatemala   Hungary   Italy   Japan   Poland   Portugal   Singapore
South Korea   Switzerland   Thailand   Turkey   Ukraine   Vietnam

Oxford is a registered trademark of Oxford University Press
in the UK and certain other countries.

10 07483723

Published in the United States of America by
Oxford University Press
198 Madison Avenue, New York, NY 10016

© Oxford University Press 1991, 1996, 2002, 2015

All rights reserved. No part of this publication may be reproduced, stored in
a retrieval system, or transmitted, in any form or by any means, without the prior
permission in writing of Oxford University Press, or as expressly permitted by law,
by license, or under terms agreed with the appropriate reproduction rights organization.
Inquiries concerning reproduction outside the scope of the above should be sent to the
Rights Department, Oxford University Press, at the address above.

You must not circulate this work in any other form
and you must impose this same condition on any acquirer.

ISBN 978-0-19-977389-3
A copy of this book's Cataloging-in-Publication
Data is on file with the Library of Congress.

9 8 7 6 5 4 3 2 1
Printed in China through Asia Pacific Offset
on acid-free paper

To Azalea, Dylan, Selma, and Theo.
You bring joy to our lives.

Contents

III. Intercellular Communication

IV. Behavior and Plasticity

Preface to the Fourth Edition

Much has changed in the field of neuroscience since the publication of the third edition of *The Neuron* in 2002. We know a great deal more now about all aspects of nervous system structure, organization, and function than we did in that dim, dark, distant past at the dawn of the twenty-first century. This is the case whether one's scientific interests focus on molecular details of the activity of individual neurons and glia or on neural systems that underlie complex behaviors. Why does neuroscience continue to advance at so dizzying a pace? In part, it is because exceptionally talented individuals choose to devote their creative efforts to understanding how the brain works, as witnessed by the fact that no fewer than 17 neuroscientists have been awarded Nobel Prizes (in Physiology or Medicine, and Chemistry) since 2002. Another important reason is that spectacular technological advances, particularly (but not exclusively) in genome sequencing and optical imaging, have provided exciting new tools with which to explore the molecules, cells, and systems of the brain in innovative ways. In any event, whatever the reasons for the astounding progress, the current state-of-the-art is such that the fourth edition of *The Neuron* is by necessity significantly different from its predecessors.

While this fourth edition continues to be aimed at students undertaking a first course in the cell and molecular biology of nerve cells, with an overall approach and organization comparable to those of prior editions, the method of presentation is substantially different. Readers of the third edition may notice that the printed book is a little slimmer than in the past. We have taken advantage of other spectacular advances, in the area of information technology and ways of delivering content to students, to move many experimental examples that populated the earlier print editions of *The Neuron* to a website that now constitutes an integral part of the book. In addition, we have progressed from the two-color figures in the past to full color for the fourth edition, which has presented the opportunity to redraw every figure in the book and rethink and redesign most of them. Finally, the website hosts a series of videos and newly designed animations that help to clarify many of the key concepts we present. The overall reader experience will be very different than in the past.

As always, we thank the numerous colleagues who contributed materials for this and the previous editions. Those contributions are acknowledged where the materials appear in the book. In addition we would like to thank Brian Rash and Elise Stanley for their advice on specific topics in the book, and Paul Forscher and Dylan Burnette for providing beautiful images for the cover. Our Oxford University Press editor, Joan Bossert, understood our psyches better than we do, and pushed our buttons at the right times to move the project along. Especially noteworthy is our partnership with Don Slish, who drew or redrew all of the figures, and designed and executed the animations. Don's enthusiastic participation and talent for illustration were invaluable for the redesign of the fourth edition. Finally, as for all the prior editions, the wonderful intellectual environment of the Marine Biological Laboratory in Woods Hole provided the backdrop for much of the work that went into preparing this book. If ever we do a fifth edition, no doubt we will do it in Woods Hole.

Philadelphia and New Haven, February 2015   I.B.L.
                                                                                 L.K.K.

Preface to the Third Edition

The passage of a mere five years since the second edition of *The Neuron* has rendered several statements that were made in the last two editions completely inaccurate. It is not that we are acknowledging any major errors of scientific fact in the past editions (although time will certainly show that some of what we now believe about the functioning of the brain is wrong). Rather, the arrival of a new millennium has made nonsense of most of the statements that began "Earlier in this century. . ." This by itself, however, was insufficient to induce us, yet again, to rent a room at the Marine Biological Laboratory in Woods Hole, to prepare a manuscript that represents a substantial revision of both the organization and content of the prior editions. It was, instead, the continuing rapid march of progress in neuroscience that provided the major incentive.

The third edition of *The Neuron* continues to provide a comprehensive first course in the cell and molecular biology of nerve cells, and the overall approach matches that of previous editions. The recent mapping of the human genome and those of other species has, however, led to the discovery of numerous new proteins that regulate the development, excitability, and other functions of neurons. These have been incorporated into the new edition in nearly all of the chapters. The first section of the book, which deals with neuronal excitability, has been reorganized substantially, so as to make it more readable for those students with less background in the physical sciences. This section has also been endowed with an additional chapter, prompted by the spectacular growth in our knowledge of the molecular nature of proteins that regulate excitability. A new chapter, on the "Birth and Death of Neurons," has also been added in the section on Behavior and Plasticity. In addition to covering new discoveries about the early development of neurons, this chapter describes the recent discovery that new neurons are continually being formed in certain parts of the adult mammalian brain. It also describes research on stem cells, which holds the therapeutic potential for the repair of damaged or diseased brain tissue. Finally, the use of imaging technologies in the study of the brain has expanded enormously in the past few years. The new edition describes some of these new approaches. Moreover, the introduction of full color plates in the new edition now allows many new images to be presented in their original form.

In addition to acknowledging again all of our colleagues and coworkers who read earlier editions and who were instrumental in providing materials for those editions, we now would also like to recognize the invaluable contributions of Thomas Kuner, Vincent Pieribone, Paul Forscher, Jim Trimmer, Matt Rasband, Phil Haydon, and Rod MacKinnon. Finally, we must say that it has been delightful, as always, to work with Fiona Stevens and Jeff House at Oxford University Press. In the last millennium, Fiona and Jeff worked for different publishers, and in fact were competitors for the first edition of *The Neuron*. It is our fantasy that it was this early competition that brought them together and that led to their eventual marriage. Even neuroscientists must be permitted the occasional romantic fantasy.

Woods Hole, Summer 2001

I.B.L.
L.K.K.

Preface to the Second Edition

There have been enormous advances in neuroscience in the five years since the first edition of *The Neuron* was published. Neuroscience remains one of the most dynamic research areas in modern biology, with cellular and molecular approaches to understanding brain function leading the way. This progress in cellular and molecular neurobiology has paralleled equally exciting advances in other areas of basic cell and molecular biology, which have been instrumental in helping us to understand how nerve cells work. Indeed, in the first edition of this book we emphasized the features that cells of the brain have in common with other cells, and these are becoming more and more apparent as time passes.

This rapid progress has led us to revise *The Neuron* thoroughly for this second edition. What was formerly the first chapter has been expanded substantially and divided into two separate chapters, to emphasize features of the cell biology of neurons and glia and their commonalities with other kinds of cells. The section on intracellular communication has also been expanded and reorganized. We now introduce the concept of ion channels as specialized membrane proteins at an early stage, making it easier to understand the idea of selective membrane permeability in terms of the properties of particular ion channel proteins. In addition, we emphasize the astonishing diversity of the voltage-dependent ion channels that has become evident in recent years, and discuss the implications of this diversity for neuronal physiology. In the section on intercellular communication, the chapter on secretion of neurotransmitters has also been rewritten to reflect the new level of understanding of secretion that has resulted from identification of many of the molecular players in vesicle fusion and exocytosis. Here again, information from many kinds of cells, including lower eukaryotes, has been instrumental in advancing our understanding of secretion. The remaining chapters in the section on intercellular communication have also been revised thoroughly, to include new information resulting from the cloning and characterization of the multitude of glutamate receptors, as well as to describe novel elements of intracellular signaling pathways in neurons and other cells. Finally, the substantial revision of the last section reflects the fact that cellular and molecular studies of development and plasticity have also been exceptionally fruitful during the last five years. As more and more of the molecular entities that are essential for neuronal development and adult plasticity are identified and characterized, phenomena that previously could be studied only at the descriptive level are now becoming understood in molecular detail.

It is an exciting time to be a neuroscientist. As was the case when we wrote the first edition of *The Neuron,* we enjoyed learning many new things from our friends and colleagues and from the burgeoning literature in cellular and molecular neurobiology. Many thanks to all who helped and encouraged us, particular our assistants Joyce Chase and Paula Shelly for dealing with a myriad of details, and our editor Jeffrey House of Oxford University Press for his constant encouragement and advice. It was fun the first time, and it was fun again this second time. We might even consider doing it again!

Summer 1996   I.B.L.
                                                                                 L.K.K.

Preface to the First Edition

A fundamental goal of neuroscience is to understand the way neurons generate animal behaviors. This requires, among other things, a knowledge of how many neurons are involved in the control of a specific behavior, where these neurons are located, and how they are connected. All this information, however, is of no avail if we do not understand the intrinsic properties of the individual neurons themselves. Just as each individual human provides his or her unique contribution to society, the properties, abilities, and "personalities" of different cells in a neuronal circuit all play roles in the output of that circuit. They also determine how that output alters with time to change the behavior of an animal.

Neuroscience has come of age. Its practitioners are no longer simply a hodge-podge of physiologists, biochemists, and anatomists who stumbled across the brain and decided to study it. Rather it is a mature discipline in its own right. This is evidenced in part by the appearance of several textbooks of neurobiology, some aimed at medical students, others designed to give undergraduates a broad overview of the field. By necessity such treatments must emphasize breadth at the expense of depth, and we have found ourselves frustrated in looking for a textbook that can be used to communicate in detail to students the principles of cellular and molecular neurobiology. Hence this book.

The text reflects our own training, research interests, and biases. It is unabashedly reductionistic, in line with our conviction that understanding the elements of the nervous system—individual neurons and the molecules that regulate neuronal activity—is essential for even the most rudimentary understanding of how the brain works. We recognize that, although understanding the elements is essential, it is not in itself sufficient for a complete description of brain function, and where appropriate we place cellular and molecular principles in the context of the nervous system in which they function. However, for the reasons given above we have preferred to emphasize depth, and this is not intended to be a comprehensive text covering all of neuroscience. Rather it is designed for a first course in cellular and molecular neurobiology, for undergraduate or beginning graduate students who already have a basic grounding in biochemistry and cell biology. Students who have mastered the material in this book can move on readily to more specialized courses in neural systems, development, behavior, and computational neurobiology.

We had fun writing this book. It gave us the opportunity to interact with and tap the expertise of knowledgeable and stimulating colleagues from whom we learned a great deal. Many of them read portions of the manuscript and pointed out (occasionally with great glee) our misconceptions and murky writing style. Eve Marder, Chris Miller, and Jimmy Schwartz undertook the heroic task of reading the entire book, and their suggestions were invaluable. Others who read several (sometimes many) chapters were Spyros Artavanis-Tsakonas, Bill Catterall, Arlene Chiu, Martha Constantine-Paton, Pietro DeCamilli, Dorothy Gallager, Scott Kasper, John Lisman, Joanne Mattessich, John Perkins, Jan Rosenbaum, Larry Squire, and Kate Turtle. We are grateful to all of them. Much of the treatment of resting and action potentials in Chapter 3 was inspired by a wonderful little book titled *Neurophysiology: A Primer*, written by Chuck Stevens many years ago (J. Wiley and Sons, New York, 1966). We are also grateful to John Dowling, Paul Forscher, Steven Hunt, Andrew Matus, Tom Reese, Bruce Schnapp, Toni Steinacker, Masatoshi Takeichi, Asa Thureson-Klein, Monte Westerfield, and especially Dennis Landis for providing us with original photographs for some of the figures. Mike Lerner gave us the idea for, and the chemical structures of, the different odorants in Figure 13–16a. Maureen Ferrari and Joyce Chase did yeoman duty at the keyboard and in dealing with endless administrative details, and Paula Shelly helped with the preparation of the index. As always it was a pleasure to interact with our editor Jeffrey House and his colleagues Edith Barry, Donna Grosso, and Susan Hannan at Oxford University Press. Finally we thank our friends and colleagues for their encouragement, for telling us over and over again that a text of cellular and molecular neurobiology is sorely needed.

October 1990   I.B.L.
                                                                 L.K.K.

How to Use the Online Textbook

Individual purchasers of the print version of *The Neuron: Cell and Molecular Biology, Fourth Edition* are also entitled to free personal access to the online edition of the textbook. This can be accessed through Oxford Medicine Online at: www.oxfordmedicine.com/theneuron4e/

The online textbook contains ancillaries meant to support your studies. Your instructor should inform you if you will be quizzed on this supplementary material.

Callouts have been placed in the text indicating the online location of this material. For videos and animations demonstrating cellular mechanisms, the relevant symbol is ⬭. Adobe Flash is required for viewing the animations. When there is more on a topic available online, further illustrating a concept with interesting findings, descriptions of experiments, tables, or figures, ❷ will appear.

The Neuron

I

Introduction

The purpose of this book is to describe the cellular and molecular properties of brain cells that allow them to carry out their assigned task in an animal. While glial cells play fundamental and essential roles in the brain, we focus in this book on the properties of individual nerve cells—the neurons—and the connections between them. We begin in Chapter 1 with an overview of neuronal signaling to emphasize that the primary function of neurons is information transfer. This includes both intracellular signaling, from one part of a neuron to another, and intercellular signaling, from one neuron to another or to a muscle cell. Chapter 2 goes on to describe features of neuronal and glial cell biology. We emphasize here that neurons share many properties with other cell types in the organism but at the same time are exquisitely specialized to carry out their signaling functions. To provide just one example, the cytoskeleton, a cellular scaffolding present in all cells, plays a particularly critical role in the organization and maintenance of neuronal form. Each of the themes introduced here will be elaborated in more detail in later chapters.

1

Signaling in the Brain

Tell me where is Fancy bred,
Or in the heart, or in the head?

lthough William Shakespeare asked this question near the end of the sixteenth century, the answer had been known, at least to some, for more than two millennia. Many ancient Greek scholars, Hippocrates and Plato among them, appreciated that there is something special about the brain and argued that the brain is responsible for behavior in humans and other animals. We now take it for granted that the brain is the organ that obtains information about the environment, processes and stores this information, and generates behavior. In addition, it is the brain that is responsible for such vaguely defined aspects of behavior as feelings, aspirations, and abstract thoughts—those qualities we consider, with more than a touch of hubris—quintessentially human. Diseases of the brain, which often are associated with aging and hence are on the increase as the human lifespan lengthens, cause untold suffering for individuals, families, and society as a whole. In essence, then, it is the brain that makes us what we are, and indeed it can be argued that all other organs are there simply to support the brain. For this reason understanding the biology of brain function is a major goal of modern science. The brain is one of the last great frontiers in the biological sciences, and the unraveling of its mysteries is comparable in complexity and intellectual challenge to the search for the elementary particles of matter or the effort to explore space.

Levels of Organization

One can study the functioning of the nervous system at a number of *levels of organization* (Fig. 1–1). Biochemists, molecular biologists, and structural biologists investigate the properties of molecules that perform tasks important for brain function. Physiologists may study the characteristics of individual nerve cells or collections of cells that communicate with one another and are functionally related. Behavioral psychologists explore patterns of behavior and its modification—learning—in experimental animals ranging from lower invertebrates to humans. Computational neuroscientists attempt to put it all together, to model behavior and other higher brain functions in terms of the known properties of molecules, cells, or collections of cells.

In this book we will focus on the top part of Figure 1–1, on the cells of the brain and the molecules that control their function. Although, where appropriate, we will discuss these cells and molecules in the context of the larger nervous system to which they belong, our major emphasis will be on *signaling*, or *information transfer*, within and between nerve cells. We shall see that such signaling is essential for an organism to (*1*) sense information about its environment, (2) import this information into its brain where it can be processed, and (3) generate a behavioral response. Chapters 3 through 7 will explore the ways in which nerve cells are specialized for *intracellular signaling*, the movement of information from one part of the cell to another. Chapters 8 through 13 will consider *intercellular signaling*, the mechanisms by which nerve cells communicate with each other and the outside world. Finally, in Chapters 14 through 19 we will address *neuronal plasticity*, changes in the properties of a neuron. This section will include cellular and molecular aspects of *development*, the processes by which nerve cell number, position, shape, and patterns of connectivity (and hence signaling) are set down during embryonic and early postnatal life. In addition, we will discuss *behavior* and its modification at the level of signaling in individual nerve cells, collections of nerve cells, and behaving animals.

Although our focus will be on cellular and molecular mechanisms of brain function, we emphasize that cellular and molecular neurobiology do not exist in a vacuum. First, many of the mechanisms we will discuss in the context of brain cells have their counterparts in other cell types—that is, there is an emerging awareness of a satisfying unity in cell biology. Second, it is evident that understanding the brain requires study at all the levels of organization depicted in Figure 1–1, from behaving human beings to single brain cells and the molecules that regulate

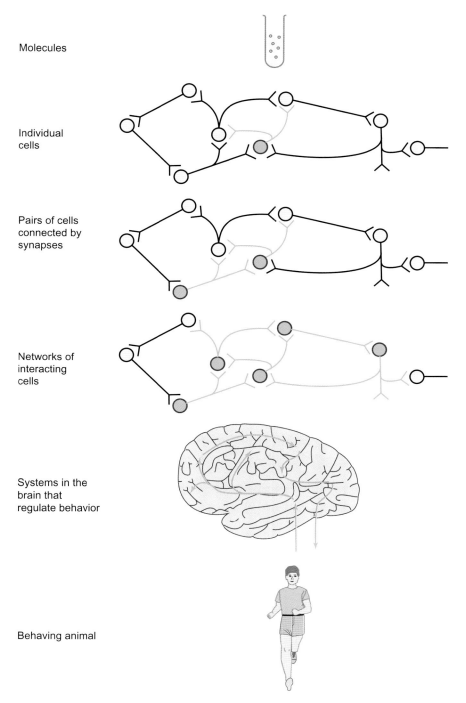

Molecules

Individual
cells

Pairs of cells
connected by
synapses

Networks of
interacting
cells

Systems in the
brain that
regulate behavior

Behaving animal

Figure 1–1. Levels of organization for studying structure and function in nervous systems. Depending on their background and training, different scientists may take different approaches to the study of the nervous system.

cellular activity. No single level is inherently more important than any other, and information from all of them will be necessary for even the most rudimentary understanding of normal and abnormal brain functions. Thus, by its very nature, the problem demands a multidisciplinary attack, one that bridges traditional scientific disciplines and facilitates collaboration among scientists with very different training and experimental approaches. History has amply shown that it is at the boundaries between disciplines where the most significant advances are likely to be made.

The Cellular Hypothesis

We have grown so accustomed to thinking of the brain as a cellular organ that it is easy to forget that this view was once the subject of intense debate. By 1840 it was evident, from the work of the anatomists Jacob Schleiden and Theodor Schwann, that discrete entities called cells are the basic architectural units of living tissues. It was, however, to be another 50 years and more before it was accepted that this principle also applies to the brain. Around the end of the nineteenth century, the great neuroanatomists Santiago Ramón y Cajal and Camillo Golgi argued passionately about whether the brain consists of enormous numbers of discrete cells or is a continuous syncytium of tissue. The answer to this question is, of course, of enormous significance for understanding how signals spread from one part of the nervous system to another. Ramón y Cajal made elegant use of a technique, which had been discovered fortuitously by Golgi, for staining tissue. For reasons that are not yet understood, this technique, known as *silver impregnation*, results in the staining of only a small subset of the neurons present in a brain section. As a result, individual neurons show up clearly (Fig. 1–2) in tissue sections that actually contain a large number of neurons. Using other staining methods that stain all the neurons, these same sections would have appeared only as "tangled thickets." Ramón y Cajal correctly identified these discrete entities as individual nerve cells, although Golgi never accepted this interpretation and continued to put forward his "reticular theory" of a continuous meshwork. The disagreement between them was so intense that Ramón y Cajal and Golgi reportedly did not talk to each other during the ceremony at which they were jointly awarded the Nobel Prize in Physiology or Medicine in 1906. In any event, the debate was definitively won by the advocates of the cellular hypothesis, and it is now universally accepted that the brain, like all other organs in the body, is

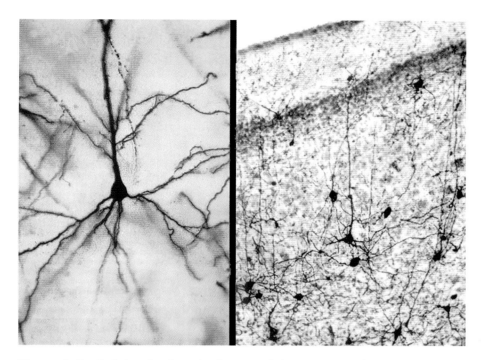

Figure 1–2. Golgi-stained cerebral pyramidal neurons are shown at two magnifications. Pyramidal neurons (and granule cells) are typical neurons of the cerebral cortex. Cell bodies of pyramidal neurons have a pyramid shape with a single apical dendrite and multiple basal dendrites. The axons are not visible in these images. (Micrographs courtesy of Thomas F. Fletcher—see http://vanat.cvm.umn.edu/neurHistAtls/pages/cns12.html.)

cellular. In retrospect it is clear that one reason for the long confusion over this issue is the complexity of brain tissue. As we shall see in this book, a large number of different neuron types exist in the brain, and many of them have a complex asymmetric, three-dimensional structure that makes it extremely difficult to ascertain where one cell ends and the next begins.

Unique Structures of Neurons

Although in the next chapter we will also discuss those structures and organelles that neurons have in common with other cells, the remainder of this chapter will emphasize the features that make the neuron unique.

It must be remembered that the essence of nervous system function is signaling, or information transfer, both intracellularly from one part of a cell to another and intercellularly between cells. It is a fundamental premise of cellular neurobiology that a great deal will be learned about how the nervous system works by investigating

1. those aspects of neuronal structure that specialize them for information transfer;
2. the mechanisms of intracellular neuronal signaling;
3. the patterns of neuronal connectivity and mechanisms of intercellular signaling;
4. the relationship of various patterns of neuronal connectivity to different behaviors; and
5. the ways in which neurons and their connections can be modified by experience.

All of these topics will be discussed in this book. We begin now with a description of three structural elements unique to neurons: the *axon*, which is specialized for intracellular information transfer; the *dendrite*, which is often the site at which information is received from other neurons (but which also has other functions); and that most highly specialized structure of all, the *synapse*, which is the point of information transfer between neurons (Fig. 1–3).

The axon. The axon is a thin, tube-like process that arises from the neuronal cell body and travels for distances ranging from micrometers to meters before terminating (at a synapse—see "The synapse," later in this chapter). As we shall see in Chapters 3 through 7, specialized proteins in the axonal plasma membrane allow the axon to transmit electrical signals rapidly along its length, from soma to terminal. The axon originates at a cone-shaped thickening on the cell body called the *axon hillock* (Fig. 1–3). It is often (but not always) unbranched until just before it terminates, but it may branch many times in its terminal region. The diameter of the axon remains more or less unchanged throughout its length. Its structure, like that of the dendrite, is formed and maintained by the *cytoskeleton*, a cellular scaffolding that is present in all cells but that exhibits certain unique properties in unusually shaped cells such as neurons. The role of the cytoskeleton in the formation and maintenance of neuronal form will be discussed briefly in Chapter 2.

The dendrite. Dendrites are neuronal processes that tend to be thicker and much shorter than axons and often are highly branched, giving rise

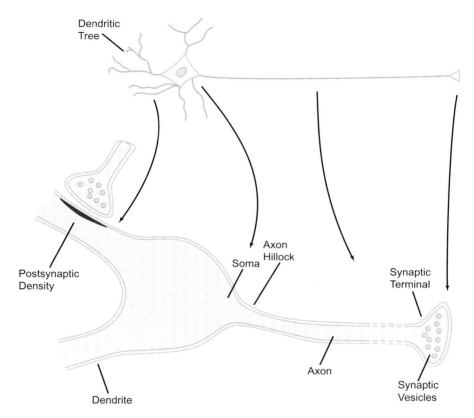

Figure 1–3. Ultrastructure of the neuron. Drawing of a typical nerve cell to show its overall shape. Structures and organelles that are found exclusively in nerve cells are labeled.

to a dense network of processes known as the *dendritic tree* (Fig. 1–3). In addition, the dendritic cytoskeleton differs from that of axons (see Chapter 2). Dendrites often originate from the cell body, but in some neurons in invertebrates they arise from the proximal regions of the axon. A three-dimensional computer reconstruction, from images taken with a confocal microscope, reveals the presence of numerous finger-like projections or thickenings on the dendrites of some neurons. These projections, called *dendritic spines*, arise from the main shaft of the dendrite (Fig. 1–4). These spines are the synaptic input sites at which the neuron receives information from another cell. Like the axonal membrane, the plasma membrane of the dendrite contains a particular set of proteins that allows the dendrite to carry out its assigned functions (see Chapters 11 and 12). In this case, the proteins

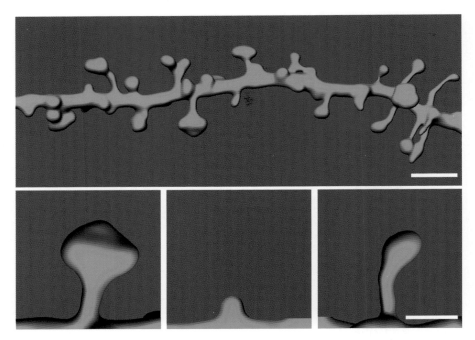

Figure 1–4. Dendritic spines. The *upper* micrograph shows a three-dimensional computer reconstruction of a tertiary portion of the basal dendrites of a CA1 pyramidal neuron in the hippocampus, labeled with a fluorescent dye called green fluorescent protein. Scale bar 2 μm. The *lower* micrographs illustrate three distinct morphological subclasses of dendritic spines. At *left* is a mushroom-shaped spine, *center* is a short, stubby spine, and *right* is a thin spine. Scale bar 1 μm. (Micrographs courtesy of Anne McKinney [McKinney, 2010]).

specialize the membrane in a way that allows the dendrite to receive and integrate information from other neurons and from glial cells. However, the role of the dendrite is not exclusively the receipt of information. Some dendrites share with axons the ability to transmit electrical signals, and in many nerve cells both information input and output occur on the same set of dendrite-like fine processes.

Interestingly, dendritic structure is not always fixed and immutable. Under some physiological conditions, the size and shape of dendritic spines can exhibit dynamic changes on a rapid timescale (Fig. 1–5). It is thought that such alterations of spine morphology may contribute to long-lasting plastic changes in neuronal properties, of the kind that we shall discuss extensively later in this book.

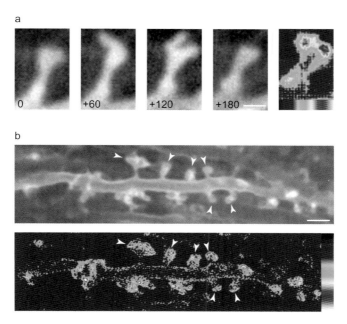

Figure 1–5. Dendritic spines are dynamic. *a*: Four frames, taken 60 seconds apart, from a live cell recording of a single dendritic spine from a transgenic animal expressing surface membrane targeted green fluorescent protein. *Right*: Pseudo-color motility image of the corresponding time-lapse recording. This composite image shows the summed morphological differences of 60 individual frames recorded 10 seconds apart. Areas with highest motility appear red while those with limited motility are dark blue. White bar = 1 μm. *b*: First frame from a 10-minute time-lapse recording of a single dendrite. Below it is its corresponding pseudo-color motility image. Note that all spines (arrowheads) have high motility, as seen by the red coloring, whereas motility of the dendritic shaft is virtually absent. White bar = 2 μm. (Images courtesy of Martijn Roelandse and Andrew Matus; see also Fischer et al., 1998.)

Intercellular Communication

The synapse. We have emphasized that intercellular communication, which results in the passage of information from one part of the nervous system to another, is the essence of nervous system function. It is this information transfer that most clearly distinguishes the brain from other organs, and thus it is not surprising that the neuron has evolved a unique and highly specialized structure, the synapse, to carry out this task

(Fig. 1–3). The term *synapse*, derived from a Greek word meaning "connect," was introduced by the British physiologist Charles Sherrington near the end of the nineteenth century. Sherrington was studying spinal cord reflexes (Fig. 1–6). Such reflexes are invariant behavioral responses to a particular kind of stimulus to the animal; they do not require the brain itself but are mediated via the spinal cord. One often-cited example is the rapid withdrawal of the arm when the fingers encounter a hot stove.

Sherrington's studies on reflexes coincided with the time when the neuron doctrine was becoming well established as a result of the work of Ramón y Cajal and his followers. He defined a set of neuronal connections, a *pathway*, responsible for the reflex, and noted that information always travels through the reflex pathway in one direction only. More specifically, input is via the *sensory* component of the pathway (yellow in Fig. 1–6), which provides information about the outside world. Output is through the *motor* component (blue in Fig. 1–6), which drives muscle contraction to provide an appropriate behavioral response to the sensory input. Sherrington, like Ramón y Cajal before him, also became convinced from his detailed anatomical and physiological studies that the pathway is not unicellular, but that there is a discontinuity between the sensory

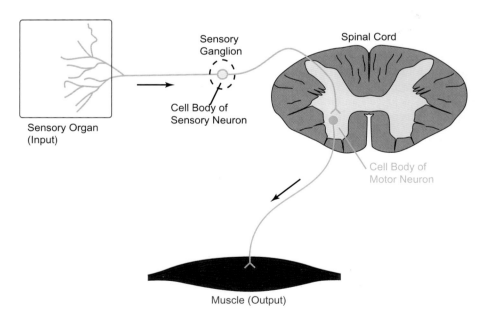

Figure 1–6. A reflex arc. The classic reflex arc, with its sensory neuron input and motor neuron output. The arrows indicate the direction of information flow in the reflex pathway.

and motor components. In many reflex arcs there may also be an additional neuron called an *interneuron* in the spinal cord, which provides the link between the sensory and motor neurons. With remarkable insight, Sherrington proposed that at the point of contact between the sensory and motor neurons there might be a structure—the synapse—that allows unidirectional information transfer between the neurons. It is important to emphasize that Sherrington's definition of the synapse was a functional one, and it was to be many years before the anatomical correlate of this functional connection was identified. Sherrington's pioneering work established the idea that understanding synaptic structure and function is essential for understanding how the brain works. This emphasis on the synapse in cellular neurobiology has been maintained into the twenty-first century. We know now that the strength of synaptic transmission—information transfer at synapses—is not fixed. This is in contrast to information transfer along the axon, which has been thought of (not entirely correctly, as we shall see) as all-or-none. Accordingly, one widely held belief in modern neurobiology is that changes in the properties of synapses underlie the plasticity of nervous system function, including learning and memory (see Part IV). In addition, it is becoming evident that malfunctioning synapses are associated with a whole range of debilitating diseases, among which are Parkinson's disease, Alzheimer's disease, amyotrophic lateral sclerosis (ALS), fragile X syndrome, bipolar disorder, and schizophrenia. As our understanding of synaptic structure and function has increased, it has become possible to think about designing rational treatments for these diseases, but the relatively limited success in this area emphasizes how far we still have to go. It is these considerations, together with the sheer intellectual excitement of trying to understand how intercellular communication works in the nervous system, that continue to motivate research on synaptic transmission.

Two kinds of synapses. Through the first half of the twentieth century there was a bitter controversy about the nature of information transfer at synapses. The school of neuropharmacologists, led by Sir Henry Dale, insisted that synaptic transmission is mediated by a chemical substance liberated from the terminal of one neuron (the *presynaptic cell*), which interacts with and influences the properties of the follower neuron or muscle cell (the *postsynaptic cell*). The first convincing piece of evidence for this idea came from the study of a neuron-to-muscle synapse in the frog. It had been known that electrical stimulation of the vagus nerve leads to slowing of the heart rate. In a classic experiment carried out in 1921, the Austrian pharmacologist Otto Loewi placed a frog heart, innervated by the vagus nerve, in a chamber containing physiological saline, and connected the chamber with another that contained a second, noninnervated heart (Fig. 1–7). The experimental arrangement allowed the saline solutions in which the two hearts were sitting to exchange freely. He then stimulated the vagus nerve, which, as expected, resulted in slowing of the first heart. Loewi noted that, after some delay, the second heart slowed as well (Fig. 1–7). He concluded correctly that some chemical released by the firing of the vagus nerve resulted in slowing of the heart rate, and that this chemical diffused through the saline solution to the second chamber, where it acted on the second heart. The chemical

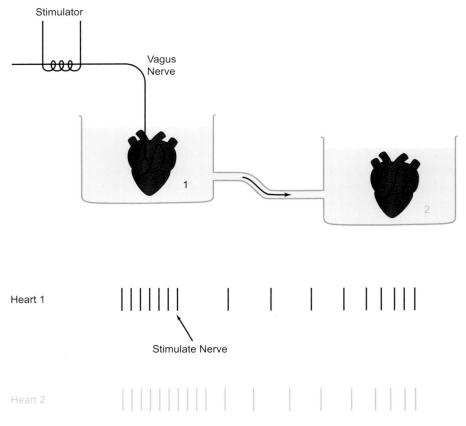

Figure 1–7. Chemical transmission at a nerve–muscle synapse. In this famous experiment performed in 1921, Otto Loewi placed an innervated (1) and a noninnervated (2) heart in two separate chambers connected by a bridge of physiological saline. At the bottom is shown the rate of beating of both hearts before and after stimulation of the vagus nerve connected to the first heart.

responsible for this phenomenon was subsequently isolated and identified as acetylcholine, the first chemical to be characterized as a *neurotransmitter*. Although this is a neuron-to-muscle synapse, it is now clear that neuron-to-neuron chemical synapses operate in the same general way, and that acetylcholine is an important neurotransmitter in the brain as well as in the peripheral nervous system.

The conflicting point of view about the fundamental nature of synaptic transmission, put forward most forcefully by the electrophysiologist Sir John Eccles, held that synaptic transmission is electrical and results from the movement of ions from one neuron to another via direct physical connections between the neurons. Again, a variety of evidence supported this contention. The arguments between the two factions were often acrimonious because each believed that there is a single global mechanism of synaptic transmission. We know now that they were wrong in this belief, and that both points of view were correct. Chemical and electrical modes of synaptic transmission exist side by side in most and probably in all nervous systems.

Properties of chemical and electrical synapses. Although both chemical and electrical synapses mediate intercellular information transfer, they do so by very different mechanisms. It is instructive to consider those features that distinguish the two types of synapses in the context of the different functional roles that they may play. We know a good deal more about chemical than electrical synapses, but this may reflect historical accident rather than the relative importance of the two kinds of synapses. Convenient experimental preparations for the investigation of chemical synaptic transmission have been available for many years, but only more recently have conceptual and technical advances allowed a more thorough examination of the properties of electrical synapses. It is worth mentioning in this context that, until quite recently, much of what we know about chemical synaptic transmission was gleaned from studies of the vertebrate *neuromuscular junction*, which is a neuron-to-muscle rather than a neuron-to-neuron synapse (see Fig. 8–10). This focus on neuromuscular synaptic transmission arose in part because the frog sciatic nerve–gastrocnemius muscle synapse is such an accessible and convenient experimental preparation—far more accessible than neuron-to-neuron synapses in the complex cellular networks of the central nervous system (an example of the latter is shown in Fig. 1–8). In addition, a number of gifted investigators, most notably Sir Bernard Katz and his collaborators in London, exploited this preparation brilliantly to elucidate the details of synaptic transmission between nerve and muscle cells.

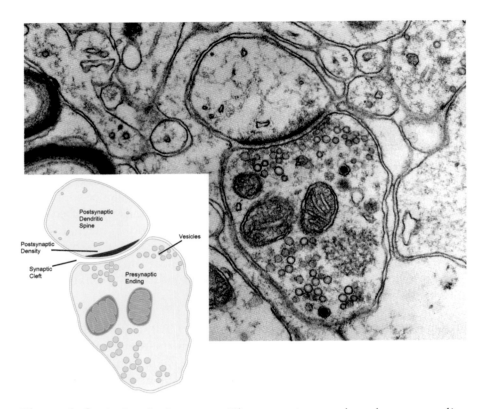

Figure 1–8. A chemical synapse. Electron micrograph and corresponding labeled drawing of a synapse onto the spine of a Purkinje cell dendrite. (Micrograph from http://medcell.med.yale.edu/histology/nervous_system_lab/synapse_em.php.)

Accordingly, for many years the picture neurobiologists held of the "typical" chemical synapse was the neuromuscular junction. However, as recent technological advances have enabled more detailed mechanistic investigation of central nervous system synapses, it has become evident that this picture is inappropriate in many ways. For example, many central nervous system chemical synapses operate on a timescale orders of magnitude slower than that of the neuromuscular junction, and the molecular mechanisms involved in the transduction of the chemical signal into an electrical response in the postsynaptic cell can be very different (see Chapters 11 and 12). Although we will be referring often to studies on the neuromuscular junction in discussing various aspects of chemical synaptic transmission in subsequent chapters—and indeed, this is reasonable because (with apologies to Gertrude Stein) a synapse is a synapse is a synapse—these caveats must be kept in mind.

Chemical synapses. The most obvious difference between chemical and electrical synapses is in their structure as seen in the electron microscope. Chemical synapses have an *asymmetrical* morphology, with distinct features found in the presynaptic and postsynaptic parts, or "elements," of a synapse (Figs. 1–3 and 1–8). The presynaptic ending is a swelling of the axon, often but not exclusively at its terminal, containing mitochondria and, most important, a variety of vesicular structures. As will be discussed in Chapter 8, the vesicles contain the neurotransmitter, for example, acetylcholine, that is released from the presynaptic terminal and produces some change in the postsynaptic cell. The vesicles are often clustered adjacent to the membrane of the presynaptic terminal at its point of closest contact with the postsynaptic cell, in so-called active zones (see Fig. 1–8). This clustering may be related to the association of vesicles with specialized cytoskeletal filaments.

The electron microscope also reveals that the pre- and postsynaptic elements of a chemical synapse are separated by a 200 to 300 Å gap, the *synaptic cleft* (Fig. 1–8). This gap is somewhat larger than the normal extracellular space between cells, and its presence emphasizes that there are not direct membrane connections between the pre- and postsynaptic cells at chemical synapses. Stains that bind to sugars reveal that the synaptic cleft is loaded with carbohydrate, presumably associated with glycoproteins in the pre- and/or postsynaptic membranes. The function of this extracellular carbohydrate is not well understood, although one idea favored by many investigators is that it provides a matrix for some molecules that are important for synapse formation (see Chapter 17).

Just as most presynaptic endings are axon terminals, most postsynaptic elements of central nervous system synapses are dendrites, giving rise to the term *axodendritic synapse*. However, like most rules, this one has exceptions. *Axosomatic synapses* (where the postsynaptic target is a neuronal cell body or soma) and (more rarely) *axoaxonic synapses* exist. Moreover, as we have already mentioned, it is becoming evident that dendrites can also act as presynaptic elements (*dendrodendritic synapses*). In fact, in some nerve cells there are fine dendrite-like processes at which both input and output of information take place.

Virtually the entire somatic and particularly the dendritic surface of most central nervous system neurons is covered with presynaptic terminals (Fig. 1–9). In other words, there can be enormous convergence, onto a single neuron, of input information from hundreds or even thousands of presynaptic cells (Fig. 1–10). As we will see in subsequent chapters,

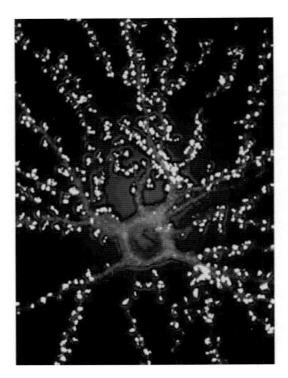

Figure 1–9. Convergent synaptic inputs (yellow) on a single neuron (gray). (Micrograph from Carlo Sala and Morgan Sheng; see http://www.brain.riken. jp/bsi-news/bsinews28/files/special0101-big.jpg.)

neurons are constantly integrating these multiple synaptic inputs, and the location of any given postsynaptic site can be very important in determining its contribution to the neuron's overall activity. When we consider, in addition, that a single presynaptic axon may branch many times and provide input to dozens or hundreds of postsynaptic targets (divergence; see Fig. 1–10), we can begin to appreciate the complexity of the computations that even the simplest of nervous systems carry out.

In contrast to the presynaptic terminal, the postsynaptic element is usually characterized by the absence of vesicles adjacent to the plasma membrane. There is often a highly electron-dense structure, the *postsynaptic density*, associated with the postsynaptic membrane immediately opposite the accumulation of vesicles on the presynaptic side (Figs. 1–3 and 1–8). The significance of this distinctive morphological specialization has been the subject of much investigation, and both the structure and function of the postsynaptic density are becoming well understood. Among its functions is to help anchor receptors for neurotransmitters in

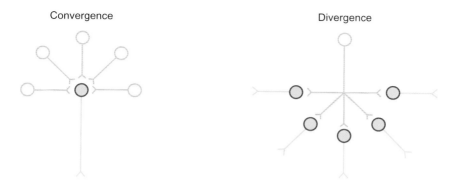

Figure 1–10. Convergence and divergence in nervous system function. *Left*: Any given nerve cell (blue) can receive convergent synaptic inputs on its dendrites from a large number—perhaps as many as thousands—of other neurons (yellow). *Right*: The axon of a single neuron (yellow) may branch many times, and thus divergent information may be provided to a large number of postsynaptic targets (blue) of this single neuron.

the postsynaptic membrane. It also contains molecules that are involved in *transduction*, the conversion of the chemical signal into an electrical response in the postsynaptic cell (see Chapters 11 and 12).

Associated with the morphological asymmetry of chemical synapses is a fundamental functional asymmetry: chemical synapses are *unidirectional*. That is, rapid transfer of information only occurs from the pre- to the postsynaptic cell (see Fig. 1–6). Like most rules, this one too has an exception; we shall see that a different kind of information does appear to flow from the post- to the presynaptic cell via retrograde messengers. However, this is very different from the more or less symmetrical *bidirectional* information transfer characteristic of most electrical synapses. In addition, at chemical synapses, there is a delay that may be a millisecond or longer between the arrival of information at the presynaptic terminal and its transfer to the postsynaptic cell. This delay reflects the several steps required for the release and action of the chemical neurotransmitter (Chapter 9). Furthermore, the response of the postsynaptic neuron may outlast the presynaptic signal that evokes it, sometimes by a very long time. The transduction mechanisms that may be responsible for such long-lasting changes in the target cell will be considered in detail in Chapters 11 and 12.

Electrical synapses. The striking asymmetry in structure and function of chemical synapses may be contrasted with the symmetrical morphology and bidirectional information transfer characteristic of electrical synapses. First, there are no morphological specializations that allow pre- and postsynaptic elements to be distinguished. Indeed, since signals can move in both directions through electrical synapses, each cell may be pre- or postsynaptic at different times. Instead of the synaptic cleft that separates the two elements of the chemical synapse, electrical synapses are characterized by an area of very close apposition between the membranes of the pre- and postsynaptic cells. Within these areas of the membrane are found *gap junctions*, cell-to-cell pores that allow ions and small molecules to pass freely from the cytoplasm of one cell to the next. Gap junctions exist in many types of cells. One test that is often used to establish whether a pair of cells is connected by gap junctions is dye coupling—that is, a low-molecular-weight dye such as Lucifer yellow or neurobiotin injected into one of the cells can spread rapidly into the cytoplasm of its synaptic partners (Fig. 1–11). As is the case for gap junctions in other cell types, molecules with a molecular weight up to about 1000–1500 can be transferred through most neuronal gap junctions. The structure and function of gap junctions will be discussed in more detail in Chapter 8. It is the movement of small ions from one cell to another through cell-to-cell gap junction channels that mediates intercellular signaling at electrical synapses. These channels provide a low-resistance pathway for ion flow between the cells without leakage to the extracellular space, and thus signals can be transmitted with little attenuation. Two important functional properties follow immediately from this mode of transmission. First, as referred to earlier, information transfer can be *bidirectional*—that is, a functional symmetry accompanies the structural symmetry. There are examples where the efficacy of electrical synaptic transmission is higher in one direction than in the other (so-called rectifying synapses); in fact, the first electrical synapse whose properties were investigated in detail, between two large axons in the crayfish, rectifies markedly. In general, however, the rule of symmetrical bidirectionality holds, and, in fact, it is an important criterion in identifying electrical synapses. The second important functional consequence of this mechanism is that electrical synapses are *fast*. There is no delay analogous to that seen with chemical synaptic transmission.

The extent of electrical synaptic connectivity in the central nervous system remains unclear. One functional role for electrical synapses that is widely accepted is in the synchronization of the electrical activity of large populations of neurons. For example, in both vertebrates and invertebrates it has been demonstrated that populations of neurosecretory neurons that synthesize and release biologically active peptide

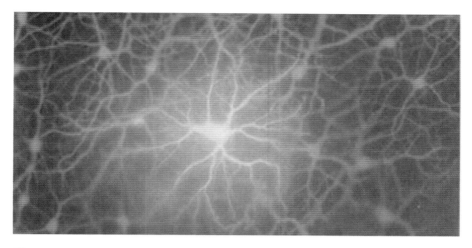

Figure 1–11. Dye coupling via electrical synapses between horizontal cells (a type of neuron) in the rabbit retina. The dye spreads via the gap junctions linking the horizontal cells to reveal the injected neuron and hundreds of its neighbors. (Modified from Mills and Massey, 1994.)

neurotransmitters and hormones are connected extensively by electrical synapses. Simultaneous recording from several neurons within such populations reveals that they all are electrically active at the same time, probably resulting in concerted release of their neurotransmitter. Synchronization via electrical synapses may also be important in various aspects of neuronal development, including the formation of chemical synapses. Large numbers of electrical synapses are also found in the retina, where they may influence and coordinate the processing of visual information.

It has been suggested that electrical synapses are less subject to alteration of their properties than chemical synapses and thus might provide an invariant mode of intercellular communication. However, it is becoming evident that the efficacy of electrical synaptic transmission can be regulated, perhaps as extensively as that of chemical synapses. Furthermore, as we shall see in Chapter 18, an examination of the properties of small networks of neurons reveals that many neuronal pairs in such networks are connected by both chemical and electrical synapses, and that the efficacy of one of the synaptic types may be modulated by the other. Thus, there is no question that more detailed information about electrical synaptic transmission will be necessary as we extend our understanding of intercellular communication in the nervous system.

Summary

The neurons are the cells of the brain that are responsible for intracellular and intercellular information transfer, or signaling. Neurons are asymmetrical cells with morphologically and functionally distinct regions that specialize them for signaling. In this chapter we have focused on the unique structural elements characteristic of neurons throughout the animal kingdom. These include the dendrite, among whose functions is the receipt of information from other neurons. The axon, in contrast, is specialized for the intracellular transfer of information over long distances. Finally, we have discussed the synapse, the highly specialized structure that mediates the transfer of information from one neuron to another. It is this intracellular and intercellular communication that is the essence of nervous system function, and that makes the brain so complex and difficult to study and yet at the same time so fascinating for the student of cell and molecular biology.

Form and Function in Cells of the Brain

I n the first chapter we focused exclusively on neurons, and particularly on those aspects of neuronal structure that specialize them for intra- and intercellular signaling. We will now consider some features of the structure and function of the various classes of *glial cells* that comprise by far the majority of cells in the brain. We will then emphasize that neurons share many structural features with other kinds of cells, including the glial cells. Finally, we will conclude with a discussion of the *cytoskeleton* and its role in the formation and maintenance of the structure of neurons.

The Brain Consists of Neurons and Glia

In the middle of the nineteenth century the German anatomist Rudolf Virchow recognized that cells in the brain could be divided into two distinct groups: (*1*) neurons, and (*2*) a far more numerous group of cells that appear to surround the neurons and fill the spaces between them. Virchow called this second category of cell the *neuroglia*, or nerve glue, the implication being that one of its functions is to hold the neurons in place. It now appears that this, indeed, is one of the many functions of glia, and certainly the name itself has stuck!

Glial cells can themselves be divided into several subclasses based on their appearance in the microscope. In the central nervous system the main types of glial cells are the *astrocytes*, the *oligodendrocytes*, and the *microglia*. As their name implies, the astrocytes have a star-like appearance, with numerous long arms radiating out from a central cell body (Fig. 2–1a). The oligodendrocytes also have a central cell body, with

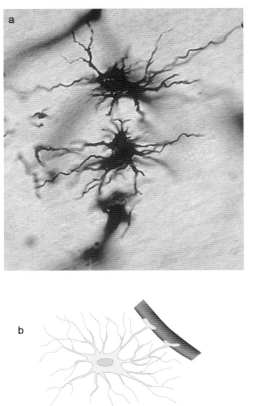

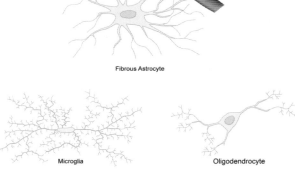

Fibrous Astrocyte

Microglia   Oligodendrocyte

Figure 2–1. Morphology of glial cells. *a*: Two fibrous astrocytes stained with the Golgi stain. *b*: Drawings to illustrate the different morphologies of astrocytes, oligodendrocytes, and microglia. (Micrograph courtesy of Thomas V. Fletcher—see http://vanat.cvm.umn.edu/neurHistAtls/pages/images/Glia2.jpg.)

radial arms that tend to be shorter and more branched than those of the astrocytes (Fig. 2–1b). As will be discussed later, the oligodendrocytes play an essential role in the functioning of neurons by forming the *myelin sheath* around axons in the central nervous system (see Fig. 2–3). In the peripheral nervous system the *Schwann cell*, another class of glial cell, is responsible for forming the myelin sheath. The microglia (Fig. 2–1b) respond to disease and injury by acting as immune cells within the central nervous system, and they also phagocytose cell debris. It has become evident more recently that the microglia also play important roles in the development and function of synapses in the healthy central nervous system.

Thus, in addition to their best known role, myelination, a number of functional roles have been ascribed to glia. For example, they are known to

1. act as a scaffolding for neuronal migration and axon outgrowth;
2. participate in the uptake and metabolism of the neurotransmitters that neurons use for intercellular communication;
3. take up and buffer ions from the extracellular environment;
4. act as scavengers to remove debris produced by dying neurons;
5. segregate groups of neurons one from another, and act as electrical insulators between neurons;
6. provide structural support for neurons, fulfilling a role played by connective tissue cells in other organs;
7. play a nurturing role, supplying metabolic components and even proteins necessary for neuronal function; and
8. participate in intercellular signaling, and thereby play a role in information handling and memory storage.

Thus it appears that glia have evolved multiple functional roles.

Glial signaling. In the past, neurobiologists relegated glia to a role secondary to that of neurons because they thought that neurons are uniquely capable of intracellular and intercellular signaling. More recent evidence, however, suggests that glia may in fact be active participants in brain signaling, and it is conceivable that neurons even form functional synapses with glial cells as their partners. Much better established is the fact that astrocytes possess on their cell surfaces the protein receptors for certain neurotransmitters that mediate synaptic signaling between neurons. Astrocytes are also capable of releasing these neurotransmitters, which may then influence nearby neurons or other glia. It is also known that astrocytes can respond to neurotransmitters by producing oscillations in their cytosolic calcium concentration that can spread from one astrocyte to another (Fig. 2–2), also producing changes in calcium concentrations in neighboring neurons (Video 2–1) ⬥. Such communication among glial cells, and between glia and neurons, implies that our traditional picture of glia as staid and stodgy cells that play only a supporting role in the

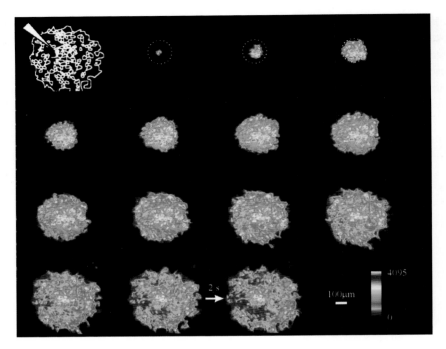

Figure 2–2. Spread of a calcium wave through astrocytes. Cultured astrocytes were loaded with a dye that emits a fluorescent signal when it binds calcium. Following stimulation of a small locus (upper left), a wave of elevated intracellular calcium (yellow and red) spreads throughout the network (for a more detailed discussion of calcium imaging, see Chapter 9). Successive images are 2 seconds apart. (From an experiment by Phil Haydon and colleagues. See also Haydon and Carmignoto, 2006; Carmignoto and Haydon, 2012; and http://www2.neuroscience.umn.edu/eanwebsite/CaWaves.htm.)

drama of brain function is inadequate. Fortunately, the need to ascribe a more central role to glia is becoming widely recognized.

The myelin sheath. Myelination is one role of glia that is well understood. The myelin sheath surrounds many, but not all, axons in the vertebrate nervous system. Although it is formed by glial cells and is not strictly a part of the axon, it is fundamental for axonal function. The sheath is formed by oligodendrocytes on central nervous system axons and by Schwann cells in the peripheral nervous system. When a myelinated axon is examined in cross section with an electron microscope, the axon is found to be surrounded by concentric circles of alternating dark and light bands (Fig. 2–3a). This structure arises by the tight wrapping of the membrane of the oligodendrocyte or Schwann cell around the axon during development (Fig. 2–3b). The cytoplasm of the glial cell is gradually squeezed out of this region as the cell wraps around the axon, so that the concentric circles represent layers of closely apposed glial plasma membrane. One can get a feel for the structure of myelinated axons by examining a roll of paper towels end-on: the central cardboard tube represents the axon, and the layers of paper, the wrapping of glial plasma membrane.

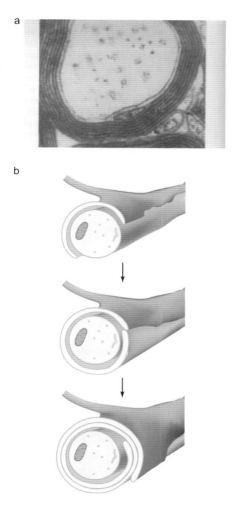

Figure 2–3. The myelin sheath. *a*: Electron micrograph of a cross section through a myelinated axon (modified from Peters et al., 1976). *b*: Formation of the myelin sheath. In the first stage (*top*), an axon (blue) is enclosed by the process of the oligodendrocyte (yellow). The oligodendrocyte then completely envelops the axon, but has not yet lost its cytoplasm (*middle and bottom*). The mature myelin sheath has cytoplasm only in the inner and outer ends of the spiraled process. (Modified from Peters et al., 1976.)

A single Schwann cell may occupy up to about 1 mm of the length of a peripheral nervous system axon. Since some axons may be up to 1 meter or more in length, the myelin sheath consists of a large number of Schwann cells, each occupying its small portion of the axon. Between adjacent Schwann cells are gaps of several micrometers, known as *nodes of Ranvier* (Fig. 2–4). The Schwann cell–covered region between the nodes of Ranvier is known as the *internode*. To carry our analogy further, the myelinated axon can be compared with a large number of rolls of paper towels lined up end-to-end, with small gaps in the paper layer (but, of course, not in the cardboard tubes) representing the nodes of Ranvier.

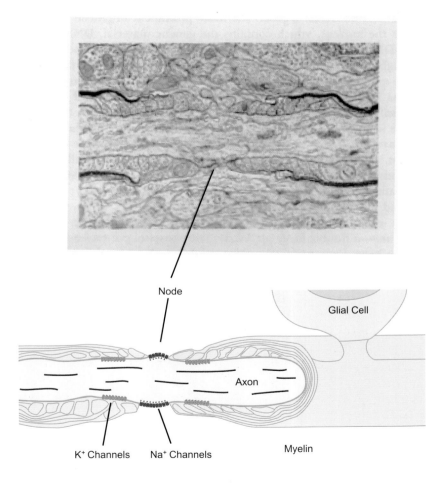

Figure 2–4. Electron micrograph and diagram of the node of Ranvier. At this region, the myelin sheath is interrupted for a short distance between adjacent glial cells, exposing the axonal plasma membrane to the extracellular space. (Micrograph courtesy of Dennis Landis.)

This multiple-membrane layer, which also happens to be unusually rich in lipid, insulates the axonal cytoplasm (the *axoplasm*) from the extracellular fluid. This means that electrical current can flow across the axonal plasma membrane only at the nodes, and, as we shall see in Part II, this has profound implications for the speed of transmission of electrical signals along the axon. For example, in those species, including humans, in which birth occurs before myelination is complete, the newborns are extremely limited in motor performance and, in fact, are quite helpless until myelination is achieved. The functional importance of myelin is also underscored by the severe impairments in motor function observed in the demyelinating diseases, such as multiple sclerosis, which are associated with extensive degeneration of the myelin sheath.

In addition to this critical role in the transmission of electrical signals, other interactions between Schwann cells and neurons have been documented. For example, neurons produce certain molecules, such as growth factors, that are necessary for Schwann cell proliferation. Schwann cells in turn produce molecules that influence the expression of neuronal proteins important for neuronal survival and differentiation. Thus the traditional picture of a purely mechanical interaction between Schwann cells and neurons must be revised to include reciprocal chemical influences as well.

The Nerve Cell Body: Neurons Are the Same as Other Cells

As Ramón y Cajal recognized, the nerve cells, or *neurons*, are the individual signaling elements of the brain. Although we will argue that the intercellular communication that is the hallmark of brain function makes the brain a unique organ, in many respects neurons (and glia) closely resemble other types of cells. Figure 2–5 is a diagram of the archetypal neuron; it is similar to that in Figure 1–3 but with additional organelles included. Note that this neuron consists of processes of different size and shape emanating from a cell body. It is this neuronal cell body, or *soma*, an enlarged drawing of which is shown in Figure 2–6, that most resembles cells in other organs. The most prominent organelle in the cell body is the *nucleus*, which contains the genetic material, DNA. The genomic DNA in neurons is identical to that in other cells in the organism (although in the so-called giant neurons in some invertebrates the genome divides many times without corresponding cell divisions, resulting in as many as 50,000 copies of the genome in the nucleus of some of these cells; the functional consequences of this are not understood). Even though the genome is no different from that in other cells, genes are regulated in specific ways that

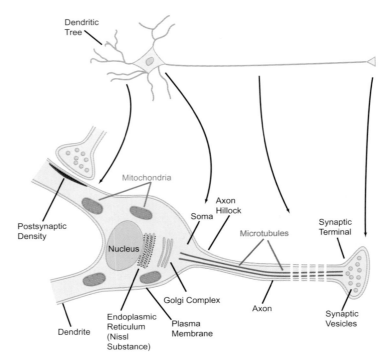

Figure 2–5. Ultrastructure of the neuron. Drawing of a typical nerve cell to show its overall shape and characteristic organelles. Structures and organelles that are common to all types of cells are shown (compare with Fig. 1–3).

result in the synthesis of a pattern of proteins specific to neurons. Of course, specific gene expression also occurs in all other tissues and accounts for the existence of specific cell types in liver, muscle, heart, and other organs, in addition to the many different types of neurons and glia found in the brain.

The entire neuron, like all other cells, is enclosed by a *plasma membrane*—a double layer (or *bilayer*) of phospholipid molecules, which acts as a barrier preventing the contents of the cell from mixing with those of the extracellular space. The plasma membrane is also an effective electrical insulator, hindering the diffusion of charged ions in and out of the cell. This is important, because signaling in nerve cells requires the controlled movement of ions across the plasma membrane, a process mediated by specialized proteins located in the membrane.

What other common organelles are found in neuronal cell bodies (Figs. 2–5 and 2–6)? The cell body (as well as the neuronal processes) contains *mitochondria* to supply the cell's energy needs. In fact, because a great deal of energy is required to maintain the transmembrane ionic

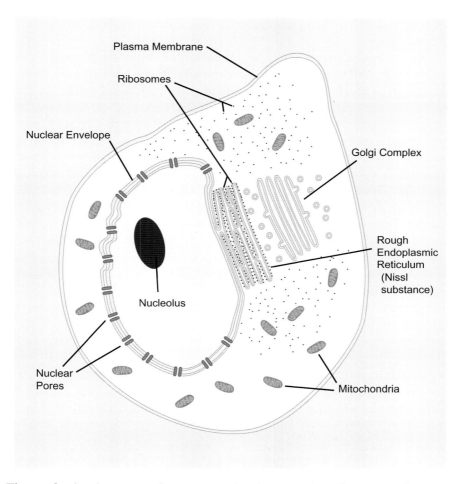

Figure 2–6. The soma of a neuron. The diagram identifies some of the key intracellular organelles.

gradients that are essential for neuronal signaling, neurons tend to be particularly rich in mitochondria. The cell body also contains *ribosomes*, which are responsible for the synthesis of proteins destined for insertion into membranes or for secretion; they are located on the membranous sacs of the *rough endoplasmic reticulum*, which is often unusually dense adjacent to the nucleus of neurons, giving rise to the structural feature called the *Nissl substance* (named after its discoverer). Other membranous components include the *smooth endoplasmic reticulum* and the *Golgi complex*, which are involved in the processing of proteins for membrane insertion or secretion (see Chapter 8), and lysosomes and other granules involved in the breakdown and disposal of cellular components.

We have been listing these organelles of the cell body at a rapid pace simply because the reader with a background in cell biology will already be familiar with their structures and functions. Although some features may be particularly characteristic of neurons, such as the presence of the Nissl substance or the density of mitochondria, these are relatively minor quantitative differences, rather than qualitative ones, between neurons and other cells. Only a trained observer would readily identify an electron micrograph as being an image of a neuronal cell body rather than of some other kind of cell.

Perhaps the most striking example of the conservation of structures and their functions among different cell types relates to molecular mechanisms of secretion. We will discuss this topic in detail in Chapter 8, but it is worth noting here that almost identical proteins and molecular mechanisms are involved in vesicle trafficking and secretion processes in organisms ranging from yeast to humans. Nothing emphasizes better the unity of cellular biological mechanisms, or that neuronal cell biology is fundamentally the same as that of other cell types.

Formation and Maintenance of Neuronal Form

When one examines a picture of a neuron with its remarkably asymmetrical structure (Figs. 1–3 and 2–5), an obvious question comes to mind concerning the mechanisms by which such neuronal polarity arises during development: How do structures such as axons and dendrites originate? In the rest of this chapter, we will introduce some of the important molecules and processes that contribute to the formation and maintenance of neuronal structure.

The cytoskeleton. The neuron, like all cells, contains a heterogeneous network of filamentous structures known collectively as the *cytoskeleton*. The major components of this network are the *microfilaments*, the *neurofilaments*, and the *microtubules*.

The function of microfilaments is best understood in skeletal muscle, where they are made up of the proteins actin and myosin. These filaments are present in highly ordered structures and interact to produce muscle contraction. Actin is also found in axons and, as we shall see in Chapter 16, is particularly prominent in the growing tips of axons, the *growth cones*, where it may contribute to the regulation of membrane movement. Actin is also thought to play a critical role in dynamic changes in dendritic spine morphology, of the sort illustrated in Figure 1–5. In many cell types, including neurons, actin accounts for an extremely high proportion of the total protein in the cell.

The neurofilaments are probably the least well understood of the cytoskeletal components. They are long filaments approximately 10 nm in diameter, intermediate in size between actin filaments (about 5 nm) and microtubules (about 20 nm). For this reason they fall into the general class of cytoskeletal components known as *intermediate* filaments in non-neural cells. Certain pathological conditions, including Alzheimer's disease, are associated with a profound disorganization of neurofilaments. Microtubules carry out a variety of functions in different cells. They play an important role in cell movement and are the major component of the *mitotic spindle*, an organelle that participates in cell division. Microtubules are also prominent inhabitants of axons and dendrites. Like the other filaments, they are polymeric structures, made up of large numbers of repeating units of two similar 50 kDa proteins known as α- and β-*tubulin* (Fig. 2–7). The polymerization of tubulin into microtubules depends on the nucleotide GTP and is promoted by *microtubule-associated proteins* (MAPs), which may also help anchor microtubules to membranes or to other cytoskeletal components (Fig. 2–7). A number of different MAPs exist, and these are often differentially associated with axons and dendrites. In addition, the amounts of the various MAPs change in characteristically different ways during neuronal development. Figure 2–8 shows a

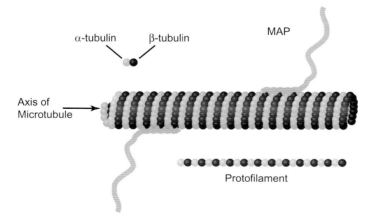

Figure 2–7. A microtubule. The rigid wall of the hollow microtubule consists of a helical array of protofilaments, each of which is a dimer formed from α- and β-tubulin. Microtubule-associated proteins (MAPs) bind to the microtubule.

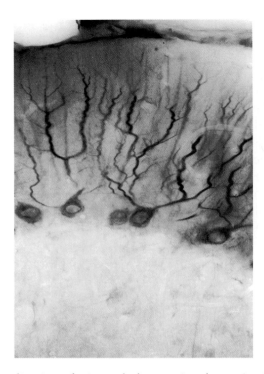

Figure 2–8. Localization of microtubule-associated proteins (MAPs) in axons and dendrites. In this experiment carried out by Andrew Matus and colleagues, a section through a rat cerebellum was stained with an antibody that specifically recognizes one particular microtubule-associated protein, MAP1. The dendrites of the large Purkinje cell neurons stain prominently with this antibody.

micrograph of a section through the rat cerebellum, stained with an antibody that recognizes one particular MAP, MAP1. The staining is particularly prominent in the dendrites of the large cerebellar Purkinje neurons. Earlier in development, however, MAP1 is restricted to the axons of these cells. Findings such as these suggest that the MAPs play a crucial role in the generation of structural differences between axons and dendrites.

Axonal transport. Another fundamental question that arises from an examination of neuronal structure is how it is maintained during the normal everyday life of the neuron. Since portions of the cell may be as far as 1 meter or more from the cell body, the neuron must have mechanisms for providing proteins and other necessary metabolic materials to its distal regions. Although it had long been thought that protein synthesis is restricted to the cell body, there is now compelling evidence for the existence of ribosomes that actively support protein synthesis in some dendrites. Furthermore, it is established that axons and their presynaptic terminals contain mRNAs encoding a large number of proteins, many of which are targeted to mitochondria. Nevertheless, other proteins required for normal membrane turnover, and enzymes necessary for metabolic functions such as neurotransmitter synthesis and degradation appear to be synthesized in the soma and must make their way down the axon to the presynaptic terminal. How do they get there?

It has long been known that the neuron has evolved a series of elegant transport systems known collectively as *axonal transport*. Of course, all cells are faced with the problem of moving cellular components from one part of the cell to another, but as pointed out earlier, this problem is particularly acute for neurons. Hence, their transport systems are highly specialized. Such systems are necessary because in the absence of an active process, a typical protein would require approximately 10 days to diffuse passively down a 1 cm axon from soma to terminal. ❽

Molecular motors. How is it that mitochondria and other large vesicular organelles are able to move long distances along microtubules to distal portions of the neuron? It is now understood that a specialized class of proteins called *molecular motors* drive axonal transport and other cell movements. The best understood of the molecular motors is *myosin*, which interacts with actin to form the basic contractile unit of muscle. In fact, it is now known that a large family of myosin molecules exists, the members of which participate in such fundamental processes as cell division, cell motility, and changes in cell shape during development and differentiation. Myosin and other molecular motors use the energy derived from ATP hydrolysis to drive movement, by mechanisms that have been studied for a long time and are becoming better understood.

It is not myosin, however, that drives axonal transport. That task falls to *kinesin*, another kind of motor protein. Studies in yeast, nematodes, and fruit flies have identified a large family of kinesin genes. The various kinesin proteins are approximately 120 kDa in molecular mass and share a common overall molecular architecture (Fig. 2–9a). The 350 or so amino acids at the amino terminal are highly conserved and constitute the motor domain, which contains the enzymatic machinery responsible for the hydrolysis of ATP (the *ATPase activity*). The remainder of the molecule is divergent in sequence (Fig. 2–9a) and is responsible for targeting the kinesin to a particular kind of organelle or region of the cell. Hence, this divergent tail determines the biological role of a particular kinesin, while the motor domain simply drives the required movement. This modular structure of the kinesins is emphasized by the use of recombinant DNA techniques to make fragments of kinesin molecules that contain only the motor domain. Such purified motor domain fragments alone can produce ATP-dependent movement of microtubules along nonbiological surfaces such as glass.

How does kinesin drive axonal transport of vesicles, mitochondria, and other organelles? Kinesin interacts with both microtubules and vesicles and uses the energy of hydrolysis of ATP to drive the movement of one relative to that of the other (Fig. 2–9b). Since in axons the microtubules are effectively fixed in place by other cytoskeletal components and cell membranes, the net effect of kinesin's action is to move vesicles in a proximal-to-distal (*orthograde*) direction along the axon. Organelles can also be transported in a distal-to-proximal direction—that is, toward the

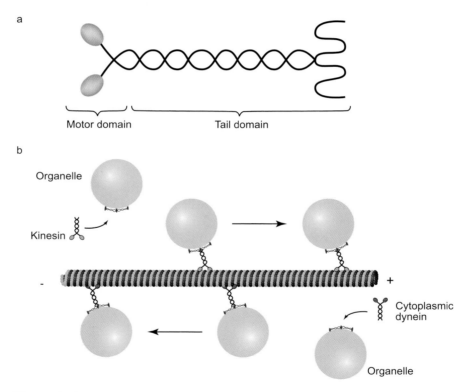

Figure 2–9. Vesicles and organelles are transported by an active process along microtubules. *a*: Molecular architecture of kinesin, a molecular motor. The functional unit is a kinesin dimer, consisting of a conserved motor domain and a divergent tail. *b*: The binding and movement of organelles such as vesicles along microtubules are mediated by kinesin and another molecular motor, cytoplasmic dynein.

cell body (*retrograde transport*). This is mediated, at least in part, by another molecular motor, a microtubule-associated protein (MAP1c), also called cytoplasmic *dynein*, that is distinct from kinesin. A single microtubule can serve as a track for transport in both the orthograde and retrograde directions, the direction of transport being determined by the nature of the motor that binds to the organelle (Fig. 2–9b).

Summary

The brain, like all other organs, is made up of vast numbers of cells. Unlike some other organs, however, the brain contains a wide variety of cell types. In addition to many different kinds of neurons, which are the main characters of our story, there are several classes of *glial cells*. The glia outnumber the neurons by a factor of 10 or more, but with the notable exception of myelination, their functions are only beginning to be understood. In addition to playing a critical role in the formation of the myelin sheath around axons, glia may be involved in immune responses, synaptic transmission, and long-distance calcium signaling in the brain.

Neurons share many features in common with other cells (including glia), but they are distinguished by their highly asymmetrical shapes. The neuronal *cytoskeleton* is essential for establishing this cell shape during development and for maintaining it in the adult. The neuron utilizes a process known as *axonal transport* for moving vesicles and other organelles to regions remote from the neuronal cell body. Proteins such as kinesin and dynein, called *molecular motors,* make use of the energy released by hydrolysis of ATP to drive axonal transport. Thus, the neuron has evolved unique mechanisms to establish and maintain the form required for its specialized signaling functions.

Electrical Properties of Neurons

We have emphasized in the introductory section that information transfer, within and between nerve cells, is an essential element of nervous system function. We have seen also that neurons are highly asymmetrical cells, with processes that often extend a considerable distance from the cell body. The basic question addressed in the next five chapters is how information is transferred intracellularly, from one part of the neuron to another. Chapter 3 deals with the basic phenomenology of electrical signaling. Neurons, like other cells, exhibit a voltage difference across their plasma membranes. Rapid changes in this transmembrane voltage, called *action potentials*, can be propagated from one part of the cell to another and are used by neurons (but not by most other cells) to encode information. Chapter 4 introduces *membrane ion channels*, a ubiquitous class of highly specialized membrane proteins that have evolved to provide exquisite control over the movement of ions across the plasma membrane. We discuss the biophysical and molecular properties of ion channels in considerable detail in this and the next three chapters, because an appreciation of how channels work is essential for understanding electrical signaling. Chapter 5 covers the molecular structure of ion channels and the methods that have been used to investigate these proteins. Molecular cloning and X-ray crystallographic methods now enable us to identify the structural features of an ion channel protein that are responsible for a particular aspect of its function. The goal of such approaches is ultimately to understand electrical signaling in terms of its underlying molecular mechanisms. Chapter 6 examines the ways in which the combined activities of different ion channels give rise to action potentials. Neurons differ from most other cells in their particular ensemble of membrane ion channels, which allow the generation and propagation of action potentials in complex temporal patterns. Finally, Chapter 7 describes the diversity of voltage-gated ion channels, with particular focus on calcium and potassium channels. This diversity of channel types allows the electrical behavior of any neuron to be adapted very precisely to its function in the brain.

3

Electrical Signaling
in Neurons

Although neurons have many features in common with other cells, they are unique because their primary functions are to receive, modify, and transmit messages. This includes information transfer from one cell to the next, as well as between different parts of the same cell. In this and the following four chapters, we will focus on the nature and mechanisms of the electrical signals that neurons use both for intracellular communication and as stimuli for the generation of intercellular messages.

Intracellular Transfer of Information: The Axon

In Chapters 1 and 2 we showed that neurons are highly asymmetrical cells and that different parts of the neuron exhibit structural features that enable them to carry out specialized tasks. We will concentrate first on the *axon*, the part of the neuron responsible for transmitting information from one part of the cell to another.

Axons are thin tube-like structures that arise from the neuronal cell body. They vary widely in size, shape, and other characteristics (see Fig. 1–3). Some axons within the central nervous system are only a few micrometers in length, not much greater than the diameter of the neuronal cell bodies from which they arise. In contrast, axons that run from the central nervous system to other parts of the body can be as long as 1 meter in humans and even longer in larger animals. It is immediately apparent that whatever the mechanism of axonal information transfer, it must be able to operate over long distances without garbling or losing messages. Axon diameter also varies, from less than 1 μm to almost 1 mm, and the diameter of any single axon may be different at different distances from the cell body. Axon diameter is an important factor in determining the speed at which information moves along the axon. Moreover, we shall see that whether or not an axon is myelinated also influences speed of transmission.

We begin here with a descriptive treatment of how information is passed along an axon, and then discuss the diverse patterns of electrical activity exhibited by different kinds of neurons. In subsequent chapters we will describe the mechanisms of these phenomena in detail. To summarize the descriptive message: There is a voltage difference across the axonal membrane, and information is carried in the axon in the form of rapid changes in this voltage difference. These voltage changes, which are generally referred to as *nerve impulses, spikes,* or (most commonly) *action potentials,* travel rapidly along the axon from the cell body toward the distal portion of the axon.

Ion Channels Underlie Electrical Signaling in Neurons

Neurons, like all other cells, exhibit a voltage difference known as the *membrane potential* across their plasma membranes. It will become evident in the next chapter that the membrane potential results from the unequal distribution of electrical charge, carried by *ions,* on the two sides of the membrane. Some ions can move across the plasma membrane

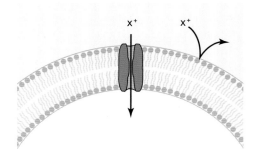

Figure 3–1. Our concept of an ion channel, drawn as a hydrophilic pore (blue) that spans the hydrophobic plasma membrane (yellow). Ions (X^+) can cross the membrane only through the aqueous pore provided by the ion channel.

more readily than others, and the movement of ions may be different under different circumstances (for example, when at rest as compared to during the action potential). Such differential permeability arises because of the presence in the plasma membrane of specialized proteins known as *ion channels*. These proteins, which will be described in detail in Chapter 4 and will reappear throughout this book, form aqueous pores across the membrane through which specific ions can flow (Fig. 3–1). It is the flow of ions through ion channels that is responsible for electrical signaling in neurons.

Resting Potential and the Passive Membrane Response

When the axon is at rest, that is, when it is not conducting nerve impulses, the value of the membrane potential is called the *resting potential*. In neurons the resting potential is usually in the range –40 to –90 mV. By convention, membrane potentials are expressed relative to the extracellular fluid—that is, negative membrane potentials indicate that the inside of the cell membrane is more negative than the outside. When the membrane potential is less negative than the resting potential, the cell is said to be *depolarized*; when it is more negative, the cell is *hyperpolarized*.

It is possible to measure the membrane potential (V_m) by inserting a measuring electrode, connected to an electrometer, into the cell (Fig. 3–2). The electrode can be either a silver wire or a fine-tipped glass pipette filled with a conducting salt solution. When the tip of the measuring electrode (M) is in the extracellular fluid, there is no voltage difference between it and the reference electrode (R), which is also in the extracellular fluid (left side of Fig. 3–2). When the measuring electrode tip is passed through the plasma membrane, there is a sudden negative voltage deflection of some 40–90 mV relative to the extracellular reference electrode (right side and bottom of Fig. 3–2), reflecting the negative resting potential (V_r).

With appropriate (and very simple) electronics, one can also inject negative (hyperpolarizing) or positive (depolarizing) current into the cell via the same electrode. With negative currents, the membrane potential changes in a hyperpolarizing direction, and the size of the change simply mirrors the amount of applied current stimulus (Fig. 3–3). Such voltage shifts in response to hyperpolarizing current injection reflect *passive membrane properties*. Similar passive responses are seen in response to

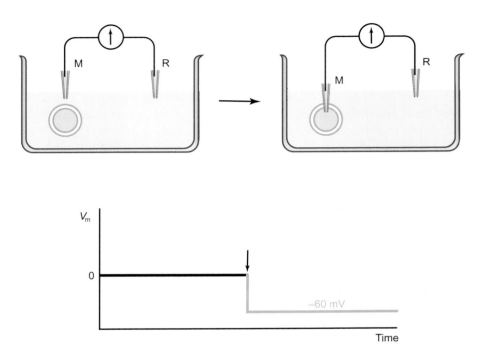

Figure 3–2. Measurement of the resting potential. The voltage difference (V_m) across the cell membrane can be determined as the voltage difference between a measuring electrode (M) inside the cell and a reference electrode (R) in the extracellular fluid.

small depolarizing stimuli (Fig. 3–3). This injection of current through use of electronic instrumentation is a useful experimental manipulation because it mimics what happens in the nervous system. We shall see that in the real world, neurons are constantly bombarded with physiologically relevant current "injections" that result from synaptic activity or sensory input.

The Plasma Membrane Is a Capacitor and a Resistor Connected in Parallel

The plasma membrane of a nerve cell or, indeed, of any cell provides a resistance to the flow of ions between the intracellular and extracellular compartments. Accordingly, it can be thought of as an electrical *resistor*, with the membrane resistance, R_m, being measured in ohms (Ω). In addition, the lipid bilayer provides an extremely thin insulating layer between two conducting solutions. This allows the membrane to act as an

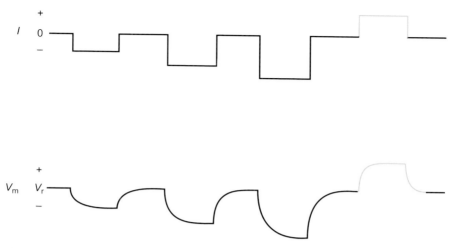

Figure 3–3. Passive response of the membrane. Current (I) can be injected into the cell via an electrode, and the resulting change in membrane potential (V_m) from that at rest (V_r) can be measured with the same or another electrode.

electrical *capacitor*, a device that is capable of separating and storing electrical charge. The membrane capacitance, C_m, is measured in farads (F). These considerations allow us to describe the electrical properties of the lipid bilayer membrane simply in terms of an *equivalent electrical circuit*, as shown in Figure 3–4a. This description is introduced not to torment the student of cell and molecular biology, but rather because it is extremely useful in understanding the electrical behavior of biological membranes under a variety of physiological conditions.

What does all this mean for changes in the voltage difference across the neuronal plasma membrane? Figure 3–3 illustrates that the time course of the change in membrane voltage does not mirror precisely that of the injected current. In fact, the rate of change of the voltage is a function of *both* the injected current (I) and the membrane capacitance (C_m) and is given by

$$\frac{dV}{dt} = \frac{I}{C_m}$$

This is because the membrane capacitance must be charged (or discharged) before the voltage can change, and this cannot occur instantaneously; rather, it takes some time determined by the membrane time constant, τ, which is equal to the product of the membrane resistance and capacitance:

$$\tau = R_m C_m$$

a

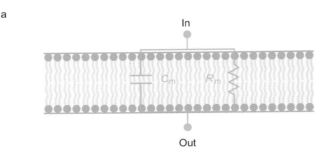

b

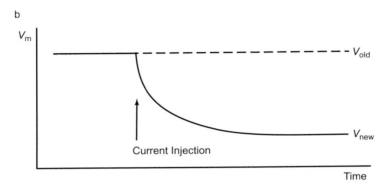

Figure 3–4. Passive electrical properties of the plasma membrane. *a:* The plasma membrane can be depicted as a resistor (R_m) and a capacitor (C_m) connected in parallel. *b:* Voltage changes do not occur instantaneously. V_m, membrane potential.

As can be seen in Figure 3–4b, the voltage changes exponentially with time (*t*), according to the equation:

$$V_t = V_{new} - \left(V_{new} - V_{old}\right)e^{-t/\tau}$$

In other words, the voltage falls to 1/*e* of its initial value in a time equal to one time constant. Time constants in biological membranes vary over a wide range, even though C_m per unit of membrane surface area is remarkably constant at about 1 μF/cm² in all membranes examined. Accordingly, the dissimilar time constants must reflect large differences in R_m from one neuron to another, and even between separate membrane regions of the same neuron. It will become evident in subsequent chapters that such differences in R_m reflect differences in the type, density, and regulation of membrane ion channels. We shall see also that different neurons exhibit distinct and often highly complex patterns of endogenous electrical activity in the absence of external stimulation. The membrane time constant, determined by the complement of ion channels, plays an important role in determining just what a neuron's endogenous activity is, and how the cell reacts to external stimuli.

The Action Potential

All characteristics of the passive membrane response described earlier apply to hyperpolarizing stimuli of any size and to small depolarizing stimuli (Fig. 3–3). The situation is very different, however, with larger depolarizing stimuli. As the strength of the depolarizing stimulus is increased, a critical stimulus strength, or *threshold*, is reached, below which only a passive response is seen and above which the response looks very different (Fig. 3–5). The actual level of the threshold will vary from neuron to neuron, and at different times within a single neuron (again depending on the complement and regulation of ion channels), but it tends to be in the range 10–20 mV depolarized from V_r. Beyond the threshold, one observes a large change in V_m several milliseconds in duration, superimposed on the passive response (Fig. 3–5). The membrane potential depolarizes very rapidly, and then there is a slightly less rapid return to the resting level. Note that V_m does not go to 0 but actually

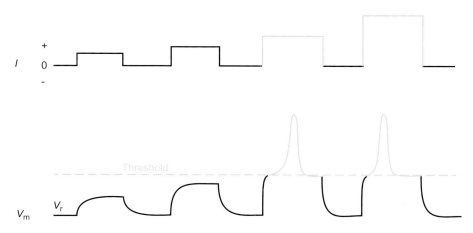

Figure 3–5. Active responses to large depolarizing stimuli. Although small depolarizing responses produce a passive membrane response, as in Figure 3–3, the voltage response to larger depolarizing currents is very different. *I*, current; V_m, membrane potential; V_r, resting potential.

becomes some 50 mV positive—that is, the inside of the neuronal membrane is briefly positive relative to the extracellular side.

This active response of the membrane when the depolarization exceeds threshold is the nerve impulse, or *action potential*. It is this signal that is responsible for the transfer of information from one part of a neuron to another. The threshold is essential to ensure that small, random depolarizations of the membrane do not generate action potentials. Only stimuli of sufficient importance (reflected by their larger amplitude) result in information transfer via action potentials in the axon. Another important property of action potentials is that they are all-or-none events; this *all-or-none law*, as it is called, is an essential feature of axonal signal transmission. The all-or-none law is illustrated in the right side of Figure 3–5, which demonstrates that any stimulus large enough to produce an action potential produces the same size action potential, regardless of stimulus strength. In other words, once the stimulus is above threshold, the amplitude of the response no longer reflects the amplitude of the stimulus. This is very important; it means that information about stimulus strength must be represented—*encoded*—in the axon in some way other than action potential amplitude.

Although the amplitude of the action potential is generally independent of stimulus intensity, many of its other properties are not. In particular, the *latency*—the time delay from the onset of the stimulus to the peak of the action potential—is a function of stimulus strength. As shown by a careful examination of the two action potentials in Figure 3–5, the stronger the stimulus, the shorter the delay between stimulus and action potential. We shall see that this *strength–latency relationship*, together with another phenomenon known as the *refractory period*, allows the encoding of stimulus strength in terms of the *frequency* of action potentials in the axon.

For several milliseconds after the firing of an action potential, it is impossible to evoke another action potential, no matter how large the depolarizing stimulus; in other words, the axon is *refractory* to stimuli during this time. This *absolute* refractory period is followed by a *relative* refractory period, during which the stimulus must be larger than normal to evoke an action potential. One useful way of thinking about the refractory period is in terms of the threshold (Fig. 3–6). During the absolute refractory period the threshold is essentially infinite, and no stimulus, no matter how large, can exceed it. During the relative refractory period the threshold is larger than normal—that is, it requires a larger than normal stimulus to exceed it. The threshold returns to the normal level with a time course shown in Figure 3–6. Because only an above-threshold stimulus will evoke an action potential, this curve describes the stimulus

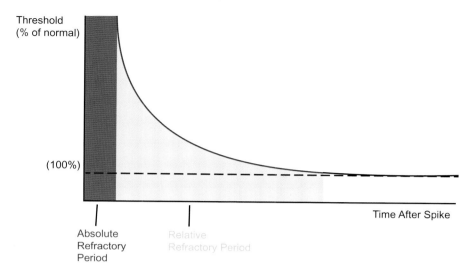

Figure 3–6. The threshold is not fixed. For a short period of time after the firing of an action potential, the threshold is much greater than normal.

strength required to generate a second action potential, as a function of time after the first action potential. We shall see in subsequent chapters that the mechanism of the refractory period can be understood in terms of the properties of the membrane ion channels that are responsible for the generation of the action potential. Let us now examine the way the refractory period contributes to neuronal information coding.

Frequency Coding

Consider the response of the axon to a sustained stimulus in light of these concepts. If the stimulus depolarizes the axon above the normal resting threshold, an action potential results. However, even if the depolarizing stimulus is maintained, a second action potential will be evoked only after the threshold has dropped back below the level of the sustained stimulus (Fig. 3–7a). This will take some time, as described in Figure 3–6. The same will be true for all subsequent action potentials during the stimulus. Thus the axon will fire action potentials as long as the above-threshold stimulus is maintained, but they will be spaced apart in time. Now consider the response of the same axon to a larger depolarizing stimulus (Fig. 3–7b). Again, the second action potential will fire only after the threshold drops back below the stimulus level, but this happens

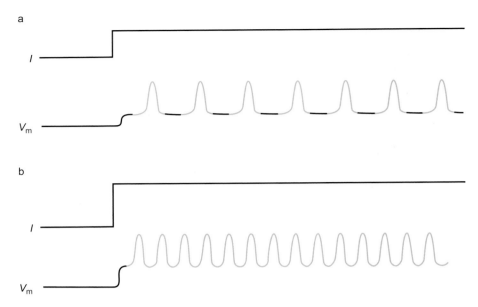

Figure 3–7. Frequency coding in axons. *a*: An above-threshold sustained depolarizing stimulus (*I*) produces action potentials at a certain frequency. *b*: When the depolarizing stimulus is larger, the frequency of action potential firing is greater. V_m, membrane potential.

more quickly because the stimulus is larger (and consequently the threshold does not have to drop quite as far). Accordingly, the second and subsequent action potentials occur with less delay. It can be seen from a comparison of Figures 3–7a and 3–7b that the larger stimulus is reflected in a higher frequency of action potentials in the axon. Thus, even though action potential amplitude obeys the all-or-none law and does not reflect stimulus intensity, the phenomena of threshold, latency, and refractory period do indeed allow the encoding of stimulus intensity as a *frequency code* in the axon.

Passive Spread and Action Potential Propagation

Everything we have discussed thus far refers to local changes in the membrane potential at a single point in the axon. However, we have also emphasized that the axon is specialized to move information from one part of the neuron to another. Thus it is time to ask how nerve impulses spread along the axon from the point of a stimulus.

Although it may seem to be a contradiction in terms, a phenomenon known as *passive spread* plays an essential role in the propagation of the active response. Let us look first at passive spread in terms of hyperpolarization of the membrane potential. Suppose an axon is penetrated by several microelectrodes some distance apart, and a hyperpolarizing current is injected through one. As can be seen in Figure 3–8a, a voltage change is observed at all the electrodes, but it is largest at the stimulating electrode and decreases in amplitude with distance away from this electrode. When the amplitude of the voltage response is plotted as a function of distance from the stimulating electrode, it can be seen to fall exponentially with distance (Fig. 3–8b). In other words, the voltage change does

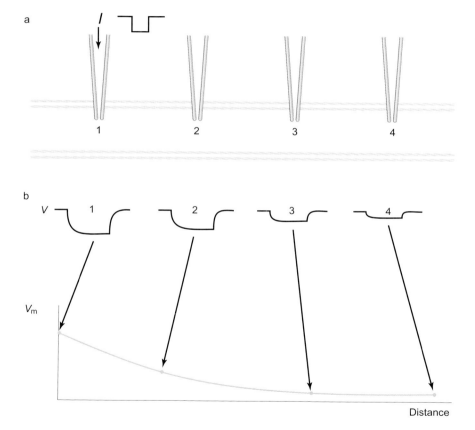

Figure 3–8. Passive spread. *a*: When hyperpolarizing current (*I*) is injected at one point (1) in the axon, a voltage change (*V*) can be measured at that point. Electrodes in other parts of the axon some distance away (2, 3, 4) measure smaller voltage changes. *b*: Plot of the voltage change (V_m) as a function of the distance from the stimulating electrode.

spread from one point to another but is attenuated with distance until it eventually becomes so small that it is essentially undetectable. This phenomenon is known as *passive spread*, because it can be seen in a dead axon or even in an electric cable with similar properties.

The extent of attenuation of the voltage change is determined by the membrane *space constant*, λ, defined as the distance at which a voltage change has fallen to 1/*e* of its initial value. The voltage V_d at some distance *d* can be described in terms of V_0, the voltage at distance 0, and the space constant:

$$V_d = V_0 e^{-d/\lambda}$$

The space constant can vary markedly from axon to axon, depending in particular on axon diameter and the molecular characteristics of the axon membrane—again, most notably its complement of ion channels.

These considerations for hyperpolarizations also serve to describe well the passive spread of a small (subthreshold) depolarizing voltage

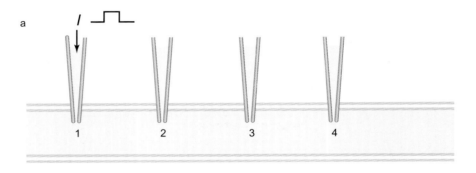

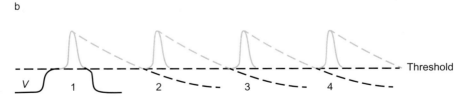

Figure 3–9. Propagation of the action potential. *a*: Electrodes are inserted at different points (1, 2, 3, 4) along the axon to measure the membrane voltage, and an action potential is elicited by the injection of depolarizing current (*I*) through the first electrode. *b*: Although the large voltage change (*V*) due to an action potential at one point decreases with distance (dashed lines), the depolarization that spreads to the adjacent region of the axon is still above threshold. Thus a full-size action potential is generated at each point in the axon.

change. When the local depolarization exceeds threshold, however, the picture changes dramatically. The above-threshold depolarization, of course, evokes a very large voltage change, the action potential. There is decrement of this large voltage change as it spreads along the axon, just as there is for smaller depolarizations or hyperpolarizations. However, because the local depolarization is so large, the passive spread is still sufficient (in spite of the attenuation with distance) to depolarize neighboring regions of the axon above threshold, and a full-size action potential is generated at a point adjacent to the original one (Fig. 3–9). This is repeated for each small region of axon until the action potential has swept over its entire length. An often-used and highly appropriate analogy is the lighting of a firecracker fuse: Ignition of one point on the fuse brings the neighboring segment above its ignition temperature, and this process continues until the fuse has burned down to its end.

To summarize this descriptive treatment of axonal information transfer, several important characteristics of axonal membranes enable action potentials to carry information faithfully from one part of the neuron to another:

1. There is a *threshold* for generation of action potentials that guarantees that small, random variations in the membrane potential are not misinterpreted as meaningful information.
2. The *all-or-none law* guarantees that once an action potential is generated, it is always full size, minimizing the possibility that information will be lost along the way.
3. The *strength–latency relationship* and the *refractory period*, together with the threshold, allow the encoding of information in the form of a *frequency code.*
4. The phenomenon of *passive spread*, which arises simply from the cable-like properties of the axonal membrane, allows the propagation of action potentials along the axon and the transfer of information over long distances within the neuron.

Action Potentials Jump Along Myelinated Axons

We mentioned in Chapter 2 that the speed of action potential propagation along the axon is determined in part by myelination. This comes about because the myelin sheath, which consists of a large number of layers of glial plasma membrane wrapped about the axon, acts as an excellent electrical insulator. The space between the axon and the myelin is not an ion-containing extracellular solution, and no current can flow across the axonal membrane in the regions of myelination. Remember, however, that the myelin sheath is interrupted periodically at the nodes of Ranvier (Fig. 2–4), and at these nodes the ionic conduction pathways that we have been discussing are indeed present. Thus action potentials can be generated only at the nodes. The passive spread of the action potential depolarization along the myelinated portion can bring the adjacent node above threshold, allowing the action potential to "jump" along the axon from node to node. The nodes in the myelinated axon are spaced some 1–2 mm apart, so that the depolarization produced at one node is still well above threshold by the time it reaches the next node. To put this another way, the space constant of the axonal membrane and the spacing between nodes are coordinated to ensure that the action potential is propagated. The high resistance of myelin forces the current to move down the axon rather than leak out across the axonal membrane, and thus the myelin sheath itself contributes to the space constant.

We shall see later that the *voltage-dependent sodium channel* is a particular kind of ion channel that plays a fundamental role in action potential generation and propagation in axons. The sodium channels are not spread evenly throughout the axonal plasma membrane but are packed together at a very high density at the nodes and are sparse in the intervening membrane under the myelin (Fig. 3–10a and b). In contrast, voltage-dependent potassium channels are present in high density in the juxtaparanodal region adjacent to the nodes and help to limit the excitability of the axonal membrane in the Schwann cell–covered internodal region (Fig. 3–10). Together these factors allow the action potential to jump from node to node, achieving conduction with the minimum use of ion channels or energy-consuming pumps.

This *saltatory conduction* (from the Latin *saltare*, meaning to "leap" or "dance") permits conduction at speeds many times faster than in non-myelinated axons of the same diameter. Certain diseases of the nervous system, the best known of which is multiple sclerosis, are characterized by loss of myelin from some myelinated axons. Associated with the loss of myelin is a redistribution of the axonal sodium and potassium channels so that they are spread more evenly throughout the axon and no longer contribute to focusing conduction at the nodes. These demyelinating diseases can have severe consequences, because they result in the slowing (and sometimes the complete blockage) of axonal conduction, with devastating effects on the neuronal pathways in which the demyelinated axons participate.

a

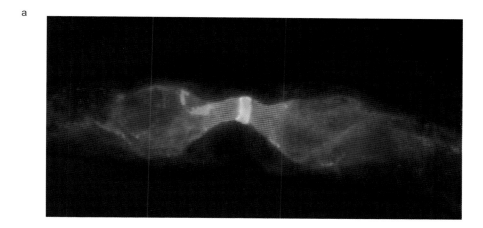

b

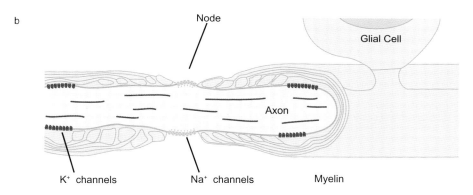

Node

Glial Cell

Axon

K⁺ channels   Na⁺ channels   Myelin

Figure 3–10. Ion channels are differentially distributed in myelinated axons. *a*: Voltage-dependent sodium channels (green) are concentrated in the node of Ranvier, while voltage-dependent potassium channels (blue) are in the juxtaparanodal region. (Micrograph from experiments by Matthew Rasband, James Trimmer and colleagues; see Ogawa et al., 2010). *b*: Drawing of a myelinated axon similar to Figure 2–4 to better illustrate ion channel distribution in and near the node of Ranvier.

Having discussed the axonal characteristics that are essential for the generation and propagation of action potentials, we will see now that neurons generally do not fire only single action potentials, or trains of action potentials at constant frequency. Rather, they may exhibit complex temporal patterns of firing that are appropriate for the particular tasks the neurons must carry out.

Different Patterns of Neuronal Electrical Activity

Cells in different regions of the nervous system are remarkably diverse in their morphology and in their electrical and biochemical properties. In the first part of this chapter we gave a description of action potential generation and propagation in a typical axon. However, even so funda-mental a phenomenon as the action potential can vary in shape and size in different neurons (Fig. 3–11). In addition, the pattern of action poten-tial firing exhibits great diversity in different neurons (Figs. 3–12 and 3–13). Again, this diversity reflects differences in membrane ion channels. Diversity in electrical and other properties should come as no surprise

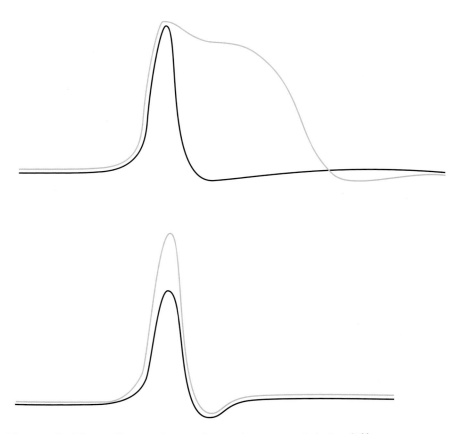

Figure 3–11. Different shapes for action potentials in different neurons. When the electrical activity of different neurons is recorded, action potentials of different amplitudes and durations are seen.

when one considers the wide variety of different behaviors and physiological functions that neurons have to control. For example, the neuronal pathways and individual neurons that control fast visual reflexes or rapid escape behaviors are very different from those that control slow behaviors, including breathing, feeding, and reproduction. In this section, we describe some of the different types of firing patterns that are encountered in nerve cells. In addition, we provide some examples of how a change in its electrical properties allows a neuron to regulate different types of behavior.

It is important to note that the relatively simple picture we have painted of axonal information transfer becomes more complicated when one moves to the neuronal cell body or dendrites. Dendrites, long thought of as passive elements that do little more than receive information from other neurons, are now known to be capable of generating action potentials and participating in complex ways in neuronal signaling. Furthermore, some cell bodies do not fire action potentials at all—they are said to be *electrically inexcitable*—and they carry out electrical signaling in more subtle ways. Even in those cell bodies that do fire action potentials, the more subtle mechanisms may still be present, leading to far more complex patterns of electrical activity than are usually observed in axons.

Silent, Beating, and Bursting Neurons

Although the generation of an action potential is fundamental to a neuron's ability to transmit information, there are many other aspects of its electrical properties that play important roles in shaping neuronal input and output. Some neurons have a steady, unchanging resting potential in the absence of external stimulation—that is, they are *silent* (Fig. 3–12a). Other neurons, however, generate a variety of endogenous electrical patterns. For example, some cells fire repetitively at constant frequency—that is, they *beat* (Fig. 3–12b). Although external stimulation can change the firing rate of the cell or inhibit it altogether, the mechanisms that drive repetitive firing are often intrinsic to the neuron itself and do not require continual synaptic activation or other external stimuli.

Some neurons that fire spontaneously in the absence of external stimulation do not fire at fixed regular intervals but instead generate regular bursts of action potentials that are separated by hyperpolarizations of the

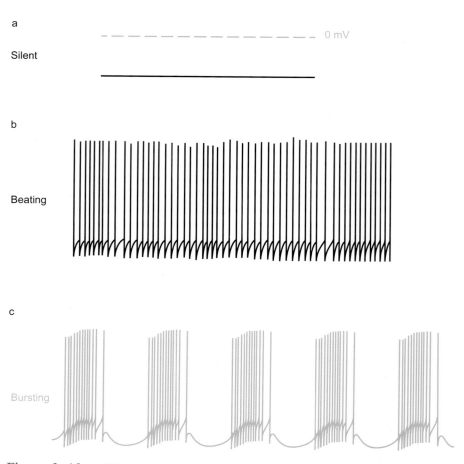

Figure 3–12. Different patterns of endogenous electrical activity. Some neurons do not fire spontaneously at all (*a*). Others may beat (*b*) or burst (*c*) in a regular manner.

membrane, as shown in Figure 3–12c. Such cells are termed *bursting neurons*. This ability of a neuron to burst repetitively is used by the nervous system in at least two different ways:

1. Bursting neurons generate rhythmic behaviors. Many fundamental behaviors, such as breathing, walking, swimming, and the chewing of food, require the continual rhythmic stimulation of a group of muscles. Numerous examples of electrical bursting can be found in neuronal circuits that generate such rhythmic motor outputs. Although in many cases the exact form and timing of the bursts, and even their generation, may be regulated by interactions among several different neurons (see Chapter 18), the ability to

generate bursts can also be intrinsic to specific neurons that continue to burst in the absence of external inputs.

2. Bursting neurons are used to secrete neurohormones. Neurons, in addition to acting directly on other neurons or on muscle cells, may secrete hormones into the circulation. Figure 3–12c shows the bursting activity of a nerve cell that secretes peptide hormones in the marine mollusc *Aplysia* (a sea hare). This neuron, like many other molluscan neurons, is large and readily identifiable on the basis of morphological, biochemical, and electrical criteria. Accordingly, these cells can be given names; this one, called *neuron R15*, has been studied extensively as a model bursting neuron. Another example of such bursting activity is found in neurons that are located in the hypothalamic region of the mammalian brain and have been termed *magnocellular neurons* (see Chapter 10). Individual magnocellular neurons contain either vasopressin or oxytocin, peptide hormones that are used in the control of water retention and lactation, respectively. For reasons that are not yet fully understood, it appears that a bursting pattern of electrical activity, such as that in the magnocellular neurons and in *Aplysia* neuron R15, is more effective than a steady pacing pattern of firing as a stimulus to the intracellular machinery that causes peptide release (see Chapter 8).

The Response to Sustained Stimulation of a Neuron

Thus far we have been discussing patterns of neuronal activity that are intrinsic to the neuron under study. However, under physiological conditions, neurons are often subjected to external stimuli, for example, a continual barrage of synaptic stimuli from other neurons. Experimentally, such continual stimulation may be mimicked by a sustained depolarization or hyperpolarization from an intracellular microelectrode. Three different ways that a neuron may respond to a depolarizing stimulus are illustrated in Figure 3–13. The cell may generate action potentials repeatedly throughout the period of stimulation, as described in the first part of this chapter, with a constant frequency that reflects the strength of the stimulus (Fig. 3–13a; see also Fig. 3–7). Alternatively, a neuron may fire only one or a few action potentials at the onset of the stimulus and remain silent thereafter (Fig. 3–13b). This response is sometimes termed *accommodation* to the stimulus. Finally, a neuron may not fire at the onset of stimulation but may generate action potentials only after a delay (Fig. 3–13c). In this case, short periods of stimulation fail to trigger any action potentials in the cell.

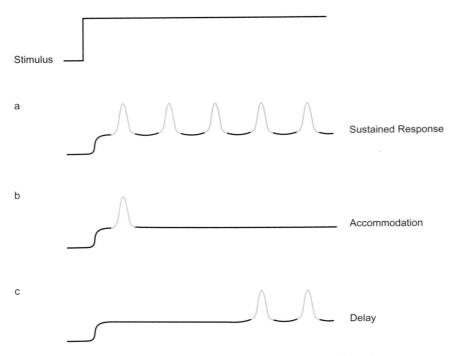

Figure 3–13. Different responses to a sustained stimulus. Some neurons respond to a sustained depolarizing stimulus with a sustained response (*a*), as shown also in Figure 3–7. Other neurons may fire one or a few action potentials and then stop responding (*b*), or may fire action potentials only after a delay (*c*).

The first two modes of response may be found in a variety of cells, for example, those that relay sensory information from the environment to the central nervous system. Accommodation in such sensory neurons would result in *behavioral habituation*, the commonly observed decrement in response during a sustained sensory stimulus (see Chapter 19). The third mode, in contrast, would be expected in cells that respond only to excess stimulation. A clear example of this is provided in motor neurons of the ink gland in *Aplysia*. Inking in *Aplysia* is a defensive response to a strong noxious stimulus, such as a mechanical stimulus that punctures the skin. It is believed that the ink that is extruded makes the surrounding water murky and provides camouflage for the animal, allowing it to hide from the predator generating the stimulus. The neurons that control the ink gland are normally not active, and they do not respond to small or transient stimuli; they begin to fire only when they receive the sustained synaptic input that is generated by a large and prolonged noxious stimulus. The mechanism of this delay can be understood in terms of the properties of the particular ion channels in the membrane of these neurons.

Neuromodulation—Changes in Neuronal Electrical Properties

There may also be long-term modulation of neuronal properties in response to more subtle external stimuli than the sustained excitation just described. Few behaviors that are controlled by the nervous system remain fixed throughout the life of an animal. For example, feeding and reproductive behaviors have to be turned on and off at appropriate times. A defensive or escape response to a tactile stimulus may be appropriate at one time and not at another. Even the characteristics of essential physiological functions such as breathing may be altered in response to external stimuli. To a large extent, such changes in the behavior of an animal occur because of changes in the electrical properties of neurons that control those behaviors, a phenomenon known as *neuromodulation*. Synaptic or hormonal stimulation may produce either short- or long-term changes in the shape of action potentials, in the endogenous pattern of firing of a neuron, or in the way the cell responds to other external stimuli (Fig. 3–14). Such modulation of neuronal electrical properties, mediated by transduction mechanisms, some of which are described in Chapters 11 and 12, not only allow the nervous system to adapt its output in the face of a continually changing environment but also are the basis for many

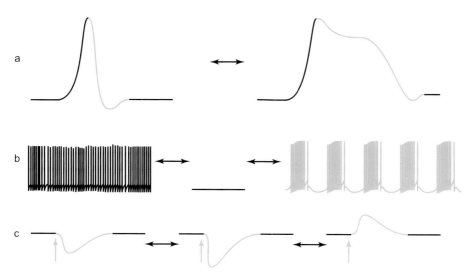

Figure 3–14. Modulation of neuronal excitability. Common modulatory changes in neuronal membrane properties include (*a*) changes in the amplitude and/or duration of action potentials, (*b*) changes in endogenous firing patterns between beating (left), silent (middle), and bursting (right) modes, and (*c*) changes in the efficacy of synaptic inputs.

long-lasting changes in behavior. Because neuromodulation underlies the choice of different patterns of behavior at different times, it is of critical importance for the proper functioning of the nervous system. We shall see that the direct phosphorylation of the ion channel proteins that underlie all neuronal excitability is a molecular mechanism mediating many, although by no means all, neuromodulatory events.

Summary

The language of intracellular signaling in nerve cells is electrical. There is a voltage difference known as the *membrane potential* across the plasma membrane of all cells. In neurons, information is carried from one part of the cell to another in the form of *action potentials*, large and rapidly reversible fluctuations in the membrane potential, that propagate along the axon. Since action potentials are all-or-none events, their amplitude carries little information about the stimulus that triggered them; instead, several fundamental membrane properties associated with the generation and propagation of action potentials allow information about stimulus strength to be encoded in the frequency of action potential firing.

Different neurons exhibit different patterns of action potential firing. Some neurons are normally silent. That is, their membrane potential remains at the resting potential unless the firing of action potentials is triggered by some external stimulus, and they return to their non-firing state when the stimulus is no longer present. However, many neurons exhibit more complex endogenous electrical activity, often firing action potentials in a regular pattern without an external stimulus. In some cases it is possible to interpret the pattern of endogenous activity in terms of the particular function that the neuron is assigned in the nervous system.

Finally, the electrical properties of a neuron are not fixed but are subject to modulation by input from the environment. This includes sensory information from the outside world, hormones released from other parts of the organism, and chemical and electrical signals from other neurons to which the neuron is functionally connected. Such modulation of neuronal properties is of fundamental significance, because it allows the animal to respond and adapt its behavior in a continually changing environment.

4

Membrane Ion Channels and Ion Currents

The electrical activity of nerve cells—indeed of all cells—depends on the movement of charge, carried by small inorganic ions, across the plasma membrane. The phenomena described in Chapter 3—the membrane potential, the firing of action potentials, and the grouping of action potentials in complex temporal patterns—all arise from such transmembrane ion flow. In addition, modulation of the endogenous electrical activity by external stimuli involves changes in transmembrane ion flow. But how is it that ions can move across the plasma membrane at all? The lipid bilayer of the plasma membrane is an excellent electrical insulator and is largely impermeable to charged species (see Fig. 3–1). It requires an enormous amount of energy to move an ion through the hydrophobic interior of the bilayer, and accordingly the cell must make special provision to allow transmembrane *ion current* to flow.

One way for ions to cross the plasma membrane is via energy-driven pumps or transporters, which use the energy from ATP to overcome the energy barrier imposed by the plasma membrane. Such pumps or transporters are proteins that pick up an ion on one side of the membrane, physically transport it across the bilayer, and release it on the other side. Because energy is expended in this process in the form of ATP hydrolysis, it is possible for such active transport processes to move ions against a concentration gradient.

Pumps and transporters are essential for many cell functions, including the establishment and maintenance of concentration gradients of various inorganic ions (most notably sodium, potassium, and calcium ions) across the plasma membrane. Some of them are also *electrogenic*—that is, their activity results in a *net* flow of ions across the membrane; hence they can influence the membrane potential. Nevertheless, pumps and transporters play only a supporting role in electrical signaling in most nerve cells. The stars of this show are the *ion channels*, a ubiquitous class of specialized membrane proteins that span the plasma membrane. These form hydrophilic pores through which ions simply flow from one side of the membrane to the other down their electrochemical gradients (see Fig. 3–1). We will now discuss in considerable detail ways of measuring the activity of ion channels and describe some of their fundamental properties that have been deduced from such measurements. In subsequent chapters we will go on to consider what is known about the molecular structures of ion channel proteins and how their function can be related to their structures.

Single Ion Channels

The possibility that ion currents might flow through hydrophilic pores in the membrane was first suggested in the mid-1950s. Although this idea became widely accepted, more than 20 years passed before the activity of ion channels could be measured directly. The breakthrough came with the advent of *single channel recordings*, methods for measuring the activity of individual ion channels either in their native membrane or after their insertion into artificial bilayer membranes constructed from phospholipids. The most important development was *patch clamp recording* (Fig. 4–1). This technique, developed by Erwin Neher, Bert Sakmann, and coworkers, allows the current passing through single ion channels in the membrane of a cell to be measured directly. The information derived from this revolutionary approach, for which Neher and Sakmann were awarded the Nobel Prize in Physiology or Medicine in 1991, has dramatically advanced our understanding of ion channel properties in neurons (and in other cells).

To carry out a patch clamp recording, a glass pipette, with an internal diameter of the order of a micrometer or so at its tip, is placed against the membrane of a cell. The application of suction to the inside of the pipette can lead to an electrical seal between the glass and the membrane. This seal becomes so tight that ions effectively are prevented from leaking out through it. Depending on the exact size of the patch of membrane under the pipette and the density of ion channels in the membrane, one or more ion channel proteins may be isolated under the pipette. Current carried by ions flowing into or out of the cell through these channels can be

a

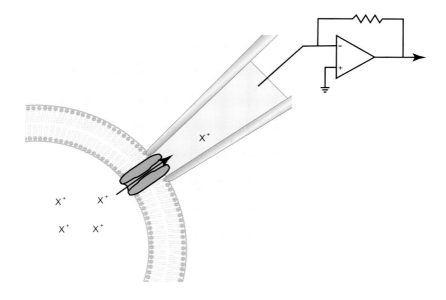

b

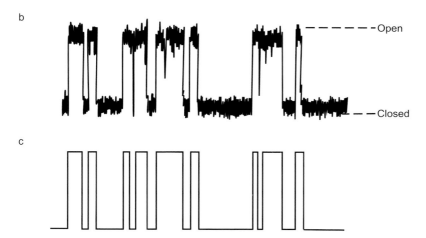

Open

Closed

c

Figure 4–1. Patch clamp recording of single ion channel activity. *a*: Illustration of the cell-attached mode of patch clamp recording, with a current-to-voltage converter that is connected to the electrode to measure the flow of ion X⁺ across the membrane. *b*: An example of recordings of single ion channel activity obtained with this method. *c*: Simple computer programs can be used to produce idealized single channel records, which reproduce faithfully the openings and closings seen in the real record.

detected by a sensitive current-to-voltage converter that is connected to the inside of the electrode (Fig. 4–1a).

When a single ion channel is isolated under the pipette, the patch clamp technique can be used to reveal abrupt transitions between an open state, during which a detectable amount of current flows through the channel, and a closed state, during which no current flows. These *functional states*, measured electrophysiologically, must be the manifestation of stable *structural conformations* of the ion channel proteins. Figure 4–1 shows an example of a real channel recording (Fig. 4–1b) together with an idealized description of its opening and closing (Fig. 4–1c). The upward

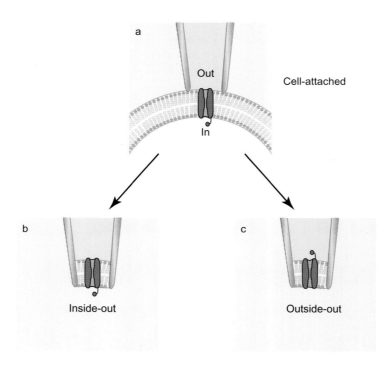

Figure 4–2. Detached patches. Because the seal between the patch electrode and the plasma membrane is mechanically stable (*a*), it is possible to pull the electrode off the cell with the patch remaining attached to the electrode. Depending on the conditions under which this is done, one can obtain either an inside-out patch (*b*), in which the former cytoplasmic portion of the channel (identified by the small ball attached to the membrane-spanning portion) is exposed to the bathing medium, or an outside-out patch (*c*), in which the former extracellular portion of the channel is exposed to the bathing medium (see Hamill et al., 1981). See Animation 4–1.⬤

transitions are openings and the downward transitions are closings of the single ion channel. The computer-generated idealized record is free of noise and hence suitable for computer-aided quantitative analysis of channel activity.

For some purposes it may be desirable to have access to both the intracellular and extracellular sides of the patch membrane. In the cell-attached patch recording technique, it is possible to alter the composition of the extracellular medium in the pipette, but there is no direct access to the inside of the patch. Fortunately, other configurations of single channel patch recording have been invented that do provide for manipulations of both the inside and the outside of the patch. Two such variants of patch clamping are termed *inside-out* and *outside-out* cell-free patch recording, illustrated in Figure 4–2. Both techniques rely on the fact that the seal between the glass pipette and the cell membrane is tight not only electrically but also mechanically. Accordingly, when a cell-attached patch pipette (Fig. 4–2a) is pulled away rapidly from a cell, the patch of membrane frequently comes away with it. In the inside-out configuration the former cytoplasmic membrane surface is exposed to the bathing medium (Fig. 4–2b), whereas in the outside-out patch the external membrane surface is accessible (Fig. 4–2c). Many ion channels can survive for a long time in such cell-free patches of membrane, and a full characterization of the properties of the channels can be carried out readily.

Ion Flow Through Ion Channels Is Fast

That such measurements of single channel currents can be made at all should not be taken for granted—it really is rather astonishing. Ion channels are proteins, and when we measure the activity, the opening and closing, of a single ion channel, we are observing the activity of a single protein molecule! Compare this with the standard enzyme assay in a test tube, where typically one is measuring the sum of the activities of some 10^{10} or more protein molecules. The ability to measure single channel activity is due in part to advances in modern electronics; current-to-voltage converters capable of measuring as little as 10^{-13} A (0.1 pA) of current are available. In other words, there is a highly sensitive assay for ion channel activity. However, this assay would not be sufficiently sensitive to measure single channel currents if the rate of ion transport through channels were not remarkably fast.

The current flowing through a single ion channel, such as that illustrated in Figure 4–1, is typically in the 1–20 pA range. This corresponds to the movement of some $0.6-12 \times 10^7$ ions per second through the channel. If we think of an ion channel as an enzyme whose job is to catalyze ion transport, then the turnover rate for this enzyme is of the order of 10^7-10^8 reactions per second. Turnover rates for most enzymes tend to be of the order of 10^2 per second, with the fastest being in the range of 10^5 per second. Active transport systems also have turnover rates in the 10^2-10^4 per second range; indeed, they have a theoretical limit of about 10^5 reactions per second because of the time it takes them to physically carry the ion across the membrane.

These uniquely high turnover rates for ion channels lead to the fundamental conclusion that the ion transport they mediate must be via diffusion through a pore. The fact that we can measure single channel events at all makes this conclusion inevitable; single carrier currents could be no larger than about 10^{-3} pA and would not be detectable with presently available techniques. This in turn has enabled us to draw a picture of an ion channel (see Fig. 3–1) as a membrane-spanning hydrophilic pore, which must be accurate in general outline if not in detail. The astonishing thing is that this could be done years ago, well before high-resolution protein structural information became available for any ion channel. The landmark determination, by X-ray crystallography, of the three-dimensional structures of voltage-dependent ion channels beginning in the late 1990s confirmed in remarkable detail many predictions that had been made on the basis of functional measurements more than 20 years earlier (see Chapter 5).

Different Kinds of Ion Channels

There are many different types of single ion channel activities, even in the membrane of a single neuron. These may be classified according to several different criteria:

1. *Single channel conductance*, a measure of the rate at which ions pass through the open channel;
2. *Ion selectivity*, the nature of the ions that are allowed to pass through the open channel;
3. *Gating*, the opening and closing of the channel under the influence of such factors as the transmembrane voltage, the binding of neurotransmitters, hormones, and other agents to sites on the outside of the channel, and the actions of certain intracellular metabolites and enzymes; and

4. *Pharmacology*, the susceptibility of the channel to various compounds that may block the pore or otherwise influence channel properties.

Single channel conductance. The voltage across the patch of membrane may be set to different levels, and the size of the current that flows through the open channels (Fig. 4–3a) can then be plotted against the voltage, as has been done in Figure 4–3b. For many channels, a straight line is obtained over a wide range of voltages. Such a plot provides two pieces of information: the *unitary* or *single channel conductance* of the channel, and the *reversal potential* for the current that flows through the channel. Knowledge of the latter allows conclusions to be drawn about the ion(s) that can permeate the channel.

The conductance of a channel is a measure of the ease of flow of current through the channel. Recall that the electrical conductance G is the inverse of electrical resistance R. Conductance is closely related to *permeability*, which is the term we have been using to describe the ease with which an ion moves across the plasma membrane. The unitary or single channel conductance (g) is the slope of the open channel current versus voltage plot (Fig. 4–3b). It is given by the equation:

$$g = \frac{\Delta i}{\Delta V}$$

This equation is simply Ohm's law, a fundamental law of physics that we will discuss in more detail in Chapter 6. When i is given in amperes and V is in volts, the unit of conductance is siemens (S) or reciprocal ohms (note that i is used to denote the current passing through a single channel, and I the *macroscopic* or total membrane current that passes through the many channels in the cell membrane; similarly, g is the single channel conductance and G the macroscopic membrane conductance). Single channel conductances are usually given in picosiemens (pS, 10^{-12} S). The conductances of channels in biological membranes that have been measured to date are generally in the range of 5–400 pS (although some channels in mitochondrial membranes can be as much as 10 times larger than this).

Ion selectivity. Another critical piece of information about an ion channel is the nature of the ions that normally flow through the channel when it is open. Channels are able to *select* for one kind of ion over another. For example, channels in biological membranes will allow either cations or anions to flow, but not both. Within these broad classes of cation and

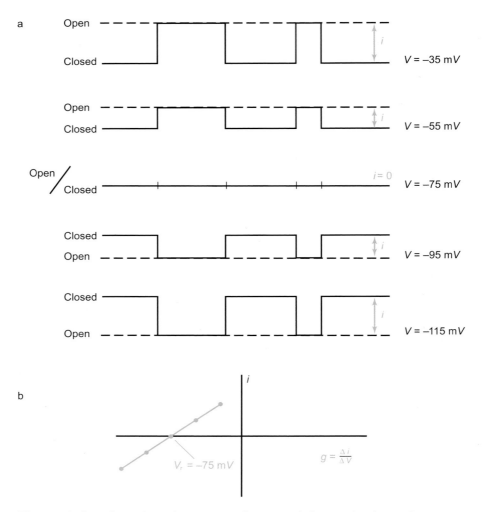

Figure 4–3. Ohm's law determines the size of the single channel current. *a*: The amplitude of the current (*i*) passing through a single open ion channel varies as the voltage across the membrane is changed. *b*: Plot of the single channel current amplitude as a function of voltage (*V*). The current is zero at the reversal potential (V_r), equivalent to the equilibrium potential for the ion that passes through the channel. The equilibrium potential will be discussed in Chapter 6. *g*, single channel conductance.

anion channels, most channels are also selective for one particular cation or anion. For example, we shall see that the action potential is dependent on the presence in axons of two distinct cation channels, one highly selective for sodium ions and the other for potassium ions. Selectivity is so fundamental a property of an ion channel that channels are usually (but not always) named according to the ions they prefer (e.g., *sodium* channels, *potassium* channels, *calcium* channels, *chloride* channels).

How do ion channels exhibit selectivity, often exquisite selectivity, for one ion over another? They do so because they are far more than simple holes in the membrane. Although a detailed discussion of channel selectivity mechanisms is well beyond the scope of this book, the X-ray determination of the structure of a potassium channel demonstrated strikingly that a potassium ion, together with its strongly bound shell of water molecules, must make a tight fit with the narrowest region of the channel protein, the *selectivity filter*, to pass through it (although there is more to selectivity than this). This will be discussed at more length when we consider ion channel structure and function in Chapter 5.

Gating. By now it will be evident that ion channels are not simply inert pores in the membrane. Rather, they are dynamic entities that can undergo extremely rapid transitions between an open state, in which they conduct ions, and a closed state, in which they do not allow ions to pass. These open/closed transitions, which are readily apparent in single channel recordings such as that in Figure 4–1, must reflect conformational changes in the structure of the channel protein.

The opening and closing of a channel is often termed *gating*, because it is convenient and instructive to modify our simple picture of the ion channel as a pore (Fig. 3–1) to include a gate, presumably an integral part of the channel protein, that can open to allow ion flow or shut to prevent it (Fig. 4–4a). These two states of the protein are in dynamic equilibrium, and the amount of time the channel spends in each state will depend on the relative values of the free energies of the two states. These free energies in turn will be reflected by easily measured quantities, the *rate constants*, for channel opening and closing (Fig. 4–4a).

Ion channel gating may be influenced by a variety of external conditions. We often say that such conditions cause channels to "open" or "close," but what we really mean that the relative free energies of the open and closed states have been changed, so that the channel is more likely to be open or closed than it was previously. This will be seen in the single channel records as a change in the rate constant for opening or closing, or sometimes for both (Fig. 4–4b).

Voltage-dependent gating. Many channels, particularly those that shape the ongoing electrical behavior of a neuron, are *voltage-dependent* channels. The frequency with which such channels open and close depends on the membrane potential. As we shall see in the next chapter, different types of channels may either increase or decrease the amount of time they

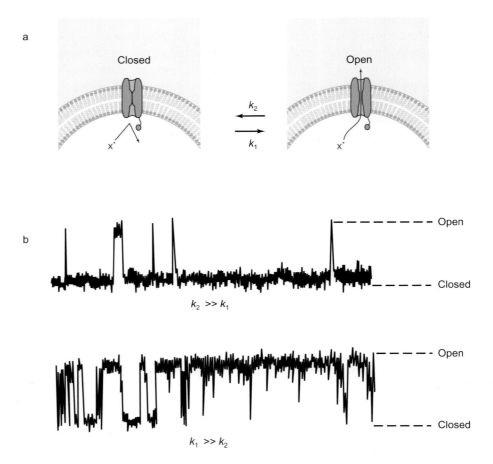

Figure 4–4. Ion channel gating. *a*: An ion channel drawn as a pore with a gate that when closed (*left*) blocks ion flow. There is a dynamic equilibrium between the closed and open (*right*) states, determined by the opening (k_1) and closing (k_2) rate constants. *b*: The proportion of time that the channel spends open depends on the relative values of k_1 and k_2. See Animation 4–2.◐

spend in the open state as the voltage across the membrane is made more positive. Figure 4–5a shows the behavior of a voltage-dependent ion channel at different membrane potentials. At negative potentials, such as the resting potential of the cell, the channel opens infrequently or not at all, and when it does open, it closes again quickly. As the potential is made positive, the channel begins to open more frequently and stay open longer until, at potentials more positive than about +20 mV, the channel is fully activated. The important point here is the amount of time the channel spends in the open and closed states. The amplitude of the open channel current also changes with voltage, in the manner described in Figure 4–3, but this is not important in the present context (it will be discussed in detail in Chapter 6). Figure 4–5b is a graph of the probability of the channel being open (open probability, P_o), as a function of voltage. The data points fall on a sigmoid curve, the steepness of which reflects the channel's sensitivity to voltage.

What is the protein structural basis for this sensitivity of channel opening and closing to voltage? The channel gate must either be coupled to or itself act as a *voltage sensor*, detecting the strength of the electric field across the membrane. It is presumed that this part of the channel protein possesses some net charge and can move under the influence of the electric field to open or close the channel pore and hence allow or prevent ion flow. In Chapter 5 we will consider the progress that has been made in identifying the voltage sensor in different kinds of voltage-dependent ion channels.

Gating controlled by neurotransmitters or intracellular messengers. The activity of many ion channels is tightly linked to the action of *neurotransmitters*, chemicals released from one neuron that influence the activity of other neurons (Chapter 10). The gating of other channels may be influenced by intracellular *messenger* molecules or ions—for example, calcium ions. The way neurotransmitter binding or intracellular messengers influence the opening and closing of the channel gate will be discussed in detail in Chapters 11 and 12.

Single channel kinetics. Further characterization of an ion channel can be carried out by measuring the mean open time and mean closed time for a large number of transitions between the open and closed states. For a voltage-dependent channel, these measurements must be made at several different voltages (as in Fig. 4–5), and the voltage at which they were measured must be stated. Histograms of the number of openings or closings of a given duration can also be plotted. Such information enables simple models of the behavior of the channel to be made. Here we will provide one example of the use of such information.

Some channels show "bursty" kinetic behavior (Fig. 4–6) that cannot be interpreted simply in terms of transitions between a single open and a single closed state. Measurement of the closed times of this channel will show that periods during which the channel is closed fall into two groups: "short" closed times, which represent the brief closings of the channel *during* one burst of openings, and "long" closed times, which represent the times between the bursts of openings. A simple model for such bursty behavior consists of two closed states, C_1 and C_2. During a

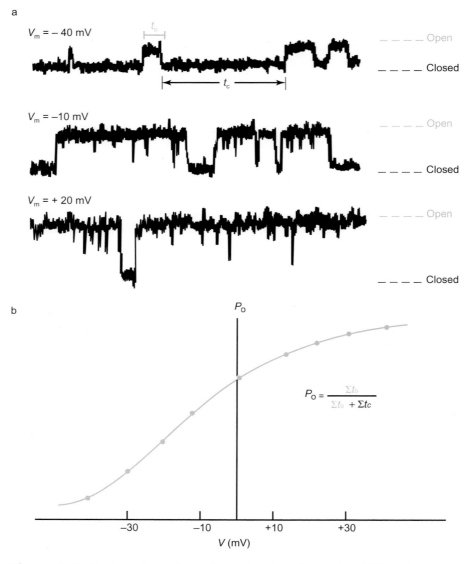

Figure 4–5. Gating of a voltage-dependent ion channel. *a*: When the membrane voltage (V_m) is varied, the amount of time that the channel spends open (t_o) and closed (t_c) changes. *b*: The channel open probability (P_o) can be plotted as a function of voltage (V). The data fit a sigmoid curve, with the open probability being highest at the most depolarized voltages. See Animation 4–3.◖

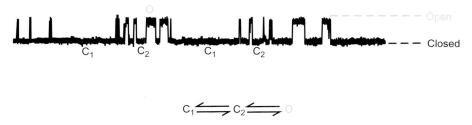

Figure 4–6. Single channel kinetics. The activity of channels cannot always be described in terms of a simple transition between one open and one closed state. Some channels exhibit bursty kinetic behavior. The behavior of the channel shown here can be explained if the channel has one open state (O) and two distinct closed states (C_1 and C_2).

burst itself, the channel flips between the open state (O) and the shorter of the closed states (C_2). Occasionally, however, the channel enters the other, longer closed state (C_1). Return from this closed state is slow, accounting for the long periods between the bursts. A simple model of this type allows the calculation of rate constants for the transitions between each of the three states of the channel. The way that these rates are affected by membrane voltage, neurotransmitters, drugs, and other parameters can then be analyzed, to provide a mechanistic description of the regulation of channel gating.

Pharmacology. Since physiologically relevant chemicals such as neurotransmitters and intracellular messengers can bind to ion channels, it will come as no surprise that nonphysiological chemicals, both naturally occurring and synthetic, can also bind to channels and influence their properties. As we will emphasize throughout this book, ion channels are proteins, and small molecules often can bind selectively and with high affinity to specific proteins. Small-molecule pharmacological probes are now available for many different kinds of ion channels, and they have proven very useful in probing ion channel structure and function. For example, *tetrodotoxin*, a drug that blocks ion flow through voltage-dependent sodium channels, has been very widely used by neurobiologists to inhibit the firing of action potentials. ❽

Activation and inactivation. One important characteristic of the kinetic behavior of a voltage-dependent ion channel is its rate of *activation*. When the membrane potential is changed abruptly, the channel open probability (P_o) may change rapidly, until a new steady-state open probability is attained for the new voltage. This increase in the channel's open probability is its activation. The rate of activation of some voltage-dependent ion channels—for example, sodium channels—is very fast, reaching a maximum within a few milliseconds after the change in membrane voltage (Fig. 4–7a). However, other channels—some potassium channels, for example—exhibit slower rates of activation (Fig. 4–7a).

As important for the electrical behavior of a neuron as the rate of activation of its different channels is their rate of *inactivation*. Some channels, once they have been induced to open by a change in voltage, maintain their new rate of opening for a prolonged period. This is the case for the slowly activating potassium channel illustrated in Figure 4–7a. Other channels, however, following their activation, undergo a progressive decrease in openings. This is illustrated in Figure 4–7a for our voltage-dependent sodium channel. The rate of loss of channel activity is termed the *rate of inactivation*. Most potassium channels also undergo inactivation, although it is generally slower than that of sodium channels. Thus channel activity

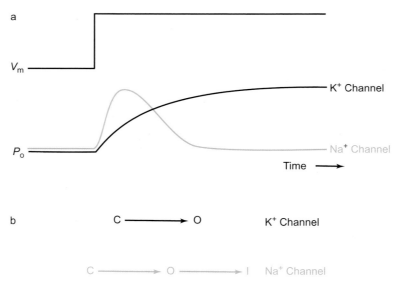

Figure 4–7. Activation and inactivation of ion channels. *a*: In response to an abrupt depolarization of the membrane voltage (V_m), the open probability (P_o) of sodium channels increases rapidly and then decreases again, even though the depolarization is maintained. In contrast, the P_o of potassium channels increases more slowly and remains increased throughout the depolarization. *b*: Kinetic schemes that can account for the gating behavior shown in (*a*). A nonconducting inactivated state (I), from which the channel recovers only very slowly, is necessary to explain the gating of the sodium channel. C, closed state; O, open state.

can be not only *voltage* dependent but also *time* dependent. At early times, voltage-dependent activation causes sodium channels to open, but at later times inactivation begins to dominate and eventually the channels are never open, even though the depolarization is maintained. This nonconducting or inactivated state is distinct from the closed state of the channel (Fig. 4–7b). As we shall see in Chapter 6, inactivation of various ion channels plays an essential role in shaping action potentials and in determining the electrical characteristics of many neurons.

Macroscopic Ion Currents Result from the Activity of Populations of Ion Channels

Let us now examine how the microscopic currents flowing through a population of ion channels combine to generate the much larger macroscopic current recorded from the whole cell. The voltage-dependent axonal sodium channels that open rapidly in response to a change from a negative to a more positive membrane voltage provide an excellent example (Fig. 4–8). When one channel is present in a patch (Fig. 4–8a), the response to depolarization is an increase in channel open probability, followed by a decrease again as the channel inactivates. When several sodium channels are present in the patch (Fig. 4–8b, c), the current record begins to resemble the whole-cell sodium current (Fig. 4–8d), measured by the techniques we will describe below. In other words, the whole-cell sodium current is the sum of the currents passing through all of the sodium channels in the plasma membrane. In the whole cell, the rapid change in the probability of opening of the individual sodium channels (Fig. 4–7a) is manifest as a rapid increase in the total sodium current (Fig. 4–8d). Note also in Figure 4–8 that the total current returns to zero even though the depolarization is maintained, as a consequence of the inactivation of the individual ion channels (again, compare with the channel open probability depicted in Fig. 4–7a). These current traces emphasize once again the fundamental fact that the activity of many ion channels is both voltage and time dependent.

   These considerations can be expressed in a more quantitative way by means of a simple yet useful equation. The macroscopic current I carried by one type of ion channel is given by

$$I = NP_oi$$

where N is the number of functional channels of that type present in the membrane, i is the current carried through a single channel when it is

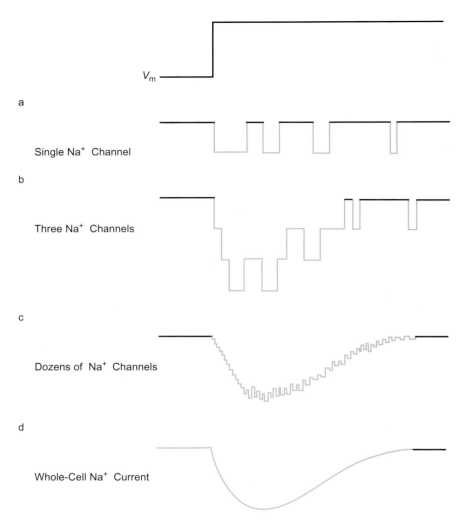

Figure 4–8. Macroscopic membrane currents result from the summed activity of individual ion channels. *a*: During a sustained depolarizing voltage pulse (V_m), single sodium channels open and then inactivate. The inward currents carried by sodium rushing into the cell through the open sodium channels are shown as downward deflections. The more channels there are in the patch (*b* and *c*), the more closely the patch current resembles the sodium current recorded from the whole cell (*d*).

open, and P_o is the probability of a channel being open. As we have seen, *i* varies with the voltage across the membrane according to Ohm's Law, and we shall see that P_o (and sometimes N) may be a function of voltage and time, and may be modulated by neurotransmitters and/or intracellular metabolic events.

A dilemma in measuring voltage-dependent ion currents. From these considerations it will be evident that I is an important parameter to measure, in order to understand channel gating and the electrical behavior of a neuron. Recall from Figure 4–5, however, that channel opening is often voltage dependent. At the same time, channel opening itself will generally result in a change in voltage, and this in turn will influence channel gating. How, then, in the face of this positive feedback loop, is one to measure I and study effectively the voltage-dependent regulation of channel gating? The answer is to devise a method to hold the membrane voltage constant, even though ion channels are opening and closing. To do this, in the 1930s K. C. Cole and colleagues invented an electronic feedback system, called the *voltage clamp*, to hold the membrane potential constant at a voltage chosen by the investigator. In its simplest form (Fig. 4–9a), the voltage clamp consists of two separate electrodes: one connected to a voltage-measuring amplifier to measure the transmembrane voltage, and the other connected to a current-passing amplifier. A negative feedback loop is created by adding a feedback amplifier, which compares a command voltage set by the experimenter (V_c) with the measured membrane voltage (V_m). The difference between these two voltages is known as the *error signal*, and the feedback system injects current through the current-passing electrode to maintain the error signal as close as possible to 0. By this means, V_m is forced to be equal to V_c; in other words, the membrane voltage is controlled by the experimenter, and the signal that is measured is the amount of current required to maintain that particular voltage. This current is, in fact, equal to the macroscopic current I flowing across the membrane at that voltage.

Other ways to measure the macroscopic membrane current. A more recent method that has been developed to measure currents in whole cells is a variant of the patch clamp technique, known as *whole-cell patch recording* (Fig. 4–9b). In this method, a conventional patch electrode is sealed to a cell as described in Figure 4–1, and the membrane under the patch is then destroyed by either a pulse of suction or a large, abrupt change in voltage. The solution in the pipette can then exchange freely with the cytoplasm of the cell, and the cell can be voltage clamped with appropriate electronics connected to the inside of the pipette. This configuration can be considered analogous to a very large outside-out patch (Fig. 4–2c), consisting of most of the cell's plasma membrane, with a very large number of ion channels contributing to the current flow across the membrane. ❷

We have emphasized techniques for measuring membrane currents under voltage clamp because of their critical importance for our understanding of neuronal membrane properties. Keep in mind that of the

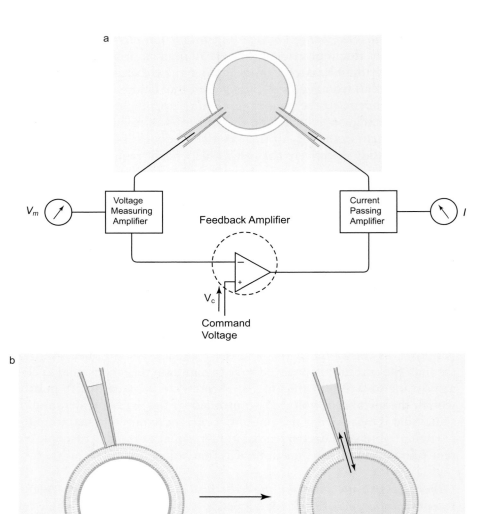

Figure 4–9. How is the whole-cell current measured? *a*: Schematic drawing of a two-electrode voltage clamp, which uses three separate amplifiers to control the membrane voltage. *b*: Whole-cell mode of the patch recording technique (see Hamill et al., 1981).

three parameters involved in regulating the transmembrane ion current—the ionic gradients, the voltage, and the gating of ion channels—the first two can be manipulated by the experimenter and, accordingly, the regulation of channel gating can be investigated in a rigorous way.

Imaging techniques have also been extremely useful in recording neuronal electrical excitability. For example, there exist *voltage-sensitive dyes*, molecules that insert themselves into the neuronal plasma membrane and change their spectral properties in response to a change in voltage across the membrane. The use of these dyes often allows the simultaneous recording of the activity of dozens or hundreds of neurons. The more recent development of genetically encoded proteins that can not only *report* on electrical activity but also be used to *control* it (so-called *optogenetics*) has triggered a revolution in this field. Optogenetics in its simplest form involves fusing a protein whose conformation can be manipulated by light to an ion channel. The first examples of this approach used the light-sensitive visual pigment rhodopsin, which is responsible for light absorption in the retina and which we shall discuss in detail in Chapter 13. When the hybrid *channel-rhodopsin* protein is expressed in a neuron and the neuron is exposed to light of an appropriate wavelength, the resulting change in conformation of the rhodopsin can cause its coupled ion channel to open (or close), thereby changing the membrane voltage.

Ion Channels, Ion Currents, and Neuronal Electrical Activity

Why should we go to great lengths to measure ion currents and to investigate the regulation of ion channel gating in a rigorous way? Recall that when ion currents flow across the plasma membrane through ion channels, the distribution of charge and hence the membrane voltage will change; this can result in action potentials, either in isolation or grouped in complex patterns, as we saw in Chapter 3. Let us re-examine the action potentials in Figure 3–11, in the context of the ion channels that are important for their generation. As shown in Figure 4–10, the depolarizing upstroke of the action potential is associated with the opening of sodium channels and the entry of positively charged sodium ions into the cell (*inward current*, which is *depolarizing*). The repolarization, in contrast, is associated with the opening of potassium channels and the exit of positively charged potassium ions from the cell (*outward or hyperpolarizing*

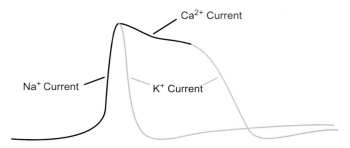

Figure 4–10. Different ion channels contribute to the action potential. Inward depolarizing currents carried through sodium and/or calcium channels are responsible for the upstroke and plateau of the action potential (black). Outward potassium current mediates repolarization (blue). The relative proportion of these currents determines the different shapes of action potentials, such as those in Figure 3–11.

current). Some cells fire longer action potentials, often associated with calcium channels that stay open longer and produce a longer-lasting inward *calcium current* (Fig. 4–10). The important points here are *(a)* multiple classes of ion channel exist, and *(b)* the combined activities of different kinds of inward and outward current channels with different kinetic properties determine the precise form of neuronal electrical activity, including the size and shape of action potentials (Fig. 3–11) and their grouping in characteristic temporal patterns (Fig. 3–12). In Chapter 6 we shall expand on this rather superficial summary and describe in more detail how voltage clamp analysis of ion currents has provided profound insights into the mechanisms of generation of action potentials and complex patterns of neuronal electrical activity. First, however, we shall discuss ion channels as membrane proteins, whose properties can be studied not only electrophysiologically but also through techniques of protein biochemistry and molecular and structural biology.

Summary

Electrical activity in neurons (and other kinds of cells) results from the movement of ions across the plasma membrane through specialized membrane proteins known as *ion channels*. Exquisitely sensitive patch clamp techniques are available to measure the current passing through single ion channels, as well as the macroscopic membrane current carried by a population of ion channels. These techniques have enabled the detailed characterization of various essential properties of ion channels, including their selectivity for particular ions, their pharmacology, and the way their activity is regulated by membrane voltage and other factors. There are many different kinds of ion channels in the neuronal plasma membrane, and their activities sum to generate action potentials and complex patterns of action potential firing.

Ion Channels Are Membrane Proteins

I n the last chapter we described methods of measuring the activity of single ion channels and the macroscopic membrane currents resulting from the activity of populations of ion channels, and discussed some of the key functional properties of channels that can be deduced from such measurements. Let us now consider ion channels as protein molecules, the amino acid sequences of which are known from molecular cloning techniques and genome sequencing. The powerful combination of electrophysiology, molecular biology, and direct structural determination has provided novel insights into how the structure of an ion channel gives rise to its critical functions.

Ion Channel Structure Inferred from Sequence

Beginning in the mid-1980s, a number of then-emerging cloning strategies were used with great success to determine the complete amino acid sequences of several different kinds of ion channels. Methods that were extremely important historically include cloning by *protein purification/ amino acid sequencing*, *positional cloning*, and cloning by *sequence homology*. These techniques have been supplanted more recently by the sequencing of entire genomes from many organisms, including humans.

Amino acid sequence of sodium channels. A cDNA encoding the full amino acid sequence of one subunit of a voltage-dependent sodium channel, the so-called α subunit, was isolated from the electric eel, a rich source of sodium channels, in the mid-1980s. The protein predicted from the cDNA sequence contains about 2000 amino acids. What can an unwieldy string of 2000 amino acids tell us about the structure of the channel? Luckily, there are a few general rules relating amino acid sequence to protein structure that can be used to provide clues as to which regions of the protein may lie within the lipid membrane itself. One of the more important of these rules is that a string of 23 or so hydrophobic amino acids can span a normal cell membrane and is likely to be organized in the form of a helix of amino acids. Such strings of hydrophobic amino acids in a protein can readily be picked out from a *hydrophobicity plot*. Each amino acid is assigned a *hydrophobicity value*, which reflects its ability to interact with water. Amino acids with nonpolar side chains interact poorly with water—they are strongly hydrophobic—and are given a high positive value. In contrast, polar amino acids are given a negative value. A running average of these values over several amino acids is then calculated around each amino acid in the protein sequence and plotted as a function of position along the protein chain. A hydrophobicity plot for the α subunit of the sodium channel is shown in Figure 5–1a. Twenty-four possible transmembrane stretches of amino acids can be found.

The apparent complexity of the structure of the sodium channel can be substantially reduced because its sequence contains four internal domains, each of which strongly resembles the others. These are marked I, II, III, and IV in Figure 5–1. Within each of these homologous regions there are six possible transmembrane segments (called *S1–S6*). With this information it is possible to construct a model for the arrangement of different parts of the protein across the plasma membrane (Fig. 5–1b). It should be emphasized that such a structure remains only a hypothetical model until direct structural determinations are made on the channel protein itself. As we shall see later, many features of the predicted structure of sodium channels have been confirmed, first by mutational analysis and later by X-ray crystallography.

Amino acid sequence of potassium channels. The first potassium channel amino acid sequence to be determined was that of the rapidly inactivating *Shaker* potassium channel in the fruit fly *Drosophila*. *Shaker* is a mutant *Drosophila* that is defective in one kind of rapidly inactivating potassium current (Fig. 5–2). When the gene at the *Shaker* locus was cloned by a positional cloning approach and its long-sought sequence finally became known, it was found to possess some quite remarkable features. The first arresting feature of the proteins encoded by the *Shaker* locus is that they are only about one-quarter the size of the sodium

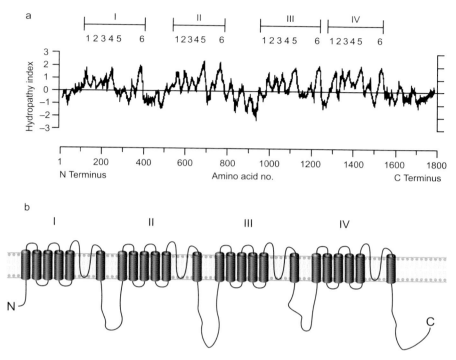

Figure 5–1. The α subunit of a sodium channel. *a*: Hydrophobicity plot for the α subunit of a sodium channel. *b*: Model for the arrangement of the sodium channel α subunit in the plasma membrane. There are six membrane-spanning segments in each of the four homologous domains (I–IV).

channel protein we described earlier. Only a single domain consisting of six putative transmembrane regions, S1–S6, is found in the protein (Fig. 5–3). This is because potassium channels are functional *tetramers*; four channel proteins have to come together to make a structure that resembles that of a sodium channel (Fig. 5–4).

The second remarkable feature is that more than one type of channel can be made from the RNA produced from the *Shaker* locus (Fig. 5–3). The DNA coding for the channel does not run continuously through the locus. Rather, regions coding for the protein (termed *exons*) are separated by stretches of noncoding DNA (termed *introns*). Because RNA is synthesized from this DNA, it initially includes these noncoding introns. During the production of the mature messenger RNA that encodes the protein, the introns are removed and the exons are *spliced* together. In the *Shaker* locus there are several different regions that can code for alternative carboxyl-terminal and amino-terminal regions of the channel protein

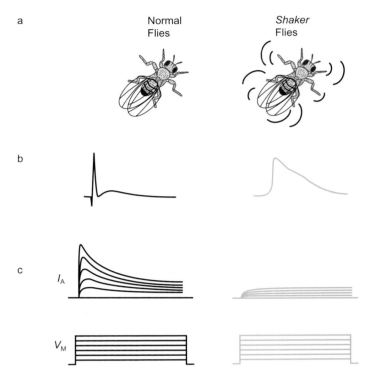

Figure 5–2. *Shaker* flies lack a rapidly inactivating potassium current. *a*: The *Shaker* mutant fly shakes its legs under ether anesthesia. This is associated with prolonged action potentials in muscle (*b*), which in turn result from the absence of a rapidly inactivating potassium current (*c*) that normally plays a prominent role in repolarization of the muscle action potential. I_A refers to A-type potassium currents, flowing through *Shaker* channels, that are evoked by depolarizations of the muscle membrane potential (V_M) under voltage clamp. (From the work of Mark Tanouye, Larry Salkoff, Bob Wyman and colleagues.)

(Fig. 5–3). Different patterns of cutting and splicing of RNA can therefore produce channels that possess the same sequence in the central region, which contains most of the hydrophobic segments, but that have different sequences at their carboxyl or amino termini. Figure 5–3 demonstrates the production of two Shaker proteins, ShA and ShB. These two proteins are identical throughout much of their sequences but differ after the fifth hydrophobic segment. We now know that many potassium channels are *heterotetramers*, in which different subunits such as ShA and ShB interact to form a mixed complex, with functional properties different from those of tetrameric channels that contain only a single type of α subunit. ❽

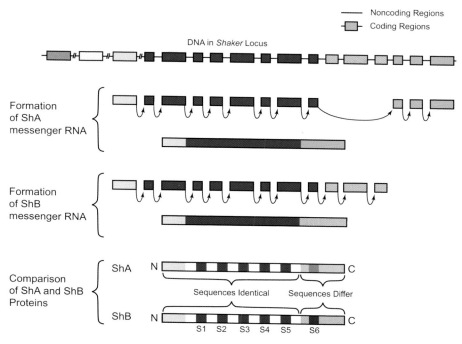

Figure 5–3. Alternative splicing in the *Shaker* locus that was cloned at about the same time by several different laboratories. Lily and Yuh Nung Jan and coworkers showed that RNA transcribed from this locus can be processed to encode one of several different *Shaker* proteins, including Shaker A (ShA) and Shaker B (ShB) (Schwarz et al., 1988).

More than 100 other genes encoding voltage-dependent potassium channels have now been identified, largely by sequence homology and whole-genome sequencing, in organisms ranging from bacteria to nematode worms to flies to humans (see Chapter 7). Some of these closely related proteins give rise to non-inactivating or slowly inactivating potassium currents rather than rapidly inactivating currents in heterologous expression systems. In fact, it appears that there is a spectrum of types of voltage-dependent potassium channels, with different functional properties, all of which fall within this family of proteins that are related structurally. Although not all of the potassium channel genes give rise to multiple protein products in the way the *Shaker* locus does, it is quite evident that the diversity of potassium channels is far greater than had been suggested from earlier electrophysiological studies. ❷

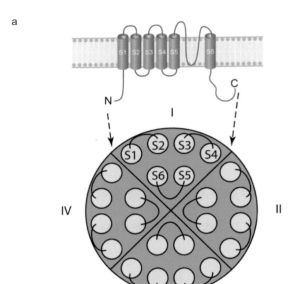

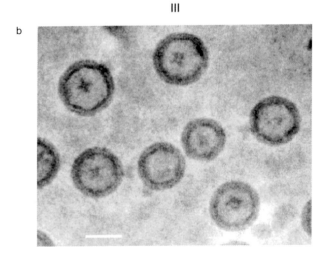

Figure 5–4. Potassium channels are functional tetramers. *a*: Each subunit of voltage-gated potassium channels is thought to have six membrane-spanning segments (S1–S6), as in each domain of sodium channels (see Fig. 5–1). Four subunits (I–IV) come together to form the functional channel. *b*: A purified voltage-dependent potassium channel, reconstituted into liposomes and imaged by cryo-electron microscopy. Scale bar 250 Å. (Micrograph courtesy of Fred Sigworth.)

The Relation of Structure to Function in the Voltage-Dependent Ion Channels

A large number of mutagenesis experiments have now been carried out on the voltage-dependent sodium and potassium channels, and many of the results from one kind of channel are generally applicable to the other. As examples of this general approach, we shall focus on three sets of studies that have characterized channel sequences involved in voltage-dependent activation, ion conduction and selectivity, and rapid inactivation in the voltage-dependent ion channels. We shall then go on to examine the high-resolution three-dimensional structures of several voltage-gated ion channels and discuss the insights from and implications of these structures for aspects of channel function.

Gating currents and the voltage sensor. It is evident that for a channel to respond to changes in voltage across the cell membrane, it must contain some charged structure that can act as a voltage sensor and actually move when the voltage across the membrane is changed. This in turn may trigger the rearrangement of the protein that allows the opening or closing of the ion channel pore (Fig. 5–5a). The charge on the voltage sensor is known as the *gating charge*.

Immediately after a change in voltage, the movement of a gating charge within an ion channel protein should itself register as a flow of current across the membrane. This gating current is very much smaller than that due to the flow of ions through the channel, and it is also very brief. Gating current flows only while the channel protein is undergoing the movement to a new conformation, and this can occur very rapidly. The ionic current, by contrast, starts to flow only after the new conformation of the protein is achieved (Fig. 5–5a). The gating currents for sodium and potassium channels in the squid giant axon (which we shall discuss in Chapter 6) and some other preparations have been measured (Fig. 5–5b). To do this it is first necessary to eliminate, either through the use of drugs or electronic wizardry, the very much larger ionic currents flowing across the lipid membrane through the ion channels. The movements of charge that can be recorded under these conditions generally match those expected for the movement of a physical "gate" that controls the entry and exit of ions through the channel. A perfect match, however, is not obtained, because not all of the charge movements within an ion channel protein lead directly to the opening or closing of the channel (Fig. 5–5c).

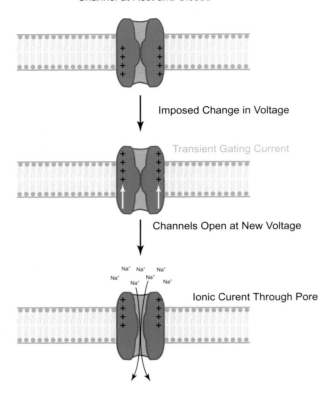

a   Channel at Rest and Closed

Imposed Change in Voltage

Transient Gating Current

Channels Open at New Voltage

Na⁺   Na⁺   Na⁺
Na⁺   Na⁺   Na⁺
    Na⁺   Na⁺

Ionic Curent Through Pore

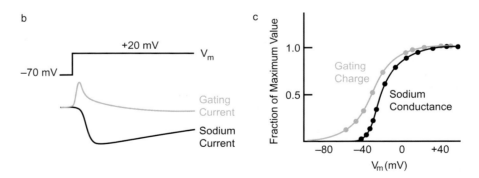

b

+20 mV

−70 mV

V_m

Gating
Current

Sodium
Current

c

Fraction of Maximum Value

1.0

Gating
Charge

0.5

Sodium
Conductance

−80   −40   0   +40

V_m (mV)

Figure 5–5. Gating currents. *a*: Generation of gating currents by movement of charged amino acids in an ion channel protein after a change in voltage. *b*: After a step in voltage, the gating current precedes the ionic current. *c*: Comparison of the voltage dependence of gating charge movement with that of the sodium conductance in an axon (this topic is covered further in a review by Armstrong, 1981). V_m, membrane voltage.

A leading contender for the voltage sensor is found in the S4 segments of the voltage-dependent ion channels. Figure 5–6a shows that this stretch of amino acids contains repeated basic residues, either arginines or lysines, in every third position. If the S4 region were to form an α-helix within the membrane, these positively charged residues would come to be arranged in a spiral form around the helix, as shown in Figure 5–6b. The positively charged residues are likely to be stabilized by negatively charged amino acids, such as aspartate or glutamate, situated on adjacent helices. It has been proposed that a change in the electric field across the α helix, such as would occur when the membrane is depolarized, leads to an uncoupling of the positive residues from their partners, followed by a displacement or rotation of the helix. This would result in the movement of charge in the direction of the imposed electric field and the establishment of a new equilibrium conformation (see Fig. 5–5a).

It is possible to produce mutant channels in which the positively charged amino acids in the S4 region are replaced by neutral or negative charges. This has been done for both sodium and potassium channels. When such mutant channels are expressed heterologously, the voltage dependence of their activation differs from that of wild-type channels in ways that are generally consistent with the idea that segment S4 is a sensor of the membrane voltage. In addition, elegant experiments with reagents that interact chemically with specific amino acids have provided compelling evidence that the S4 segment actually moves in a direction perpendicular to the plasma membrane when the transmembrane voltage is changed.

A ball-and-chain mechanism for rapid channel inactivation. What channel structures contribute to inactivation in the voltage-dependent ion channels? A classic experiment performed by Clay Armstrong, Francisco Bezanilla, and Eduardo Rojas in the 1970s demonstrated that activation on the one hand, and inactivation on the other, must involve different parts of the channel protein. They perfused the inside of the squid giant axon with pronase, a heterogeneous mixture of proteolytic enzymes, and found that this treatment eliminates inactivation but does not affect activation/deactivation of the axonal sodium current (Fig. 5–7a). This experiment demonstrated unequivocally that the channel component responsible for inactivation, the *inactivation gate*, is a protein domain that must be accessible from the cytoplasmic face of the membrane. Armstrong and Bezanilla also demonstrated that the activation gate is not accessible to pronase from the cytoplasmic side and hence must involve a different domain of the protein. This finding, which has been instrumental in shaping our ideas about how sodium channels work, led Armstrong

a

Sodium I

Val-Ser-Ala-Leu-Arg-Thr-Phe-Arg-Val-Leu-Arg-Ala-Leu-Lys-Thr-Ile-Ser-Val-Ile-

Sodium II

Leu-Ser-Val-Leu-Arg-Ser-Phe-Arg-Leu-Leu-Arg-Val-Phe-Lys-Leu-Ala-Lys-Ser-Trp-

Calcium I

Val-Lys-Ala-Leu-Arg-Ala-Phe-Arg-Val-Leu-Arg-Pro-Leu-Arg-Leu-Val-Ser-Gly-Val-

Calcium II

Ile-Ser-Val-Leu-Arg-Cys-Ile-Arg-Leu-Leu-Arg-Leu-Phe-Lys-Ile-Thr-Lys-Tyr-Trp-

Shaker

Leu-Arg-Val-Ile-Arg-Leu-Val-Arg-Val-Phe-Arg-Ile-Phe-Lys-Leu-Ser-Arg-His-Ser-

Rat Kv1.1

Leu-Arg-Val-Ile-Arg-Leu-Val-Arg-Val-Phe-Arg-Ile-Phe-Lys-Leu-Ser-Arg-His-Ser-

b

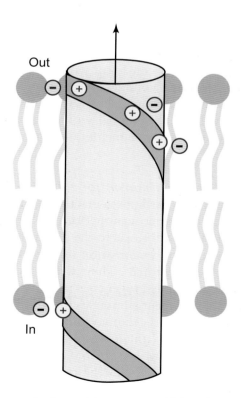

Figure 5–6. The S4 region. *a*: Amino acid sequence of S4 regions in domains I and II of several different kinds of sodium and calcium channels and in two voltage-dependent potassium channels, *Shaker* A and Kv1.1. *b*: Possible arrangement of basic residues in S4 regions in a helix that spans the membrane. (Modified from Catterall, 2010.)

and Bezanilla to propose a *ball-and-chain* model for sodium channel inactivation (Fig. 5–7b). According to this scheme, the inactivation gate (the ball) is a portion of the channel protein, tethered to the remainder of the channel by another stretch of amino acids (the chain). When the channel activates, current flows through the open channel, but only until the ball occupies and blocks its internal mouth; thus, according to this scheme, inactivation is a special form of channel block (Fig. 5–7b).

In a series of important experiments carried out in 1990, Richard Aldrich and his colleagues demonstrated that the ball-and-chain model accurately describes the rapid inactivation of *Shaker* potassium channels. To do this, they injected RNAs encoding *Shaker* channels into toad (*Xenopus*) oocytes, forcing these cells to make Shaker channels in their plasma membrane. ❷ They then applied trypsin, a more specific proteolytic enzyme, to the cytoplasmic face of detached membrane patches containing these *Shaker* channels and found that rapid inactivation was removed, mirroring the earlier pronase experiment with squid axon sodium channels. They then mutated *Shaker* to remove 20 amino acids from the amino terminal of the protein and found that the resulting channels activated normally but did not inactivate (Fig. 5–7c), suggesting that this amino-terminal region constitutes the ball. Shortening or lengthening that portion of the protein connecting the ball to segment S1 changed the rate at which the channels inactivate, which is consistent with the idea that this connecting portion constitutes the chain. Finally, when the mutated *Shaker*, lacking the 20 terminal amino acids (and hence non-inactivating), was expressed in oocytes, a synthetic peptide corresponding in sequence to these 20 amino acids was able to block the channel when it was added to the cytoplasmic face of detached membrane patches (Fig. 5–7c). Taken together, these elegant experiments provide strong evidence that the rapid inactivation of potassium channels reflects channel block by a tethered blocking particle that is an integral part of the channel protein. They also emphasize the power of carefully designed molecular mutagenesis experiments in assigning specific channel functions to particular parts of the channel protein. Interestingly, it is *not* the amino terminus that is responsible for the rapid inactivation of cloned sodium channels. Instead, mutagenesis experiments have demonstrated that the intracellular loop between the third and fourth homologous channel domains ❷ is critical for inactivation, by a mechanism that appears to be more complex than the classic ball-and-chain model. ❷

The channel pore. Another clear example of assigning function to particular channel sequences is the identification of the region that forms the ion-selective conduction pore in voltage-dependent potassium channels.

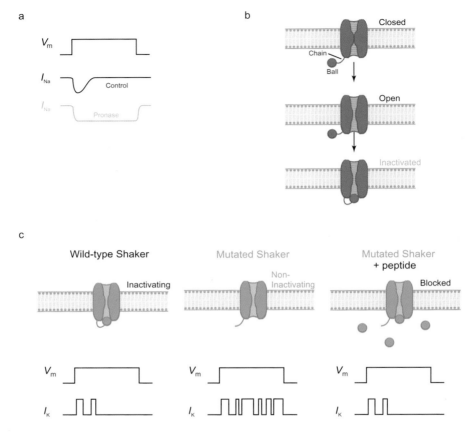

Figure 5–7. Ball-and-chain model for inactivation of sodium and potassium channels. *a*: Sodium channel inactivation before (black) and after (blue) intracellular treatment with pronase. I_{Na}, sodium current. *b*: It was hypothesized that inactivation results from block of the open channel by an intracellular portion of the channel protein. *c*: Removal of the amino-terminal 20 amino acids of *Shaker* by mutagenesis eliminates rapid inactivation. Inactivation of the mutated *Shaker* can be restored by application of a synthetic peptide, whose sequence corresponds to the amino-terminal 20 amino acids, to the cytoplasmic side of the channel. I_K, potassium current; V_m, membrane voltage. See Animation 5–1.◐

Mutagenesis experiments have demonstrated that the scorpion toxin *charybdotoxin* (CTX), a blocker of the external pore of some potassium channels, interacts with amino acids in the region between transmembrane segments S5 and S6 of *Shaker* (Fig. 5–8a). It could also be inferred that the S5–S6 linker must span the membrane, because some amino acids in this linker are accessible to the potassium channel blocker tetraethylammonium (TEA), applied from the *inside* (Fig. 5–8a). This finding

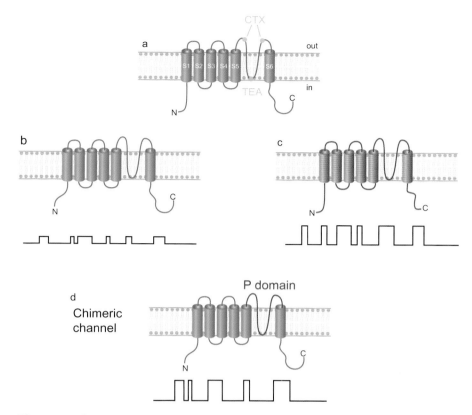

Figure 5–8. The pore domain of potassium channels. *a*: The amino acids marked in blue are critical for binding the pore blockers tetraethylammonium (TEA) from the inside of the channel and charybdotoxin (CTX) from the outside. *b–d*: The region between membrane-spanning segments S5 and S6 determines the conduction and selectivity properties of chimeric channels.

was a surprise, because the S5–S6 linker had originally been modeled as an extracellular loop based on hydrophobicity plots; it serves as a reminder that cartoons of inferred channel structure are subject to revision in the face of real data.

Further evidence that the S5–S6 linker (now called the *P domain*) contributes importantly to the channel pore comes from the finding that mutations in this region can change the ion selectivity and conduction properties of *Shaker* channels. Finally, when the P domains of two different potassium channels with different conduction properties are swapped, the size of the single channel currents of the resulting *chimeric* channels is determined by the small P domain, rather than by the remaining approximately 500 amino acids of the channel sequence (Fig. 5–8b–d).

This experiment provides an unusually clear demonstration of the critical role of the P domain in potassium channel conduction and selectivity; it has been confirmed quite spectacularly by direct structural determination.

The Three-Dimensional Structures of Voltage-Gated Ion Channels

A word of caution is in order concerning all these conclusions from mutational analysis. Site-directed mutagenesis, of course, demonstrates the relation of *sequence*, rather than *structure*, to channel function. Previously it was only possible to infer structure from the sequence information, but now that the real three-dimensional structures of some voltage-gated ion channels have become available, it is interesting to reflect on how accurate some of our earlier cartoons of channel structure were. Nevertheless, there also have been some surprises that require us to modify our picture of channel structure.

A prokaryotic potassium channel. The amino acid sequence of the P domain is remarkably invariant among different potassium-selective channels, even those that may differ substantially in their gating or other properties. Mutations within a critical region of the P domain, the so-called potassium channel *signature sequence*, disrupt the ability of the channel to discriminate between potassium and sodium ions, thereby confirming the essential role of this region in channel conduction and selectivity. Thus it came as a surprise to find such a signature sequence in a protein from the bacterium *Streptomyces lividans*. This protein is predicted from its sequence to have two membrane-spanning domains that flank a P domain, which is different from *Shaker* with its six predicted membrane-spanning domains (Fig. 5–4), but reminiscent of a large family of eukaryotic potassium channels that we shall describe in Chapter 7. This protein, nicknamed *KcsA*, forms a tetramer that can mediate the flux of potassium (but not sodium) when it is reconstituted in lipid vesicles, confirming that it is indeed a potassium-selective ion channel.

The advantage of using a bacterial channel is that it can be purified in large amounts for structural determinations. Roderick MacKinnon and colleagues were able to obtain protein crystals of the purified KcsA protein and determine its three-dimensional structure at high resolution (3.2 Å, which allows one to see atomic detail) using standard techniques of *X-ray crystallography*. This was a remarkable tour de force, because although the structures of many cytosolic proteins had been solved by these techniques, membrane proteins tend to lose their ordered structures when they are removed from the lipid bilayer and have been very difficult to crystallize. In 2003, MacKinnon shared the Nobel Prize in Chemistry for this seminal contribution to understanding ion channel structure and function.

The structure elucidated by MacKinnon's laboratory (Fig. 5–9) is very pleasing to channel aficionados. The four subunits of the tetramer are arranged to form an "inverted teepee," the apex of which points toward the intracellular side of the membrane (Fig. 5–9a). The extracellular entry

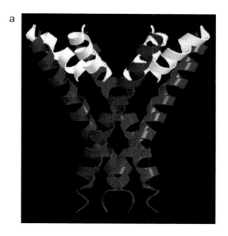

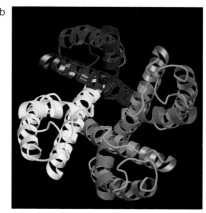

Figure 5–9. Ribbon diagrams to illustrate the three-dimensional structure of the KcsA potassium channel that was solved by Rod MacKinnon and colleagues. *a*: View from the side. Note the "inverted teepee" shape. *b*: View looking down on the entry to the channel from the extracellular side. (Modified from Doyle et al., 1998.)

to the central pore is evident from a view looking down on the membrane (Fig. 5–9b). As had been predicted many years earlier from biophysical measurements, the portion of the channel that selects for potassium over other ions (the selectivity filter) is a narrow region toward the extracellular surface of the membrane (Fig. 5–9b). Two potassium ions can occupy the selectivity filter simultaneously, with a third in a water-filled cavity deeper in the pore, again in accord with predictions from long ago.

The high-resolution three-dimensional structure of KcsA confirms beautifully many of the predictions from mutational analysis about potassium channel conduction and selectivity. Indeed, this structure goes considerably further, providing an elegant and satisfying molecular explanation for the high conduction coupled with exquisite selectivity that is a hallmark of potassium channels. Although questions about the structural changes associated with channel gating were not addressed by the KcsA studies, the structure of another voltage-gated potassium channel, a large conductance voltage- and calcium-dependent potassium channel that we shall discuss in Chapter 7, has been solved in both its closed and partially opened states (Fig. 5–10). A comparison of the different states reveals the intramolecular transitions that the channel protein undergoes in going from the closed to the open state and back again (see Video 5–1) ◐.

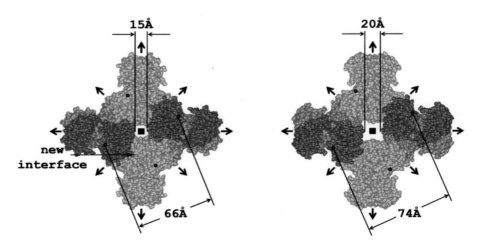

Figure 5–10. Structure of a calcium-activated potassium channel in the open and closed states. Calcium binding results in a change in the conformation of the channel protein. Note the different dimensions of the central pore in the closed (*left*) and open (*right*) states. (Modified from Ye et al., 2006; see also Jiang et al., 2002.) See Video 5–1. ◐

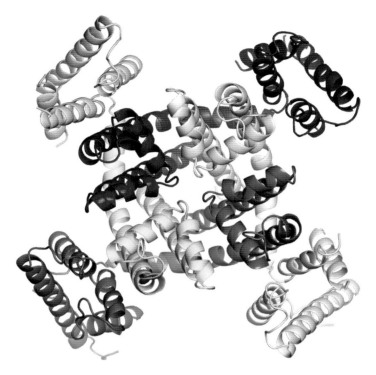

Figure 5–11. Ribbon diagram to illustrate the three-dimensional structure of a bacterial sodium channel. Compare with the potassium channel structure in Figure 5–9. Image provided by Nieng Yan. (Modified from Zhang et al, 2012; see also Payandeh et al., 2011.)

Structure of a voltage-gated sodium channel. The idea of searching for bacterial ion channels that can be purified in large quantities has proven fruitful for more than potassium channels. Several laboratories have determined the structures of bacterial voltage-gated sodium channels. The sodium channel structure, also very beautiful (Fig. 5–11), bears some similarities to potassium channel structures, but it also reveals differences that help to explain the unique properties of sodium channels (compare with Fig. 5–9). Other investigators have used the same bacterial protein purification strategy to determine the structures of chloride and other ion channels.

Summary

Advances in channel molecular biology and direct structural determination have made it clear that there are notable similarities in the molecular structures of voltage-gated sodium and potassium channels. Both have 24 membrane-spanning segments, in addition to a domain that contributes to the channel pore. In sodium channels, this overall structure is achieved with a single primary subunit that contains four homologous domains, each with six membrane-spanning segments. In contrast, the primary subunit of potassium channels is much smaller and resembles one of the four homologous domains in the sodium channel; four of these primary subunits come together to form a functional potassium channel. A variety of experiments have revealed channel protein regions involved in such functions as voltage-dependent activation, inactivation, and ion selectivity. An interesting finding from these experiments is that at least some functions of voltage-dependent channels do not involve global changes in protein conformation but can be assigned to discrete structural modules in the channel protein. Many of the inferences from mutational analysis, particularly those related to channel gating, conduction, and selectivity, have been confirmed and extended in spectacular fashion by the elucidation of the three-dimensional structures of several voltage-dependent ion channels.

6

Ion Channels, Membrane Ion Currents, and the Action Potential

A great deal of what we know about ion channel function, including information about selectivity and gating of different types of channels, was already understood long before single channel recording and the application of molecular and structural approaches revolutionized the field. This understanding developed because, for more than 60 years, the voltage clamp invented by K. C. Cole allowed us to record the macroscopic membrane current in some cells. As we discussed in Chapter 4, the *macroscopic current* is the combined current flowing simultaneously through all the active ion channels in the cell. In this chapter, we shall expand on this topic, to describe in detail just how such voltage clamp measurements of ion currents have led to a detailed mechanistic understanding of the action potential. Let us begin with a review of the fundamental physicochemical concept of the *equilibrium potential*, which is essential for understanding all electrical phenomena in biological membranes. We will then go on to consider how voltage- and time-dependent sodium and potassium currents combine to give rise to the action potential. It is important to keep in mind throughout that there is nothing mysterious about the phenomena we are discussing here—they arise logically from the properties of just a few kinds of membrane proteins, the sodium and potassium channel proteins, which (as we have seen in Chapter 5) are becoming increasingly understood in molecular and structural detail.

Ionic Equilibria and Nernst Potentials

The rate of flow of an ion across the plasma membrane is determined by

1. the *concentration gradient*, the difference in the concentrations of the ion on the two sides of the plasma membrane;
2. the *voltage difference* across the plasma membrane; and
3. the *conductance* of the ion channels, the ease with which ions move through the ion channels across the plasma membrane.

The simplest example. Consider the case (Fig. 6–1) of a plasma membrane that separates two aqueous solutions, representing the inside and outside of a cell, each solution containing only the generic ions X^+ and Y^-. This hypothetical membrane contains ion channels selective for X^+, but none for Y^-, so only X^+ can cross the membrane. Let us suppose further that the concentration of X^+ and Y^- on one side of the membrane (*inside* the cell) is 10 times as high as it is on the other side (*outside* the cell). In other words, $[X^+]_i = 10[X^+]_o$ and $[Y^-]_i = 10[Y^-]_o$. Suppose, in addition, that initially there is no voltage difference across the membrane (that is, $V_m = 0$). Furthermore, there is no net charge on either side of the membrane, because the charges on X^+ and Y^- cancel each other. In the first instant after this condition is set up (left side of Fig. 6–1), there will be a tendency for X^+ to diffuse down its concentration gradient from inside to outside the cell, via its selective ion channel. This will, of course, cause a redistribution of charge across the membrane; the inside of the membrane has lost some of its positive charge and the outside has gained some, so there will now be a voltage gradient across the membrane with the inside negative relative to the outside. This voltage gradient will tend to slow the diffusion of X^+, since the positive ion does not want to leave the region of negative charge. This continues over time, with the flow of X^+ (the ion current across the membrane) becoming slower and slower, until eventually the voltage gradient becomes large enough to oppose the concentration gradient. At this point there is no longer any net flow of X^+, and hence the voltage is no longer changing (right side of Fig. 6–1). It is important to note that the number of ions that flows across the membrane to give rise to the voltage difference is very small relative to the total number of ions in the intracellular and extracellular compartments. In other words, the voltage difference across the membrane is established without any significant change in the ion concentration gradient.

The voltage required to exactly oppose the flow of any given ion X is called the *equilibrium potential* (E_X) for that particular ion (Fig. 6–1). It

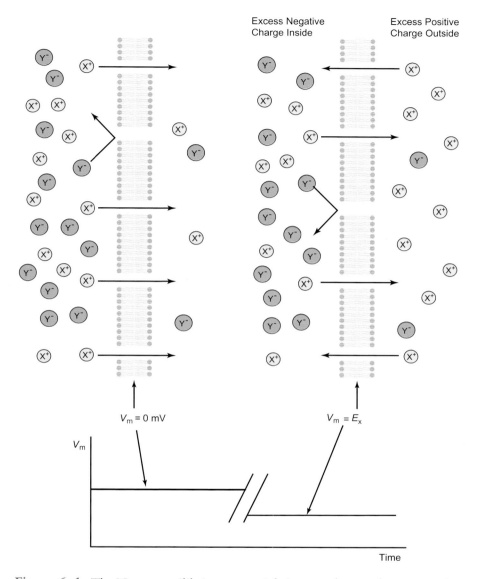

Figure 6–1. The Nernst equilibrium potential. Assume the membrane contains ion channels selective for a single charged ion, X^+, which is asymmetrically distributed on the two sides of the membrane. The counterion, Y^-, cannot cross the membrane. X^+ will flow across the membrane from the side of high concentration to that of low concentration (*left*), until the buildup of charge is sufficient to oppose net ion flow (*right*). The transmembrane voltage (V_m) is 0 before X^+ begins to flow. At equilibrium $V_m = E_x$.

will at once be evident that E_X is entirely dependent on the transmembrane concentration difference. Near the end of the nineteenth century, the physical chemist Walther Nernst described the relationship of E_X to [X⁺]i and [X⁺]o with the equation that bears his name:

$$E_X = \frac{RT}{zF} \log_e \frac{\left[X^+\right]_o}{\left[X^+\right]_i}$$

where R is the gas constant, T is the temperature in degrees Kelvin, z is the charge on the ion, and F is the Faraday (the amount of charge in coulombs carried by a mole of monovalent ions). For monovalent ions at room temperature (approximately 20°C), the Nernst equation reduces to

$$E_X = 58 \log_{10} \frac{\left[X^+\right]_o}{\left[X^+\right]_i}$$

For the 10-fold inside-to-outside concentration gradient we have been discussing here, it is evident that E_X will be −58 mV; in other words, when the inside of the cell is 58 mV more negative than the outside, the voltage gradient will balance the concentration gradient and there will be no net ion flow. As long as X⁺ is the only ion able to cross the membrane, the voltage across the membrane, V_m, will be equal to E_X. A change in the concentration gradient will, of course, lead to a shift in E_X and a corresponding shift in V_m.

A real membrane. Let us now return to the real world and consider what happens when there are several ions with different concentration gradients across the membrane, and several kinds of ion channels selective for these different ions. The example we shall use is the membrane of the giant axon of a marine mollusc, the squid *Loligo*, which has been of enormous value in elucidating the physicochemical mechanisms underlying action potential generation and propagation. The extracellular medium for the squid giant axon contains ions at concentrations very similar to those in seawater, and the following relationships are *approximately* true for the ionic gradients across this membrane:

1. $[K^+]_i = 20[K^+]_o$
2. $[Na^+]_o = 10[Na^+]_i$
3. $[Cl^-]_o = [Cl^-]_i$ (This is not strictly correct, since chloride ion concentration is somewhat higher outside than in; however, negative

charges on intracellular organic macromolecules more or less balance out the excess extracellular chloride.)

There are, of course, other ions both in seawater and inside the cell, but these are either present at low concentrations or do not have ion channels that permit them to cross the membrane. Accordingly, they contribute very little to the transmembrane flow of ion current, and we may restrict our discussion to the major charge carriers, potassium, sodium, and chloride.

The axonal membrane also contains sodium-, potassium-, and chloride-selective ion channels, and for the squid axon (and most other axons) at rest the following relationships also hold:

4. The permeability of the membrane to chloride (P_{Cl}) is essentially 0 (that is, $P_{Cl} = 0$). In other words, the chloride channels are always closed.
5. The permeability to sodium (P_{Na}) is very low. In other words, the sodium channels are mostly closed.
6. The permeability to potassium (P_K) is relatively high. In other words, the potassium channels are often open.

How then can we determine the membrane potential given these relationships? The Nernst equation can describe the membrane potential when only a single ion X^+ can flow, since under these conditions $V_m = E_X$. Many years ago David Goldman, and independently Alan Hodgkin and Bernard Katz, derived the following equation to describe the membrane potential in terms of the concentration gradient and permeation properties of several different permeant ions:

$$V_m = \frac{RT}{zF} \log_e \frac{K_o + [P_{Na}/P_K]Na_o + [P_{Cl}/P_K]Cl_i}{K_i + [P_{Na}/P_K]Na_i + [P_{Cl}/P_K]Cl_o}$$

When the chloride and sodium channels are closed (that is, when P_{Cl} and P_{Na} are zero) and the membrane is permeable only to potassium, the permeability terms in both the top and bottom of the Goldman-Hodgkin-Katz equation are zero. Under these conditions, the equation reduces to the Nernst equation and the membrane potential is determined only by the single permeant ion, potassium. When V_m is measured experimentally in the squid axon (and in many other nerve cells), it is usually found to be very close to, but slightly less negative than, the equilibrium potential for potassium. In the particular case of the squid axon, for example, V_m is usually about −70 mV, whereas the E_K calculated from the potassium concentration gradient by the Nernst equation is −75 mV. The fact that the measured V_m is usually slightly less negative than E_K reflects the fact that the resting sodium permeability, although small, is not zero. That is, the sodium channels may be open occasionally at rest, although much less so than the potassium channels. Thus, to a limited extent, sodium also contributes to setting the resting membrane potential.

This resting potential, then, is exactly large enough to balance the ion flow caused by the various permeant ions with their different concentration gradients and membrane permeabilities. At this voltage the net charge movement is zero. It is important to note that when only a single type of ion can cross the membrane, the system comes to *equilibrium* and there is no net flow of that ion (Fig. 6–1). In the squid axon and other real cells, however, V_m is not exactly equal to the equilibrium potential for any of the permeant ions. No individual ion is at equilibrium, and each will continually flow down its own concentration gradient. Thus there will be some current (I) carried by each ion. In this case, the membrane is at a steady state rather than equilibrium. The total membrane current (I_m), which must be zero because the voltage is not changing, is the sum of the currents carried by the individual ions. In other words, I_m is given by the following:

$$I_m = I_1 + I_2 + I_3 + \cdots + I_n = 0$$

where I_1, I_2, and so on are the currents carried by n different ions. It can be seen that for this sum to be equal to zero, different currents must have different signs. By convention, as we have mentioned in Chapter 4, the flow of positive ions across the membrane into the cell (inward current) is considered to be negative, and the flow of positive ions out of the cell (outward current) is positive. The opposite holds for the flow of negative ions.

If the total I_m is zero, and only sodium and potassium can flow, the currents carried by sodium and potassium (I_{Na} and I_K, respectively) must be equal and opposite. That is, $I_{Na} = -I_K$. How can this be, when P_K is so much greater than P_{Na} (that is, the potassium channels are open so much more than the sodium channels)? The answer is that I for any given ion is dependent on more than just whether its channels are open or closed. From Ohm's law, which we introduced briefly in Chapter 4, we know that the current flow between two points depends on the voltage difference (V) and resistance to current flow (R) between those points:

$$I = \frac{V}{R}$$

For the flow of an ion X$^+$ across a membrane, the relevant voltage difference is $(V_m - E_X)$ and is called its *driving force*. We can consider that R is equivalent to the *inverse* of the permeability for that ion (intuitively we can see that *permeability*, a measure of the *ease* of ion flow, is inversely related to the *resistance* to ion flow). In reality, R is actually the inverse of the *electrical conductance* (G), a measure of the ease of ion flow that is similar to, but not identical with, permeability. For our purposes, however, the terms *conductance* and *permeability* both reflect the extent to which ion channels are open and may be used interchangeably. Accordingly, we can rewrite Ohm's law to describe the current carried by any ion as follows:

$$I_X = (V_m - E_X)G_X$$

Since $I_{Na} = -I_K$ at rest, it must follow that

$$(V_m - E_{Na})G_{Na} = -(V_m - E_K)G_K$$

Now, remember that V_m is very close to E_K, so the driving force for potassium flow $(V_m - E_K)$ is very small. By contrast, the driving force for sodium $(V_m - E_{Na})$ is large enough to generate an inward sodium current equal to the outward current carried by potassium, in spite of the very much lower sodium conductance. This point, that the current carried by a given ion is dependent on *both* the membrane conductance and the driving force for that ion, is fundamental. We will return to it later in the context of the currents that flow during the action potential.

Let us now suppose that G_{Na} suddenly becomes much higher than G_K and is maintained at this high level (Fig. 6–2a). Initially, sodium ions will rush into the cell down their concentration gradient and the ion current carried by sodium (I_{Na}) will increase (Fig. 6–2b). At this time there is a net current, and the system is no longer at a steady state. As positively charged sodium ions build up inside the cell, the membrane depolarizes (Fig. 6–2c). This depolarization brings the membrane potential, V_m, closer to E_{Na}, thereby decreasing the driving force for sodium, and so I_{Na} will begin to decrease again (Fig. 6–2b). At the same time because the V_m is farther from E_K, I_K increases as a result of the increased driving force for potassium. These changes in the sodium and potassium currents combine to slow the rate of change of V_m (Fig. 6–2c). Eventually, a new steady state is reached at a different voltage. I_K and I_{Na} are again equal and opposite but are larger than they were before, reflecting the increase in total membrane conductance.

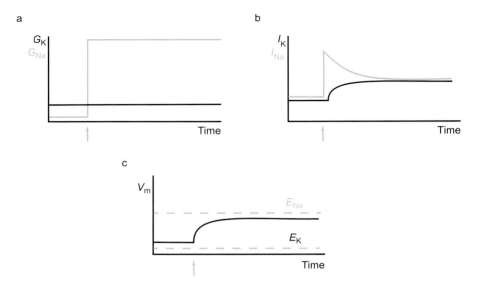

Figure 6–2. Conductance changes lead to voltage changes. Effect of a change in sodium conductance (G_{Na}) (**a**) on membrane currents carried by sodium (I_{Na}) and potassium (I_K) (**b**) and on the membrane voltage (V_m) (**c**). The time at which the sodium conductance is changed is indicated with blue arrows. E_K, potassium equilibrium potential; E_{Na}, sodium equilibrium potential; G_K, potassium conductance.

From these considerations we can see intuitively what the Goldman-Hodgkin-Katz equation tells us mathematically. When the ion concentrations are kept constant, V_m depends on the relative values of G_K and G_{Na}. Figure 6–3 is a simple graph that describes the V_m when G_K is fixed and G_{Na} is varied from much smaller than to much greater than G_K. The two extremes are the limiting cases of the Goldman-Hodgkin-Katz equation, where it reduces to the Nernst equation and V_m is equal to either E_K or E_{Na}; intermediate conductance ratios lead to intermediate values for V_m.

Since V_m is a function only of ion conductance and concentration gradient, in theory a change in either or both of these parameters could be used to alter V_m. However, it seems likely that changes in concentration gradients sufficiently large to produce significant changes in V_m would be very slow and would disrupt many other cellular functions. On the other hand, the opening and closing of ion channels in the plasma membrane can be modulated very rapidly by a variety of mechanisms. We shall see throughout this book that a selective change in the activity of one or another ion channel, with a consequent change in the membrane

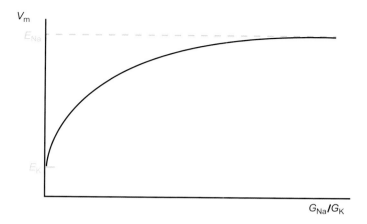

Figure 6–3. The membrane potential depends on the ratio of the membrane conductances for sodium and potassium. When the sodium conductance (G_{Na}) is very low, the membrane potential (V_m) approaches the potassium equilibrium potential (E_K); in contrast, when the sodium conductance is much higher than the potassium conductance (G_K), the membrane potential approaches the sodium equilibrium potential (E_{Na}).

conductance for the ion that flows through that channel, is used routinely by nerve cells as a means of changing V_m and rapidly producing meaningful electrical signals.

Ion selectivity revisited. Let us return to the question of ion selectivity of channels, which we introduced in Chapter 4, in the context of this concept of the Nernstian equilibrium potential. An important clue to the ion selectivity of a channel is provided by the *reversal potential* (V_r) for the single channel current, the voltage at which no current flows through the channel even when it is open. Recall that for the hypothetical channel illustrated in Figure 4–3, this occurs at −75 mV. At this potential the channel may still open and close, but no net flow of ions occurs during the openings. From our discussion of Nernstian equilibrium potentials it will be evident that the reversal potential for the current through a channel is equal to the equilibrium potential for the ion that passes through the channel. Thus, we may infer that current through the channel in Figure 4–3 is probably carried by potassium ions. Because E_K is very close to −75 mV, there will be no net flow of potassium at this voltage, even when the channel is open. It can be seen that these considerations for single channels are identical to those for total membrane currents, which result from the summed activity of large numbers of single channels.

As with the total membrane current, the Nernst equation provides a ready test for the ion selectivity of a single channel. For example, for a potassium channel, increasing the extracellular concentration of potassium (that in the patch pipette) by a factor of 10 should shift the reversal potential by 58 mV to a more positive potential. Altering potassium concentration, however, should not alter the reversal potential of channels selective for other ions, such as sodium ions, if no change has been made in sodium concentration. Some channels, however, can allow more than one species of ion to cross the membrane. To determine the extent to which a potassium channel selects for potassium over other ions, these manipulations should be made in the presence of other potential permeant ions, for example, sodium ions. A channel that strongly prefers potassium over sodium will always exhibit a reversal potential at E_K, whatever the sodium concentration on either side of the membrane. If, however, the reversal potential deviates from E_K in the presence of sodium ions, this indicates that the channel is not completely selective and allows some sodium to flow through it.

Again, we may compare this with the total membrane current in a neuron at its resting potential. This current is equal to zero (that is, it reverses its sign) near but not precisely at E_K, because the membrane is permeable mostly but not exclusively to potassium. One possible interpretation of this finding is that there is one class of ion channels, permeable mostly to potassium and slightly to sodium, that is responsible for the total membrane current. Such channels, called *nonselective cation channels*, do, in fact, exist. We know, however, that most channels in neurons are selectively permeable to one or another ion. The combination of a high potassium and low sodium permeability comes about because under resting conditions the potassium channels are sometimes open (and allow current to pass), whereas the sodium channels are closed virtually all of the time.

One difficulty in the interpretation of whole-cell macroscopic current recordings is that the membrane of a neuron has many different types of ion channels, which are selective for different ions and whose gating is influenced in different ways by voltage and neurotransmitters. Is it indeed possible to measure *selectively* the whole-cell sodium current, as we have shown in Figure 4–8? One advantage of heterologous expression of sodium (or other) channels in low-background cell types (see online Figs. 5A–1 and 5A–3) ❽ is that the expressed channels are generally the dominant channels in the plasma membrane and can be studied without contamination by other channel types. In real neurons, although it is not always easy, a careful choice of voltages and the use of drugs that block other channels often allow one to record currents that represent the opening and closing of a single class of ion channel. This will become evident in our following discussion of the ion currents responsible for the action potential and for determining patterns of neuronal firing.

Ionic Mechanisms of the Action Potential

Changes in ion channel activity such as those just described are exactly what happens during the nerve impulse. We will begin by summarizing the sequence of changes during an action potential (Fig. 6–4); then we will discuss in depth the experimental evidence for this sequence of events.

When an axon is depolarized beyond the action potential threshold, the depolarization itself causes large numbers of voltage-dependent sodium channels to open. This is seen as a rapid increase in G_{Na} (Fig. 6–4a), which quickly rises to a level very much higher than G_K (as in Fig. 6–2). As a result, inward sodium current increases (Fig. 6–4b), the membrane depolarizes further, and V_m approaches E_{Na} (Fig. 6–4c). In contrast to the situation described in Figure 6–2, however, a further series of changes occurs *as a result of the depolarization*. By the peak of the action potential there is (1) sodium channel inactivation, and hence a rapid decrease in G_{Na} back to its resting level, and (2) a slower increase in G_K. As a result of (1), the inward I_{Na}, which had been transiently very large, begins to drop and, somewhat more slowly, as a result of (2), I_K begins to increase. As the outward I_K becomes larger than the inward I_{Na}, the net current flow is now outward (hyperpolarizing), and it begins to drive V_m back toward the resting level. Notice that the increase in G_K is prolonged compared to that in G_{Na} (Fig. 6–4a), and thus for some period after the spike, the V_m may actually be more negative than the normal resting potential (Fig. 6–4c), that is, closer to E_K. This phenomenon is called the *afterhyperpolarization* of the spike. Finally, G_K begins to decrease again, and at some time after the end of the action potential the membrane conductances have returned to their normal resting levels. The time course of these changes (Fig. 6–4) can vary from cell to cell, but in general, axonal action potentials are very fast. In the squid giant axon, for example, this entire sequence of events is over in a few milliseconds. In other cells, for example, in certain neurons in the pituitary gland of vertebrates, action potentials may last for tens of milliseconds; in cardiac muscle cells they may be hundreds of milliseconds long.

Voltage clamp studies of the squid giant axon. The sequence of events described in Figure 6–4 is of fundamental importance, and the reader is encouraged to examine this figure carefully before proceeding further. The

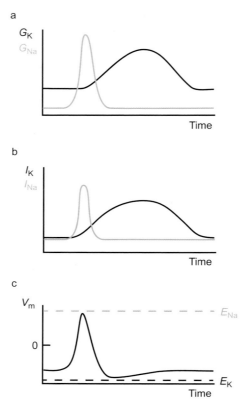

Figure 6–4. Membrane conductances, currents, and voltage during an action potential. *a*: The first change during an action potential is an increase in the sodium conductance (G_{Na}) to a level much greater than at rest. This is followed by a slower and longer-lasting increase in the potassium conductance (G_K). *b*: The sodium (I_{Na}) and potassium (I_K) currents change in accordance with the changes in conductance. *c*: The result of these changes is the characteristically shaped action potential. E_K, potassium equilibrium potential; E_{Na}, sodium equilibrium potential; V_m, membrane potential. See Hodgkin and Huxley (1952a,b,c) and Hodgkin et al. (1952). A sophisticated quantitative analysis is presented by Hodgkin and Huxley (1952d).

studies that led to the elucidation of these events represented a partnership between several brilliant investigators on the one hand and a magnificently well-suited experimental preparation on the other. The work of J. Z. Young, K. C. Cole, Alan Hodgkin, and Andrew Huxley on the squid giant axon ranks among the great success stories of twentieth-century biology. Hodgkin and Huxley received the 1963 Nobel Prize in Physiology or Medicine for their studies, and some have remarked that it is unfortunate that the Atlantic squid *Loligo* cannot be similarly honored.

Young, a British zoologist, found in the mid-1930s that the mantle of the squid is innervated by a giant axon up to 1 mm in diameter. The giant axon arises from the fusion of a large number of smaller neurons. It can be removed from the animal, and the axoplasm can be extruded and replaced by saline solutions of defined ionic composition; in other words, the transmembrane ion gradients can be manipulated by the experimenter. The axonal plasma membrane is surprisingly robust and survives this maltreatment with its electrical properties intact. The large size of the axon makes it easy to place electrodes both inside and outside the membrane to measure (and control) the transmembrane voltage. Recall from Chapter 4 that the importance of controlling as well as measuring the voltage arises from the fact that the sodium and potassium conductances themselves change as a function of voltage, and the membrane is not at steady state during the action potential. It will be evident that depolarization produces an increase in G_{Na}, which then produces further depolarization. This in turn further increases G_{Na}, and an unstable positive feedback loop results that gives rise to the *regenerative* all-or-none action potential. Accordingly, the only way to study the regulation of membrane conductance effectively is to measure the conductance properties at fixed voltages, by means of K. C. Cole's voltage clamp that we described near the end of Chapter 4.

Voltage- and time-dependent ion currents. Hodgkin and Huxley (as well as Cole) immediately recognized the importance of controlling the membrane voltage, and the experimental convenience of working with a large and manipulable axon. They carried out a series of seminal experiments (interrupted by World War II) on voltage-clamped squid giant axons. The results of these studies were published in a classic series of papers by Hodgkin and Huxley (one of them in collaboration with Bernard Katz) in 1952. They are wonderfully insightful papers, not easy reading, but essential for the serious student of neurophysiology and membrane biophysics.

Hodgkin and Huxley asked what happens when one voltage clamps the axon near the resting potential, and either hyperpolarizes it or depolarizes it before returning to the original voltage (the *holding potential*). As shown in Figure 6–5a, small hyperpolarizing or depolarizing pulses of the same size produce small inward or outward currents, respectively. These *leak currents* are not time dependent (they reach their maximum amplitude in a time that is too short to be resolved by the recording instrumentation), and they are the same size in inward and outward directions for equal-amplitude hyperpolarizations and depolarizations. When the current amplitude is plotted against the voltage during the pulse (the

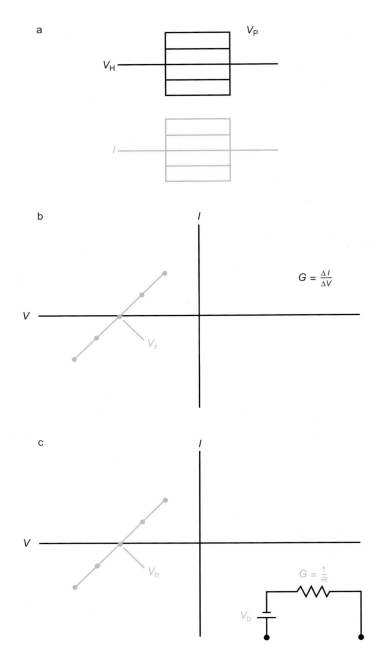

Figure 6–5. Leak currents. *a*: When small depolarizing or hyperpolarizing voltage clamp steps are made from the holding potential (V_H) to a pulse potential
(V_p), small time-independent currents (I) are seen. *b*: A plot of these currents as a
function of voltage (V). V_r, reversal potential. *c*: A similar current–voltage relationship is seen for a linear resistor in a nonbiological electrical circuit, containing
a battery of voltage (V_b).

pulse or command potential) for these small hyperpolarizations and depolarizations, the resulting *current–voltage (I–V) relationship* is a straight line (Fig. 6–5b). Such a straight line *I–V* relationship is also seen for a linear resistor in a nonbiological electrical circuit (Fig. 6–5c). Time- and voltage-independent ion channels contribute to the leak current. It can be seen that the *I–V* curve intersects the zero current axis at V_r. This is not surprising, since V_r is defined as the voltage at which the net current flow is zero.

This linear leak current is all that is seen for hyperpolarizing voltage clamp pulses, whatever their amplitude. What happens when a larger pulse, one that normally exceeds the threshold for action potential generation, is given in the depolarizing direction? Remember that the membrane voltage is held constant by the voltage clamp, so no action potential is permitted to occur. It becomes immediately obvious that the membrane current is *not* at steady state during these larger depolarizing pulses, and time-dependent currents flow (Fig. 6–6a). There is an inward-going (negative by convention) current during the first millisecond or two after the beginning of the depolarizing pulse, and then the current reverses sign and becomes outward or positive during the remainder of the pulse. When a series of such pulses is given to different depolarizing voltages, a family of curves is generated, as shown in Figure 6–6b. Note that as the amplitude of the voltage pulse increases, the early inward component of the current (blue) first increases, and then begins to decrease until it reverses sign and becomes outward at very large depolarizations. In contrast, the later outward component of the current (black) remains outward and continues to increase in amplitude with larger depolarizations.

How can one interpret these complex voltage- and time-dependent ionic currents? The relationship between the imposed membrane voltage and the membrane current that flows at that voltage can be investigated in more detail by constructing *I–V* curves. The early and late components of the current can be examined separately by measuring the current at different times. The peak inward current (usually after about 1–2 msec in the squid axon) is taken as a measure of the early component, and the current near the end of the pulse is the late component (also called by Hodgkin and Huxley the *delayed outward current*). As we shall see later in this chapter, the early and late components can also be separated on the basis of other criteria, confirming that it is appropriate to make this distinction on the basis of their kinetic properties. The *I–V* curves in Figure 6–6c extend the conclusions drawn from an inspection of the current traces themselves. Both components of the current exhibit markedly non-linear *I–V* curves, the early inward component first increasing and then decreasing in amplitude, then reversing in sign. The late outward

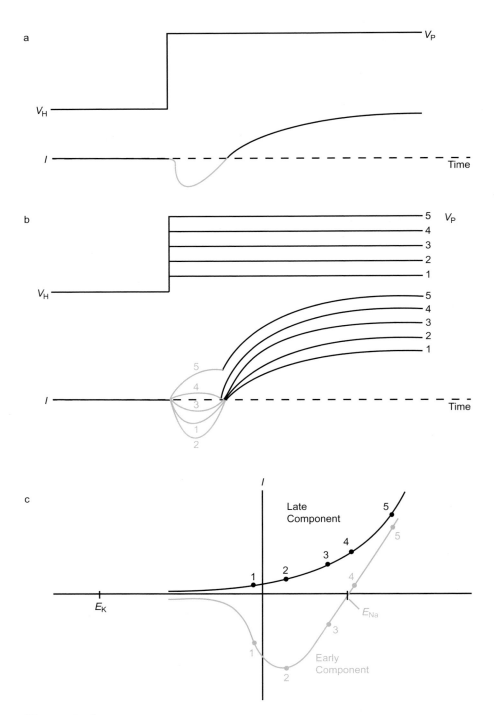

Figure 6–6.

component continues to increase (with increasing slope) over the entire voltage range examined. For these large depolarizations, then, the membrane is no longer behaving like a linear resistor; rather, it *rectifies*, and the membrane conductance exhibits voltage dependence in this voltage range. As we know from our consideration of voltage-dependent ion channel gating in Chapter 4, these conductance changes result from voltage-dependent changes in the open probability of the channels responsible for the membrane currents.

Sodium and potassium carry the inward and outward membrane currents. Hodgkin and Huxley noted that the early inward current has the right sign, amplitude, voltage dependence, and kinetics to be responsible for the upstroke of the action potential. For example, the fact that the inward current turns on only at voltages depolarized from rest provides an explanation for the phenomenon of a threshold for action potential generation. Similarly, the delayed outward current has the right characteristics to be responsible for the repolarization. They subsequently went on to ask which ions are the charge carriers for the inward and outward currents, using a series of *ion substitution* experiments. When the sodium was removed from the extracellular medium and replaced by an equivalent amount of some nonpermeant monovalent cation (for example, choline), the late component of the current was not affected, but the early component was outward over the entire voltage range examined (Fig. 6–7a). This is because the sodium concentration gradient is now reversed, and when the sodium channels open, sodium leaves rather than enters through the channels. When the extracellular sodium concentration is varied over a wide range, so as to systematically vary E_{Na}, it can be seen that the V_r for the early current is always equal to E_{Na} (Fig. 6–7b). This result confirms that this component of the membrane current is carried entirely by sodium and there is no significant contribution by any other ion.

The delayed current is not affected by the extracellular sodium concentration. Hodgkin and Huxley suspected that the delayed current was

Figure 6–6. Nonlinear voltage-dependent currents. *a*: When the depolarizing pulses are made much larger than those in Figure 6–5, time-dependent currents (*I*) are seen to flow. During a sustained depolarization, there is an early component of the current that is inward (blue) and a later component that is outward (black). V_H, holding potential; V_p, pulse potential. *b*: When a series of pulses is given to different depolarizing pulse potentials, the currents change with voltage in a characteristic way. *c*: Plot of the early and late peak currents (*I*) as a function of voltage (*V*) (see Hodgkin and Huxley, 1952a,b,c; Hodgkin et al., 1952). E_K, potassium equilibrium potential; E_{Na}, sodium equilibrium potential.

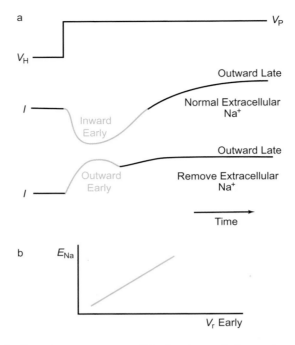

Figure 6–7. Sodium ions are responsible for the early inward current. *a*: When sodium ions are removed from the extracellular solution, the early phase of the current *I* (blue) changes from inward to outward because of the change in the sodium equilibrium potential. The late component of the current, on the other hand, is not affected by manipulation of the sodium concentration. V_H, holding potential; V_p, pulse potential. *b*: Plot of the reversal potential (V_r) of the early current as a function of the sodium equilibrium potential (E_{Na}). The relationship is exactly that predicted by the Nernst equation for a current carried exclusively by sodium ions.

carried by the outward flow of potassium ions, but they were unable to confirm this directly because they had difficulty changing the intracellular potassium concentration without damaging the axons. However, subsequent ion substitution experiments, on squid axon and other cell types, confirmed that V_r for the delayed current is always equal to E_K, demonstrating that potassium, and only potassium, is the charge carrier for this current component. Because of its kinetics, and the voltage-dependent gating that gives rise to rectification in the *I–V* curve, this current is often called the *delayed rectifier potassium current*.

Pharmacological tools have also proven to be extremely useful in separating the two components of axonal current. For example, we have seen that the Japanese puffer fish toxin tetrodotoxin (TTX) is a potent and

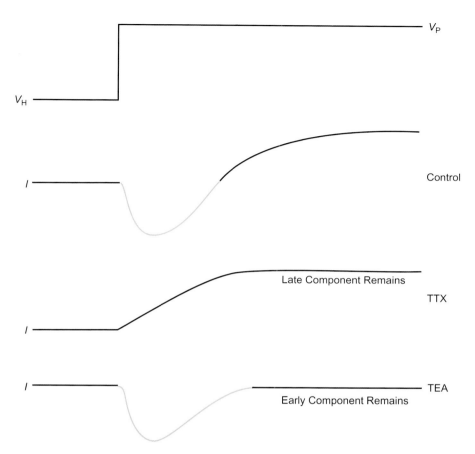

V_P

V_H

Control

I

Late Component Remains

TTX

I

TEA

I

Early Component Remains

Figure 6–8. Pharmacological agents selectively block the early and late components of the current (I). Tetrodotoxin (TTX) blocks the early component of the current, whereas tetraethylammonium (TEA) blocks only the late component. V_H, holding potential; V_p, pulse potential.

selective blocker of sodium channels. In the presence of TTX, the early sodium component of the current is eliminated, and the delayed potassium component can be studied in isolation (Fig. 6–8).

It is interesting that the puffer fish is a delicacy in Japan, and chefs are specially trained to remove the TTX-containing organs for the preparation of this dish; however, these organs are not removed entirely, because the tingle one gets from a small dose of TTX is apparently one of the major reasons this dish is so popular. In spite of the undoubted skill of these highly trained chefs, mistakes sometimes are made, and there are still occasional deaths in Japan from TTX poisoning.

The story of potassium channel blockers has (until recently) been less colorful. Several organic compounds with quaternary ammonium groups, the most useful of which is tetraethylammonium (TEA; see Fig. 5–8), are selective blockers of the delayed rectifier potassium current in squid axon, allowing the early sodium current to be examined in isolation (Fig. 6–8). As we shall see in Chapter 7, there can be many kinds of potassium currents in nerve cells, and these blockers do not affect all potassium currents to the same extent. The pharmacology of potassium channels has also expanded with the discovery that certain toxins from bees, snakes, scorpions, and other creatures can selectively block certain kinds of potassium channels. This rich pharmacopeia has proven extremely useful for the biochemical purification of potassium channels, and for probing their function and structures.

Sodium current inactivation. An essential feature of an action potential is the activation of the sodium current by membrane depolarization. But as we have seen, depolarization not only makes the sodium current turn on, it also causes it to turn off very soon thereafter. This *inactivation* occurs during the depolarizing pulse. It is distinct from *deactivation* (or reversal of activation), which occurs after the end of the pulse as a result of the return of the membrane voltage to the hyperpolarized holding potential. Inactivation of the sodium current, of course, reflects the inactivation of individual sodium channels (Fig. 4–8). An examination of Figure 6–8 makes it clear that the switch from net inward to net outward current during a depolarizing pulse is due not only to the slow turning on of the delayed outward potassium current but also to the fact that the opposing inward sodium current has inactivated and gone to sleep.

Figure 6–9 summarizes in cartoon form the sequential activation, inactivation, and recovery from inactivation of sodium channels. Under resting conditions (Fig. 6–9, upper left) the activation gate is closed and no current flows. When the membrane is depolarized (right side of Fig. 6–9), this activation gate undergoes voltage-dependent opening. As shown on the upper right of Figure 6–9, the channel is now open and current can flow. The same depolarization that opens the activation gate also leads to slower closing of the *intracellular* inactivation gate (Fig. 6–9, lower right). Although the membrane is still depolarized and the activation gate is still open, no current can flow, because the inactivation gate is closed. After the end of the depolarizing pulse, deactivation, the reversal of activation, occurs and the activation gate closes. Again, no current can flow (Fig. 6–9, lower left). At this time, a depolarization cannot evoke any current, because even though the activation gate will open as a result of the depolarization, the inactivation gate remains closed for some time following

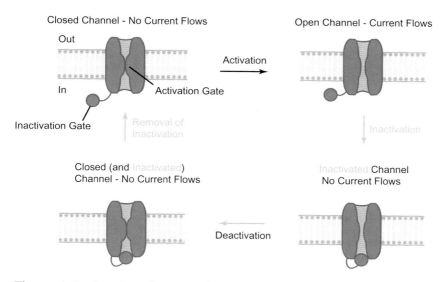

Figure 6–9. Opening, closing, and inactivation transitions in voltage-dependent sodium channels. The gating transitions in sodium channels during and following a depolarizing pulse are summarized. See text for details. See Animation 5–1.◔

the pulse. Only after inactivation is removed is the channel back in its resting state (Fig. 6–9, upper left) and available to be opened by another depolarization. ❷

The functional implications of the long time course of recovery from sodium channel inactivation are profound. Since the peak of the action potential is a voltage at which inactivation will be complete (Fig. 6A–1c) ❷, sodium current will be inactivated for some period of time, and the cell will be unable to fire a second action potential until there is sufficient recovery from this inactivation. It will be evident, then, that *sodium channel inactivation is the ionic mechanism underlying the refractory period*. The elevated threshold during the refractory period reflects sodium channel inactivation, and the return of the threshold to normal (Fig. 3–6) corresponds to removal of inactivation. Effectively, sodium channel inactivation sets the upper limit of action potential frequency in an axon. We shall see, however, that a variety of potassium channels also participate in determining the actual rate at which a neuron fires.

It is important to emphasize that there is nothing magic about inactivation gating (or any of the other gating transitions) in ion channels. With our understanding of ion channels as protein molecules, gating can be thought of as a rapid change in the three-dimensional structure of the protein, from a conformation that allows conduction to one that does not, and vice-versa. The closed and inactivated states of the channel reflect different protein conformations, neither of which conducts ions (recall the cartoon of the inactivating *Shaker* potassium channel in Fig. 5–7, and compare with Fig. 6–9). Many of the gating transitions (and hence protein conformational changes) that we have described are voltage dependent, implying that they result from the movement of some charged region of the ion channel protein, in the electric field across the plasma membrane.

Ion Pumps Maintain the Ion Concentration Gradients

It is evident that neurons cannot continue to fire action potentials forever. When membrane currents flow, ions are moving down their concentration gradients. The currents are relatively small at steady state, but are much larger during action potentials. If action potential firing continues, eventually the sodium and potassium concentrations on the two sides of the membrane will be equal, the membrane potential will be zero, and the state of the cell can be well described by the word *dead*. It may take a long time to run down the ionic concentration gradients in an axon as large as the squid giant axon, but in smaller cells significant changes may occur after relatively few action potentials. Fortunately, there exist energy-driven ion pumps (also called *transporters*) that come to the rescue before any damage is done.

A particular active ion transporter, the *sodium–potassium–ATPase* or *sodium–potassium pump*, mediates the pumping of sodium out of and potassium into the cell to maintain the ion concentration gradients. Another way to think of this is to say that the pump is responsible for charging up the membrane battery. The pump is an enzyme that hydrolyzes ATP and uses the energy to move each ion against its concentration gradient (Fig. 6–10). This transport activity may involve some sort of movement or rotation of the pump in the membrane. The stoichiometry of the sodium–potassium–ATPase, the transport ratio for sodium and potassium, is not 1:1 but instead 3:2. In other words, three sodium ions are transported outward for every two potassium ions transported inward and, as a result, the pump produces a net outward current. This kind of pump is said to be *electrogenic*, because its activity causes the cell to hyperpolarize, and it contributes (although usually only to a limited extent) to the setting of the resting potential.

Although we have de-emphasized their contribution to neuronal excitability, we can see that ion pumps do indeed play an essential role.

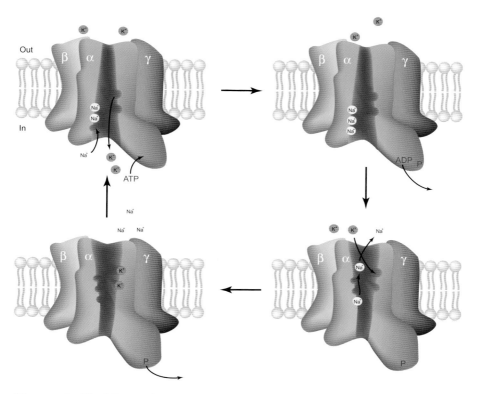

Figure 6–10. The sodium–potassium–ATPase. Depicted is the ATPase, a plasma membrane pump consisting of several protein subunits (α, β, and γ) that uses the energy of hydrolysis of ATP to move three sodium ions from the inside of the cell to the outside and, at the same time, move two potassium ions from the outside to the inside. The α subunit is responsible for ion transport and also contains the ATPase activity.

They may not be quite as flashy as the channels that allow for such rapid flow of ions, but they are always there in the background, working away quietly to ensure that the ion concentration gradients that are essential for electrical signaling are maintained.

Summary

The flow of ions down their electrochemical gradients, through populations of ion channels in the neuronal plasma membrane, gives rise to transmembrane ion currents. It is the sum of the various currents flowing at any point in time that determines the neuron's membrane potential.

Thus the normal firing pattern of a neuron, and its response to different kinds of stimulation, can be seen as a play of interactions among the currents flowing through the different kinds of ion channels in its membrane. The activities of the sodium and potassium channels responsible for axonal action potentials are themselves dependent on voltage. Voltage clamp studies, which allow the measurement of the current flowing through these channels at fixed voltage, have provided a detailed understanding of the sequence of changes in sodium and potassium channel activity that give rise to action potentials.

7

Diversity in the Structure and Function of Ion Channels

T hus far we have been discussing the axonal membrane as if it had only two kinds of ion currents, a voltage-dependent potassium current that is responsible for the axon's electrical activity (or rather its lack of activity) at rest, and a voltage-dependent sodium current that underlies the large and rapid membrane depolarization during the action potential. Although the work of Hodgkin and Huxley demonstrated that these two currents could provide a reasonably accurate picture of the electrical activity of the squid giant axon, it is now evident that there is a considerable diversity of ion currents in axons, and even greater diversity in neuronal cell bodies and dendrites.

The electrical behavior of cell bodies and dendrites certainly tends to be more varied than that of axons. Some neuronal cell bodies do not fire action potentials, and are said to be *electrically inexcitable* because they lack the rapid voltage-dependent sodium current. Other neurons produce beating, pacing, bursting, and the wide range of responses to external stimuli described in Chapter 3. The cell bodies of such neurons may exhibit many other ion currents in addition to or in place of the classic action potential currents. These other currents, which may be regulated by (*1*) voltage, (*2*) neurotransmitters that bind to receptor sites on the extracellular side of the channel, (*3*) intracellular calcium, (*4*) intracellular metabolic modulators or binding proteins, or (*5*) some combination of these factors, interact to generate complex patterns of neuronal electrical activity such as that exhibited by *Aplysia* neuron R15 (see Fig. 3–12c).

Table 7–1 Examples of the Major Classes of Voltage-Dependent Ion Channels

Channel Type	Activation Voltage Range	Physiological Function
Axonal sodium channels	−30 to +20 mV	Upstroke of action potential
Calcium channels	Variable	Calcium action potentials; calcium-mediated intracellular events, including neurotransmitter release
Potassium channels	Extremely variable	Action potential repolarization; spacing of action potentials; regulation of resting potential

Since ion currents reflect the activity of ion channels, the diversity in ion currents must be matched by an equivalent diversity in the ion channel proteins that underlie these currents. Single channel recording and molecular cloning techniques have been used to demonstrate that the diversity is even greater than had been imagined; in many cases these techniques reveal that currents that were thought previously to be carried by a single population of ion channels in fact reflect the activity of several different kinds of ion channels. Only through an understanding of the molecular basis of this diversity can we begin to understand its contribution to the electrical behavior of the neuron.

This chapter will focus on the diversity of the major classes of *voltage-dependent* ion channels (see Table 7–1). We do not mean to give short shrift to the *neurotransmitter-gated* ion channels, which are essential for chemical synaptic transmission and for the modulation of neuronal electrical properties. This group, which includes a variety of potassium-, calcium-, and chloride-selective channels, will be considered in detail in later chapters. Some of the neurotransmitter-gated channels are also voltage dependent, making the distinction between these two channel classes somewhat blurred. Thus the choice of channels to be discussed here rather than in later chapters is by necessity somewhat arbitrary.

Calcium Channels

In axons, the most important channels for the generation of inward currents are the voltage-dependent sodium channels, whose properties and structure were discussed in the last two chapters. However, there are several other classes of channels that contribute substantially to inward current flow in many neuronal cell bodies (and in some axons and dendrites). We focus now on calcium channels, which play a critical role in the lives of neurons (and other cells).

Calcium currents. In most neurons, a depolarizing voltage clamp step elicits an inward current with kinetics very different from those seen in the squid axon. The current may rise to its peak more slowly, and inactivate only partially and far more slowly, than in the squid axon (Fig. 7–1a). When pharmacological treatments and/or ion replacement are used to eliminate any sodium current, an inward current that rises slowly to its peak and inactivates only partially (if at all) can still be elicited by the depolarizing pulse (Fig. 7–1a). In fact, when the neuron is released from voltage clamp under these conditions, it is often found that it can still fire action potentials *even in the complete absence of sodium current*, although the shape and duration of these action potentials can be very different from those observed when sodium current is present (Fig. 7–1b; see also Fig. 3–11). Further ion substitution experiments reveal that most neurons exhibit a substantial voltage-dependent calcium current (I_{Ca}). In some cases, this is responsible for much or all of the regenerative depolarization during the rising phase of the action potential.

Interestingly, the very brilliance of the studies of Hodgkin and Huxley, which defined the sodium and potassium currents as both necessary and sufficient to account for action potentials in the squid axon, led to skepticism in assessing the work of early pioneers in the calcium current field. Although the calcium current experiments stood up to critical scrutiny, there was a reluctance on the part of many neurophysiologists to complicate with another ion current what had been a satisfying and apparently complete picture of membrane excitability. The neurophysiology community did not suspect in the 1960s just how drastically this simple picture was to be modified in the years to come.

Diversity of calcium channels. In some neurons, a plot of the peak calcium current as a function of voltage (Fig. 7–1c) closely resembles that for the sodium current (Fig. 6–6c), except that the current approaches zero at very depolarized voltages, reflecting the more depolarized value for the calcium reversal potential, E_{Ca}. In other neurons, this curve appears more complex, indicating that more than a single population of calcium channels gives rise to the current–voltage (I–V) relationship. In fact, on closer inspection with single channel recordings, even the simpler I–V relationships can turn out to be generated by more than one species of calcium channel. It now appears that there are multiple distinct categories of calcium channels in many neurons, which can be distinguished on the basis

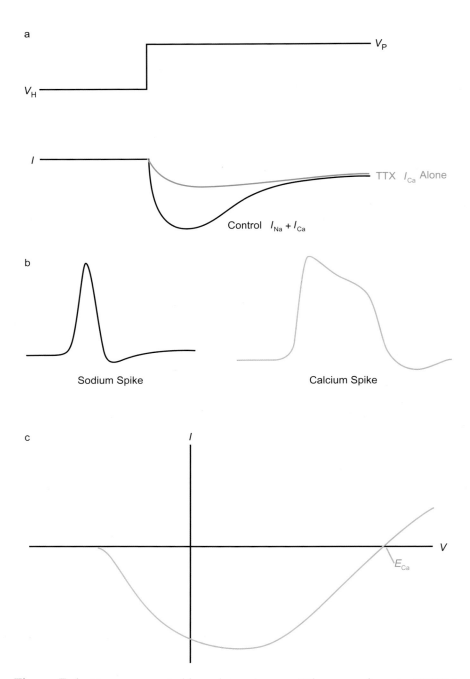

Figure 7–1. Currents carried by calcium ions. *a*: When tetrodotoxin (TTX) is used to block the sodium component (I_{Na}) of inward current (I), a calcium current component (I_{Ca}) remains. V_H, holding potential; V_P, pulse potential. *b*: The sodium action potential is contrasted with the action potential in cells in which the inward current is carried predominantly by calcium. *c*: Current–voltage (*I–V*) relationship for the calcium current. E_{Ca}, calcium equilibrium potential.

of their voltage dependence, kinetics, single channel conductance, pharmacology, and molecular structures, as well as by their location in the neuron. ❷

Purification and cloning of calcium channel subunits. Purified calcium channels, like sodium channels, include a large (175–230 kDa) subunit, termed α_1, that by itself is sufficient to form a functional voltage-dependent calcium channel, and one or more smaller subunits (β, δ, α_2, and δ; see Fig. 7–2a). Although the smaller subunits are not *required* for channel activity, they do interact with the α_1 subunits and *modulate* the properties of the channels. As we shall see later in this chapter, such auxiliary subunits also exist for potassium (and sodium) channels, and they influence not only channel functional properties but also channel targeting and localization in the plasma membrane.

Molecular cloning, based on protein sequence, has been used to study the structural basis for the functional and pharmacological diversity of calcium currents. Multiple distinct genes for both the α_1 and β subunits have been identified (Fig. 7–2b). Furthermore, all of the α_1 and most of the β subunit messenger RNA transcripts can undergo alternative splicing, a way of generating several distinct messenger RNAs and hence several different protein products from a single gene, as we saw for the *Shaker* potassium channel. The bottom line is that many flavors of the calcium channel α_1 subunit can be expressed in the brain (and other tissues), and when their interactions with β and other subunits is considered, the number of combinatorial possibilities is enormous. The diversity hinted at previously by calcium current kinetics and pharmacology pales in the face of this remarkable molecular heterogeneity.

Molecular organization of calcium channels. What have we learned about the structure and function of calcium channels from molecular cloning approaches? First, as emphasized earlier, heterologous expression tells us that there is a single polypeptide, the α_1 subunit, that is capable of forming a functional channel, reminiscent of the picture of sodium channels described in Chapter 5. Even more strikingly similar to sodium channels is the membrane organization of the calcium channel protein, predicted from the α_1 subunit amino acid sequence. Hydrophobicity plots predict 24 transmembrane segments, which can be divided into four homologous domains, each consisting of six transmembrane segments. This organization is so similar to that of sodium channels that we need not provide a new diagram here, but simply refer the reader to the diagram of sodium channel structure in Figure 5–1. Recall that this overall structural homology is recapitulated for voltage-dependent potassium channels as well (Fig. 5–4).

a

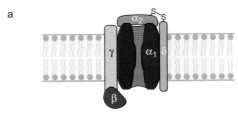

b Gene	α Subunit Subtype	Functional Channel Subtype
Ca$_v$ 1.1	α$_{1S}$	L
1.2	α$_{1C}$	L
1.3	α$_{1D}$	L
1.4	α$_{1F}$	L
Ca$_v$ 2.1	α$_{1A}$	P/Q
2.2	α$_{1B}$	N
2.3	α$_{1E}$	R
Ca$_v$ 3.1	α$_{1G}$	T
3.2	α$_{1H}$	T
3.3	α$_{1H}$	T

Figure 7–2. Subunits of calcium channels. *a*: Purified calcium channels consist of as many as five types of subunits. *b*: Multiple genes contribute to calcium channel diversity (see Catterall, 2011).

Calcium channels are special. Calcium channels are of particular interest, because calcium is far more than simply a charge carrier across the plasma membrane. As essential as calcium ions are in contributing to action potentials and other aspects of neuronal electrical activity, this role may be secondary to the intracellular messenger actions of calcium. Calcium that enters the cell interacts with calcium-binding proteins to regulate a variety of intracellular enzymes. Furthermore, intracellular calcium ions regulate the gating of several types of ion channel and can even feed back and participate in the inactivation of some of their own channels. In addition, an essential characteristic of neuronal signaling, the release of chemical neurotransmitters at synapses, is controlled directly by intracellular calcium. In this sense, calcium can be thought of as the transducer of an electrical signal, depolarization, into chemical signals inside the cell.

All of these features set calcium apart. It will thus come as no surprise that the activity of calcium channels themselves is subject to intricate modulatory influences, in cardiac and skeletal muscle as well as in neurons. We shall be hearing much more about the regulation and cell biological consequences of calcium channel activity throughout this book.

Channels That Carry Outward Current: The Potassium Channels

Even more impressive than the diversity of calcium channels is that exhibited by the potassium channels. Some half dozen or more voltage-dependent potassium currents were first identified on the basis of voltage clamp experiments. This number has expanded dramatically as single channel and molecular biological approaches have been used to study potassium channels (Fig. 7–3). It is now evident from genome sequencing that, astonishingly, there exist more than 70 genes encoding potassium-selective ion channels in organisms as diverse as humans and nematode worms. This dwarfs the number of genes (approximately 10 each) that encode the α subunits of sodium and calcium channels.

As in the case of other channels, kinetics, voltage dependence, pharmacology, single channel properties, and ultimately molecular structure determination have been used to characterize the various potassium channels. We will summarize here the physiological properties of some of these channels in the context of their cloning and molecular structures. By necessity, we can only touch on their diversity in the limited space available.

Voltage-dependent potassium channels. Once *Shaker* had been characterized, sequence homology cloning using *Shaker* sequences uncovered the large family of voltage-dependent potassium channels in a wide variety of organisms, from bacteria to humans. In mammals, there are eight genes closely related to the fruit fly *Shaker* gene, termed *Kv1.1* through *Kv1.8* (Fig. 7–3). In addition, there are 11 other subfamilies, with multiple channel genes in each subfamily. This may seem like overkill for a channel

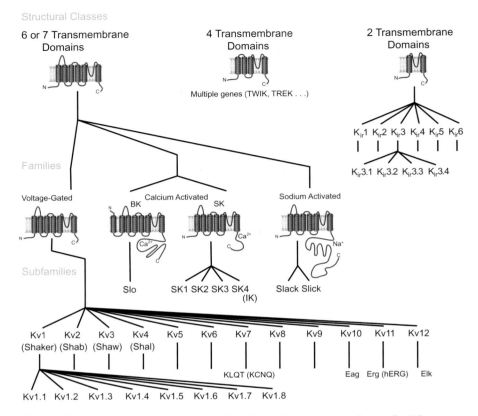

Figure 7.3 Overview of the vast diversity of potassium channels. There are three basic structural types of potassium channel subunits. The genes for potassium channels can be divided into multiple families and subfamilies. Only a few members of the subfamilies are shown for illustration (see Jan and Jan, 2012).

whose main function has been considered simply to help repolarize action potentials. Many of the channels were originally given different names or are also known by the name their gene has been given in humans. Some of these alternate names are shown in Figure 7–3. For example, $Kv7.2$ and $Kv7.3$ are also known as *KCNQ2* and *KCNQ3*, respectively, and these two subunits come together to make up the heteromeric channel responsible for the *M-current*. This current was so named because it can be regulated by the neurotransmitter agonist Muscarine; we will discuss this channel further in the context of channel modulation in Chapter 12.

Among the fundamental questions that remain to be answered is why neuronal membranes require so many distinct conductance pathways for a single ion, potassium, and how these distinct pathways evolved as organisms required an increasing diversity and flexibility in their neuronal firing patterns. In terms of their effects on neuronal excitability, however, voltage-dependent potassium channels can be divided into two general groups: non-inactivating channels such as those already discussed in terms of action potential repolarization, and *transient* or *A-current* channels that inactivate via the ball-and-chain mechanism described in Chapter 5.

Transient potassium A-currents. In considering the repolarization of action potentials, we have been discussing potassium currents that inactivate very little, if at all, during a long depolarizing pulse. As we saw in the case of *Shaker*, however, there is also a transient potassium current in many neurons, often known as *A-current*, that activates rapidly and then inactivates in a manner analogous to the sodium current. In mammalian neurons, Kv1.4 and the Kv4 subfamily of channels inactivate by an N-terminal ball-and-chain mechanism. We will encounter the Kv4.2 member of the Kv4 subfamily, an important channel in the dendrites of many neurons, again in Chapter 18.

To measure transient currents, the membrane potential must first be set to a very negative holding potential for several hundred milliseconds to remove the voltage-dependent steady-state inactivation. When the membrane is depolarized from this very negative holding potential, an outward A-current is seen (Fig. 7–4a) that mirrors the inward sodium current (Figs. 4–8 and 6–8), albeit with a more prolonged time course. The voltage dependence of A-current inactivation is such that, in neurons that have a relatively positive resting potential (more positive than about –45 mV), inactivation is more or less complete at the resting potential V_r (Fig. 7–4b). This is the reason that, in such cells, the steady-state inactivation must first be removed by hyperpolarization in order to examine this current.

The A-current is active in the subthreshold region of membrane potential and helps to determine the frequency of repetitive firing in neurons. Although it is largely inactivated near the resting potential and completely inactivated during action potentials, some portion of the inactivation is removed by the afterhyperpolarization that normally follows an action potential. Hence the A-current is active for a short while after an action potential and slows the return of the membrane potential toward the spike threshold. This in turn slows the firing frequency in a repetitively firing neuron.

Another role for A-current is to allow a delay to occur between an excitatory stimulus and the onset of action potentials. This occurs in neurons with a relatively negative resting potential, in which there is little steady-state inactivation. When such a neuron is depolarized, the A-current is activated and tends to oppose the change in membrane potential toward the threshold. As the A-current inactivates during the depolarization,

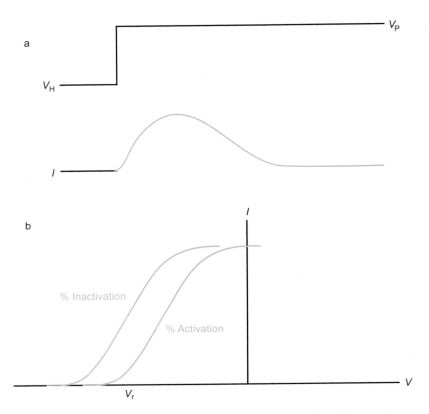

Figure 7–4. An inactivating potassium current. *a*: When the membrane is depolarized (V_p) from very negative holding potentials (V_H), a rapidly inactivating potassium current (*I*) can be observed. This potassium current has been termed the *A-current*. *b*: Extent of activation and inactivation of the A-current as a function of voltage (*V*). V_r, reversal potential.

however, the neuron begins to depolarize more rapidly. Following a delay set by the kinetics of inactivation, the neuron finally reaches threshold. An example of this is found in the *Aplysia* ink gland motor neurons, which were depicted in Figure 3–13c. The prolonged noxious stimulus causes a depolarization that leads to progressive inactivation of A-current. Only when the A-current has undergone inactivation do the neurons fire the action potentials that trigger ink release.

Again, voltage clamp experiments have shown that there are multiple types of A-currents in different cells. *Shaker* and its many cousins, the cloning and structure of which we described in Chapter 5, are rapidly inactivating A-current channels. These distinct A-current channels with different kinetic properties and voltage dependence, encoded by different genes or produced by alternative splicing from a single gene (as in the case of *Shaker*), contribute in different ways to the regulation of neuronal firing rates.

Calcium-dependent potassium currents. In many neurons, the total outward current carried by potassium exhibits a steady-state $I–V$ relationship that is very different from that seen in the squid axon (*steady state* refers here to the sustained non-inactivating current, measured many tens or hundreds of milliseconds after the onset of a depolarizing pulse). The $I–V$ curve has a characteristic N shape in the range of depolarized voltages (Fig. 7–5a), because it is the sum of several distinct current components (Fig. 7–5b, c). When the cell is injected with an agent that binds tightly to calcium, such as EGTA, or calcium entry is prevented during the depolarizations by pharmacological block of calcium channels, the resulting $I–V$ curve (black in Fig. 7–5b) looks identical to that of the delayed rectifier potassium current in squid axon (compare with Fig. 6–6c). When its kinetics, voltage dependence, and pharmacology are examined, this current can be seen to exhibit the properties of a classic delayed rectifier, whose role in action potential repolarization was discussed in Chapter 6.

The other outward current component that contributes to the shape of the steady-state $I–V$ curve at positive voltages is the one blocked by preventing calcium entry or binding intracellular calcium—a *calcium-dependent potassium current*. Its $I–V$ curve (blue in Fig. 7–5b) is obtained by subtracting the delayed rectifier component from the total outward current. This current is activated not only by the depolarization per se but also by the calcium that enters during the depolarizing voltage pulse (Fig. 7–5c). A comparison of the $I–V$ curve in Figure 7–5b with that in Figure 7–1c shows that the voltage dependence of the calcium-activated potassium current mirrors that of the calcium current; this arises from the requirement for calcium entry, through voltage-dependent calcium channels, to contribute to the activation of this potassium current. As the voltage approaches E_{Ca}, the driving force for calcium entry decreases, and hence there is less activation of the calcium-dependent potassium current (other calcium-dependent intracellular processes often exhibit a similar voltage dependence). This requirement for intracellular calcium also explains why the current is eliminated by blocking calcium entry during the depolarization. However, this current and other calcium-dependent intracellular events may still be evoked by intracellular injection of calcium or by physiological treatments that cause the release of calcium from intracellular stores (see Chapter 12).

Kinetic and pharmacological studies of voltage clamp currents had suggested that there might be some heterogeneity of the calcium-dependent

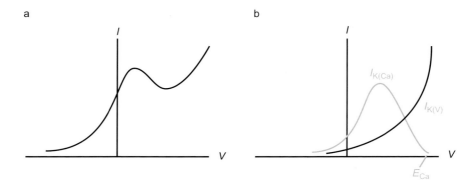

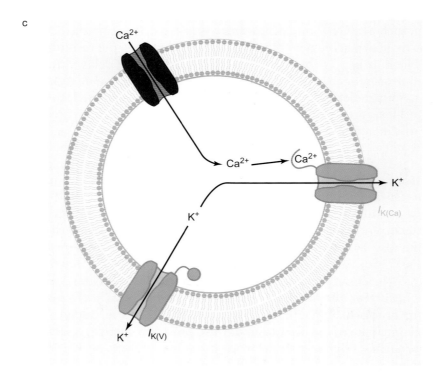

Figure 7–5. Different components of outward current. *a*: In most neuronal cell bodies, when the inward current is blocked, the current–voltage (*I–V*) relationship for the peak outward current has a characteristic *N* shape (see Meech and Standen, 1975). *b*: Two components of outward current. E_{Ca}, calcium equilibrium potential. *c*: The activation of two classes of potassium channels ($I_{K(V)}$, $I_{K(Ca)}$) is illustrated.

potassium current, but the extent of this heterogeneity became apparent only from single channel experiments and, more recently, from molecular cloning. Many cell types contain a large-conductance, calcium-dependent potassium channel (a *maxi* or *BK* channel), but there exist intermediate (*IK*) and small (*SK*) conductance ones as well (Fig. 7–3). Although the BK class exhibits voltage-dependent gating and indeed can be thought of as voltage-dependent channels whose voltage dependence is influenced by calcium, voltage is not involved in the gating of the SK class. There is even heterogeneity within the BK class. For example, in rat brain plasma membrane preparations, there are at least two separate BK channels that can be distinguished on the basis of their gating kinetics (Fig. 7–6) and pharmacology. It appears that more than one type of calcium-dependent potassium

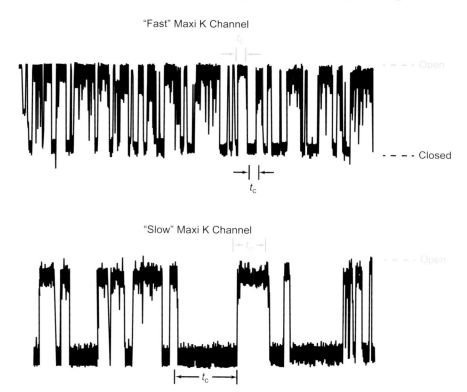

Figure 7–6. Different classes of large-conductance calcium-dependent potassium channels. At least four distinct calcium-dependent potassium channels can be seen in plasma membrane fractions from mammalian brain. Two of them are shown here. They have very similar single channel conductances (240 pS), but differ in their gating kinetics (note the differences in open time [t_o] and closed time [t_c]). These records were obtained following reconstitution of calcium-dependent potassium channels into artificial phospholipid bilayer membranes (see Reinhart et al., 1989).

channel can be present in a single cell, but the functional significance of this heterogeneity remains to be determined. Since calcium-dependent potassium currents contribute (with the delayed rectifier) to action potential repolarization as well as to interspike currents that help to control the frequency of repetitive firing, it is possible that the different kinetic properties and voltage sensitivities of different calcium-dependent potassium channels enable them to undertake distinct functional roles. As we shall discuss in Chapter 13, a particularly well-understood example is the contribution of kinetically distinct calcium-dependent potassium channels to frequency tuning in cochlear hair cells of the inner ear.

The Slowpoke *locus encodes BK calcium-dependent potassium channels.* In contrast to other voltage-dependent potassium channels, the molecular identity of the calcium-dependent potassium channels described could not be found by searching for genes similar to *Shaker* channels. Here again, *Drosophila* came to the rescue. A *Drosophila* mutant, called *Slowpoke*, lacks calcium-dependent potassium current in neurons and muscle cells. Positional cloning of the *Slowpoke* locus showed that it encodes a BK calcium-dependent potassium channel that shares many structural features with its *Shaker* family cousins, but is also different. The sequence of the *Slowpoke* cDNA predicted that the protein possesses, like *Shaker*, the six membrane-spanning segments (Fig. 7–7) that resemble one of the four homologous domains of the sodium and calcium channels (Fig. 5–3). There is in addition a seventh membrane-spanning segment, S0, that places the amino terminus of the channel on the extracellular side of the membrane (Fig. 7–7). Furthermore, instead of terminating soon after the end of transmembrane segment S6, *Slowpoke* has an additional long stretch of amino acids that constitutes about two-thirds of the channel protein (Fig. 7–7). This extended carboxyl-terminal domain contains sequences that confer calcium sensitivity on channel activity. This carboxyl-terminal domain is also one region of the channel protein that interacts with certain kinds of auxiliary subunits that influence channel properties (see discussion later in chapter).

Like *Shaker*, the *Slowpoke* messenger RNA can undergo alternative splicing to produce a number of different gene products. In fact, in the case of *Drosophila Slowpoke*, the number is extraordinarily large—its sequence predicts a total of 144 possible splice variants arising from the single *Slowpoke* gene. Furthermore, as with *Shaker*, the functional calcium-dependent potassium channel is formed by a complex of four *Slowpoke* subunits. Thus the total number of possible channels that can be derived from this single gene is 144^4 (almost 430,000,000)! Although it is not known how many of these channels actually exist in neurons, and

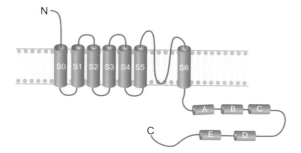

Figure 7–7. The *Slowpoke* calcium-dependent potassium channel cloned from *Drosophila*. In addition to the six membrane-spanning segments (S1–S6) also present in voltage-dependent potassium channels, *Slowpoke* possesses a seventh membrane-spanning segment, S0, and also contains a long carboxyl-terminal tail. The boxes marked A to E are cassettes of amino acids that differ among different splice variants of the channel.

only a very small number of the splice variants have been analyzed in any detail, the possibilities for functional heterogeneity are nevertheless staggering. Modulation by different combinations of auxiliary subunits or by post-translational modification of the channel protein (see later discussion) expands these possibilities even further. In addition, the IK and SK calcium-dependent potassium channels are encoded by genes distinct from *Slowpoke* (Fig. 7–3), and they also contribute substantially to the functional diversity of the calcium-dependent potassium currents in neurons.

Slack, Slick, and sodium-activated potassium channels. It has been known for some time that large conductance potassium channels activated by intracellular sodium ions exist in cardiac myocytes and many types of neurons. The currents carried by these channels contribute importantly to the regulation of neuronal firing patterns, and the channels themselves are subject to complex modulation. These sodium-activated potassium channels are encoded by the *Slick* and *Slack* genes; they resemble in their overall membrane topology (although not in their amino acid sequences) the *Shaker* channel subunits. They differ from *Shaker* channels, however, in that they possess very long, extended cytoplasmic carboxyl-terminal domains that resemble (again in structure but not sequence) those of the BK channels. Moreover, although the opening of *Slack* and *Slick* channels is increased by depolarization, they lack charged residues in the S4 segment that is the major voltage-sensor in *Shaker* channels. In excised patches, the concentration of intracellular sodium required to activate *Slack* and *Slick* channels is rather high, in the range of several tens of millimolar; nevertheless, in neurons, these channels are open even at resting levels of internal sodium ions and are further activated under physiological conditions. Neuronal excitability is modulated rather dramatically when these channels are knocked down with small inhibitory (si) RNAs. In addition, mutations in the human *Slack* gene that produce even modest (two- to three-fold) changes in *Slack* currents produce extremely severe intellectual impairment, as well as childhood epilepsies.

Inward rectifiers; potassium currents activated by hyperpolarization. All the inward and outward currents we have discussed thus far are activated by depolarization, either because they are gated directly by voltage, or because depolarization-induced calcium entry is required for their activation, or both. This gives rise to *outward rectification* in the I–V relationship, an increase in the slope of the curve with depolarization. However, in many cells, there also exist potassium currents that are activated by hyperpolarization. This causes a *decrease* in the slope of the I–V curve with depolarization (Fig. 7–8), a phenomenon known as *anomalous* or *inward rectification.*

    This all seems rather strange. Why would a cell bother with a channel that passes only inward but very little outward potassium current? This question becomes particularly pressing when we remember that under normal conditions a neuron's V_m can never be more negative than E_K, since potassium is the charge carrier with the most negative reversal potential. Thus, only under artificial hyperpolarizations imposed by the

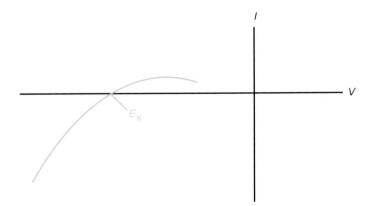

Figure 7–8. A potassium current that activates with hyperpolarization. Shown is the current–voltage (I–V) relationship for the inwardly rectifying potassium current, which displays a decrease in slope conductance with depolarization. E_K, potassium equilibrium potential.

voltage clamp will the inward flow of potassium ever occur. One clue to the role of the anomalously rectifying potassium channels is the fact that they are not perfect rectifiers and can pass some outward current in the voltage range up to about 30 mV depolarized from E_K (Fig. 7–8). Although the amount of current is not large, very few other membrane currents are active in this voltage range; accordingly, this current may play an important role in regulating the resting level of neuronal activity.

Cloning of the inward rectifier potassium channels. The inwardly rectifying potassium channels, like the calcium-dependent potassium channels, remained uncloned during the extensive homology screens that followed the initial characterization of *Shaker*. They finally fell to an approach called *expression cloning*, which we will describe in later chapters, and subsequent homology screens and genome sequencing have uncovered a large family of inward rectifiers that clearly is distinct from the *Shaker*-like family of voltage-dependent potassium channels. The structure of the inward rectifier potassium channels is interesting (Fig. 7–9). They contain only two putative membrane-spanning segments, flanking a pore domain that is very similar in sequence to that of the *Shaker*-like potassium channels; in this feature they resemble the bacterial KcsA channel (Fig. 5–9). Inward rectifier potassium channels are subject to complex forms of modulation, which we shall discuss in Chapter 12. Structural determination has contributed to a detailed molecular understanding of inward rectifier channel modulation.

"Two-pore" potassium channels. We have by no means exhausted the complement of potassium channels with the descriptions of the classes just discussed. An interesting and surprising discovery was that some potassium channels actually contain two P domains within the amino acid sequence of a single subunit (Fig. 7–9). These channels, which contribute

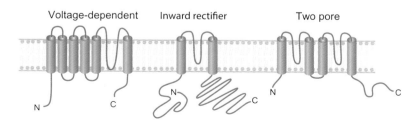

Figure 7–9. Predicted structures of other potassium channels. The inward rectifier and two pore potassium channels share some structural features with the six transmembrane voltage-dependent potassium channels.

to the resting membrane potential, form functional dimers and thus contain a total of four P domains, like the other classes of potassium channels.

Auxiliary Subunits of Potassium Channels

As mentioned previously, it was evident from early protein purification experiments that sodium and calcium channel pore-forming α subunits co-purify with a variety of auxiliary subunits, which are not essential for but often modulate channel function. Potassium channel purification, by contrast, has traditionally been extremely difficult because of the dearth of tissue sources containing sufficient potassium channel protein—hence the need to clone *Shaker* by a positional cloning approach, which provided no information about associated subunits. It is now apparent, however, that association with auxiliary subunits is another characteristic that potassium channel α subunits share with the other voltage-dependent ion channels. Indeed, there is a long and growing list of potassium channel–associated proteins (Table 7–2) that influence channel properties in a variety of ways.

Auxiliary subunits modulate potassium channel function. One potassium channel auxiliary subunit, Kvβ1, can confer rapid ball-and-chain type inactivation on a non-inactivating α subunit of a voltage-dependent potassium channel (Fig. 7–10) by providing the amino-terminal inactivation domain that this particular α subunit lacks. Roderick MacKinnon and colleagues have solved the three-dimensional crystal structure of a Kvβ subunit, in association with a portion of the α subunit. The structure

Table 7–2 Selected Examples of Auxiliary Subunits Associated with α Subunits of Potassium Channels

α Subunit	Auxiliary Subunit	Physiological Function
Kv family members	Kvβ1–Kvβn	Confer inactivation, influence membrane targeting
	Caspr2	Membrane targeting
Slowpoke family members	$K_{Ca}\beta1$–$K_{Ca}\beta n$	Modulate voltage and calcium dependence
	Slob, Slip	Modulate voltage and calcium dependence, influence membrane targeting

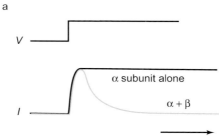

a

V

α subunit alone

α + β

I

Time

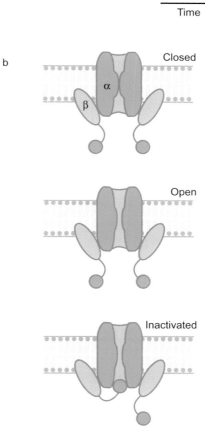

b

Closed

α

β

Open

Inactivated

Figure 7–10. Potassium channel β subunits can contribute to rapid channel inactivation. *a*: A potassium channel α subunit that lacks an inactivating ball domain (Fig. 5–7) does not inactivate when expressed alone (black), but does when it is co-expressed with β subunit (blue). *I*, current; *V*, voltage. *b*: The β subunit provides the ball-and-chain structure that confers inactivation (compare with Fig. 5–7).

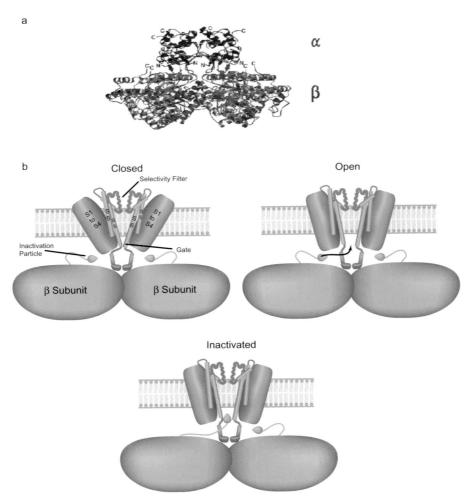

Figure 7–11. Structure of a potassium channel α–β complex. *a*: Rod MacKinnon and colleagues solved the crystal structure of a β subunit (blue) together with a portion of an α subunit (red). *b*: Among the interesting features revealed by the structure is the path through a linker region that the β subunit inactivation particle must navigate to reach and block the pore of the channel. See Animation 7–1.◐ (Based on Gulbis et al., 2000.)

has several interesting features (Fig. 7–11a), including the surprising finding that the β subunit inactivation domain must find its way through a somewhat restricted lateral opening in the α subunit in order to gain access to and block the pore (Fig. 7–11b). As was already known from the Kvβ amino acid sequence and was confirmed by MacKinnon's structural studies, this β subunit is an oxidoreductase enzyme whose precise function remains poorly understood. Other enzymes, most notably protein kinases and phosphatases that can participate in the modulation of channel function, also associate intimately with many ion channels.

The calcium-dependent potassium channels have their own complement of auxiliary subunits (Table 7–2) that differ in both structure and function from those associated with the voltage-dependent potassium channel family. As in the case of the calcium channels, the number of combinatorial possibilities and consequent functional heterogeneity afforded by multiple α-subunit genes, alternative splicing, and associated proteins is simply enormous.

Auxiliary subunits influence channel cell surface expression and membrane localization. Not all potassium channel auxiliary subunits modulate channel function directly. Instead, they may act as molecular chaperones to increase the insertion of active α subunits in the neuronal plasma membrane (a similar role has been demonstrated for some calcium channel β subunits), thereby increasing the whole-cell potassium current.

It has been known for many years that ion channels are not evenly distributed in the neuronal plasma membrane. For example, sodium channels are present at relatively low density in neuronal cell bodies but at much higher density in the axon hillock, which as a result is often the site of action potential generation. How does this differential distribution of ion channels come about? It appears that many channels are targeted via their association with auxiliary subunits, which in turn may interact with local cytoskeletal elements. A particularly striking example of this is in myelinated axons. Recall from Chapter 3 that sodium channels are present at high density at the nodes of Ranvier and are sparse in the internodal region; potassium channels, in contrast, are concentrated adjacent to the nodes in the juxtaparanodal region. This channel distribution, which can be visualized with channel-specific antibodies coupled to fluorescent markers (see Fig. 3–10a), is what allows rapid saltatory conduction of action potentials in myelinated axons. A protein called Caspr2 colocalizes with the potassium channels, but not at all with the sodium channels. Biochemical experiments demonstrate that Caspr2 binds via an adaptor protein to the potassium channel α subunit; this suggests that it might act as an auxiliary subunit that targets the channel to the juxtaparanodal region of the membrane. The signals that tell Caspr2 itself to concentrate in this region and the mechanisms by which the sodium channels are targeted to the nodes remain to be determined.

TRP Channels

The TRP (for **T**ransient **R**eceptor **P**otential) channels comprise a large and diverse family of nonselective cation channels that resemble in their overall membrane topology the six-transmembrane domain voltage-dependent potassium channels. They were first discovered in a *Drosophila* mutant in which the electroretinogram response to a steady light is transient, in contrast to the sustained retinal receptor potential seen in wild-type flies. In humans, more than 25 genes encode TRP family members, which are divided into six or seven subfamilies. They are expressed in a wide variety of cell types and play prominent roles in neurons, especially sensory neurons, where they act as transducers of sensory stimuli, including temperature and painful or noxious stimuli (see Chapter 13).

Diseases Associated with Ion Channel Dysfunction

With the detailed molecular and functional characterization of ion channels has come the realization that ion channel dysfunction is often associated with serious disorders (*channelopathies*) in many cell types, including neurons. Such dysfunction may result from mutations in either the pore-forming α subunits or the modulatory auxiliary subunits of voltage-dependent ion channels. One form of human epilepsy, for example, is caused by a mutation in a sodium channel β subunit that normally speeds up channel inactivation. When this β subunit is defective, channel inactivation is slowed, sodium current is increased, and the neuronal membrane is thereby rendered hyperexcitable. Another way to generate hyperexcitability and epilepsy is by disrupting the function of potassium channels (recall that the phenotype of the *Shaker* mutant fly is hyperexcitability). Certain types of seizures in humans are associated with one of several mutations in different portions of the Kv1.1, Kv7.2 (KCNQ2), or *Slack* potassium channel α subunits (Fig. 7–12). Indeed, as has been found for the *Slack* channels, mutations in different parts of the channel can produce very different diseases, with different ages of onset, different types of seizures, and different effects on intellectual function. In view of the remarkable ion channel heterogeneity that has become apparent and that we have stressed throughout this chapter, it is striking that a mutation in a single channel subtype can have such profound consequences for neuronal excitability. This strongly suggests that there is little functional redundancy among the multiple subtypes, and that each must play its unique and essential role in the physiology of the neuron.

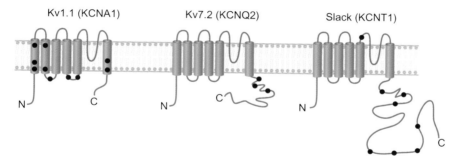

Figure 7–12. Mutations in potassium channels can cause epilepsy. The sites of naturally occurring mutations in the Kv1.1, Kv7.2 (KCNQ2), and *Slack* channels that are responsible for certain cases of epilepsy are marked in blue.

Summary

Voltage clamp and patch clamp techniques have been used to reveal heterogeneity of ion currents carried through voltage-dependent sodium, calcium, and potassium channels. Advances in channel molecular biology have made it clear that the diversity of ion channels is even greater than was suspected from these electrophysiological measurements. This diversity is achieved by several different mechanisms, including the existence of multiple genes for the pore-forming α subunits of ion channels, alternative splicing of the messenger RNA transcribed from each individual gene, formation of heterotetramers containing different α subunits of potassium channels, and modulation of channel properties by auxiliary subunits that may themselves comprise a large and diverse family of proteins. Moreover, potassium channels can be further categorized into voltage-dependent, calcium-dependent, sodium-dependent, two-pore, and inward rectifier channels. The importance of this diversity for neuronal physiology is emphasized by the emerging evidence that many human diseases are associated with dysfunction of individual classes of ion channels in neurons.

III

Intercellular
Communication

The previous section of this book described the membrane specializations that permit the transfer of information from one part of a
neuron to another. The next seven chapters address another fundamental aspect of nervous system function, *intercellular signaling*. This
includes the mechanisms that neurons use to communicate with one
another and with the outside world. Chapter 8 compares two fundamentally different modes of interneuronal communication. The first is the
direct transfer of ions and small molecules from one neuron to another
via *electrical synapses* (*gap junctions*). The second involves the release or
secretion from one cell of some chemical, a *neurohormone* or *neurotransmitter*, that diffuses to and affects the activity of a target cell. Often the
membranes of the secreting and target cells are immediately adjacent to
one another at the highly specialized structure known as the *chemical
synapse*. As described in Chapter 9, mechanisms of release of neurotransmitters at chemical synapses are best understood from the study of three
highly specialized synapses: the *nerve-muscle* synapse in vertebrates, the
giant synapse in the stellate ganglion of the squid, and the *calyx of Held*
synapse in the auditory brainstem of mammals. The various classes of
neurotransmitters and neurohormones that have been found in nervous
systems and details of their synthesis and metabolism are presented in
Chapter 10. The following two chapters describe the neurotransmitter
and neurohormone *receptors*, which are specialized membrane proteins
that recognize and bind signaling molecules and *transduce* the extracellular chemical signal into an electrical response in the target cell. Two
distinct classes of transduction mechanism are considered. Chapter 11

discusses receptors in which the binding site is part of the same molecule or macromolecular complex as the ion channel whose activity is regulated by the neurotransmitter—the *directly coupled* receptor/ion channel systems. These are contrasted in Chapter 12 with the *indirectly coupled* systems, in which the occupation of the receptor by neurotransmitter sets in motion a chain of biochemical events. These events lead ultimately to a change in the activity of an ion channel that is not intimately associated with the receptor. This chapter also addresses the concept of *neuromodulation*, the long-term alteration of neuronal electrical properties as a result of neurotransmitter or neurohormone action, and the intracellular biochemical mechanisms that are responsible for these alterations. Finally, Chapter 13 describes cells that act as *sensory receptors* by converting information from the outside world into electrical signals that can be passed on to other neurons in the brain. Sensory receptor cells use the mechanisms discussed in Chapters 11 and 12 to convert sensory information into a change in the properties of membrane ion channels.

How Neurons Communicate: Gap Junctions and Neurosecretion

W ithin any organism, cells must be able to communicate. This is of course important in tissues other than the brain, but is essential for proper nervous system function. There are three general ways in which cells talk to each other:

1. Direct transfer of molecules and ions from the cytoplasm of one cell into that of another. As we mentioned in Chapter 1, this is mediated by gap junctions.
2. The release of a chemical that diffuses to, and acts on, another cell. This release process is termed *secretion*.
3. Direct physical contact. A cell can be influenced profoundly by events triggered when molecules in its plasma membrane interact with the membranes of adjacent cells.

This chapter will deal with the first two of these modes of communication, modes that neurons use on a day-to-day basis to generate specific behaviors. The third pattern of communication plays a very important role in the development of neurons and their connections, and we shall discuss it in more detail in Chapters 14 through 17.

Gap Junctions, Connexins, and Electrical Synapses

Intercellular communication through gap junctions is conceptually the simplest form of cell-to-cell interaction. Small molecules and ions in one cell diffuse through pores in the plasma membrane directly into the cytoplasm of a neighboring cell (Fig. 8–1). These pores can be visualized in the electron microscope and are found in clusters that sometimes have a crystalline appearance. Such clusters were captured in the electron micrograph of Figure 8–2 by the technique of *freeze fracture*. In the preparation of tissues for freeze fracture, plasma membrane that has been frozen is allowed to break within the plane of the membrane itself. The exposed halves of the lipid bilayers are then coated with platinum and carbon to produce a form of bas-relief view of the inner plane of the membrane. These methods enable ready visualization of intramembranous particles, which represent integral membrane proteins such as receptors and ion channels. Figure 8–2 ❽ shows flattened sheets of membranes with arrays of gap junction particles at an electrical synapse between two neurons. These particles contain the proteins that form the cell-to-cell pore.

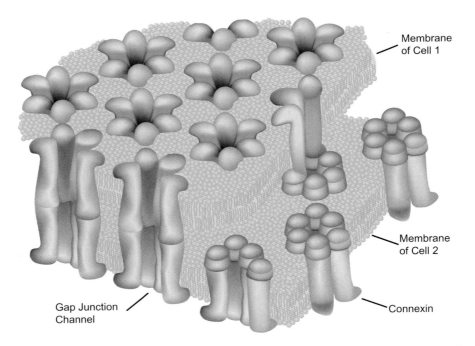

Membrane of Cell 1

Membrane of Cell 2

Gap Junction Channel

Connexin

Figure 8–1. Gap junctions. Pores spanning two cell membranes are made of connexin proteins. This picture evolved from the X-ray diffraction work of Makowski et al. (1977).

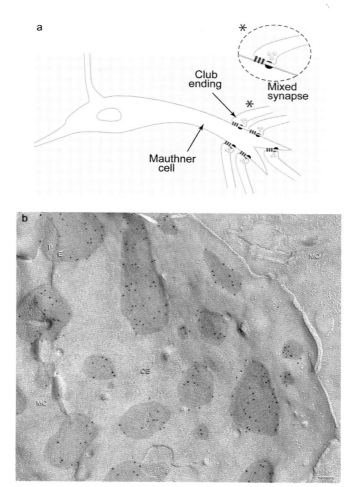

Figure 8–2. Freeze fracture replica of gap junction particles. *a*: Diagram representing the electrical synapse between presynaptic auditory neurons and a large neuron termed the *Mauthner cell* in the fish nervous system. The presynaptic terminal is called a *club ending*. This synapse also uses chemical neurotransmission and is therefore termed a *mixed synapse*. *b*: Freeze fracture electron micrograph of the membranes between the club ending (CE) and the Mauthner cell (MC). Multiple arrays of gap junction particles are shown in pink. While connexins constitute the bulk of the gap junctions, a small subset of these connexins have been labeled as black dots in this experiment (from Pereda et al., 2003).

Gap junctions are abundant in tissues such as the liver and the lens, which have been used as sources for the purification of gap junction proteins. Most of these proteins belong to the *connexin* family of proteins. A pore is believed to be formed by a complex of six connexin proteins in the membrane. To link the pore to the cytoplasm of an adjacent cell, the

connexins bind to another hexameric complex of connexins in that cell
(Fig. 8–1). The major connexin protein in the nervous system has a molecular weight of 36,000, and has accordingly been given the name CX36.
Two other connexins, CX45 and CX57, appear, however, to constitute the
gap junctions in subsets of neurons. In addition, there is a second family
of proteins, the *pannexins*, which have structural similarity to connexins,
and two members of this family, PX1 and PX2, also form gap junctions
between some neurons. This diversity may reflect the fact that the properties of gap junctions, and the way these properties are altered by neurotransmitters or hormones, vary in different cells.

Surprisingly, the electrical properties of gap junctions appear similar
in many respects to those of the ion channels we have encountered in
previous chapters. Despite the fact that gap junctions allow the passage
of relatively large molecules, with molecular weights up to about 1000,
recordings of the opening and closing of gap junctions resemble the gating
of channels that allow specific ions to cross the plasma membrane. This
can be measured by whole-cell patch clamp of a pair of cells that are coupled by only a small number of gap junctions (Fig. 8–3). When the membrane potential of one cell is maintained more negative than that of its
partner, current flows across the pore only while the cell-to-cell channel
is open.

Gap junctions are likely to function in the embryonic development of
tissues, a time when many gap junctions are formed and then broken
again. One of their roles may be to allow the transfer between cells of
small molecules that are important for development and second messenger molecules involved in intracellular signaling (see Chapter 12). This in
turn allows a group of cells to act as one functional unit. In some cases,
an assembly of gap junction proteins in the plasma membrane may open
even when it is unconnected to a second cell. When this happens, the
assembly is termed a *hemichannel*, and it may serve to release molecules
from the cytoplasm of the cell to the outside medium rather than to a
neighboring cell.

In the nervous system and other excitable tissues, gap junctions take
on a special significance. Connections between nerve cells via gap junctions are often called *electrical synapses*, in recognition of the fact that
they are involved in rapid electrical signaling and information transfer.
For example, they allow groups of neurons to synchronize their electrical
activity. However, electrical synapses are also found between pairs of neurons that do not always fire in synchrony. In such cases, their role may be
to allow synaptic inputs into one neuron to be registered in a neighboring
neuron. Later in the book (see Chapter 18), we shall encounter specific
examples of neurons that couple electrically through gap junctions.

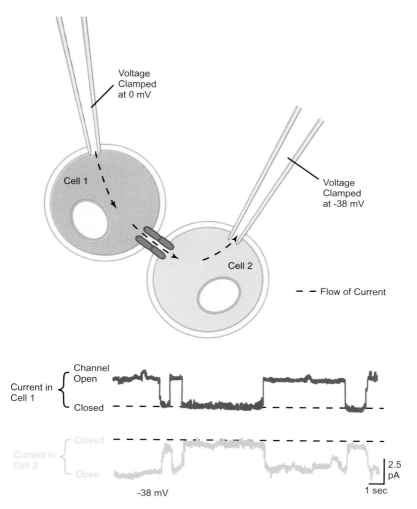

Figure 8–3. Single channel recordings of gap junctions. In an experiment carried out by Burt and Spray (1988), two cells are voltage clamped at different potentials. When the gap junction channel opens, the current that flows between the cells is recorded as outward current in Cell 1 and inward current in Cell 2.

Neurosecretion

Although gap junctions constitute a very basic and rapid form of communication, they are limited to bidirectional interactions between neighboring cells. When information has to be transferred from one cell to a distant cell, the usual mechanism is for the first cell to release a chemical into the extracellular space. This chemical then either diffuses, or is transported to the target cell. A large number of different chemicals may act as extracellular signals. These include amino acids and other small organic compounds, lipids, small peptides, and large protein complexes. Chemicals that are released from a neuron and alter the excitability of another neuron or a muscle cell are termed *neurotransmitters*.

Figure 8–4 shows some of the mechanisms that allow molecules to leave a cell. Lipid molecules diffuse readily across a cell membrane. Once synthesized, they can leave the cell without requiring any specialized machinery (Fig. 8–4a). For certain other, more hydrophilic, molecules, carriers or pores exist in the plasma membrane to enable them to cross the membrane (Fig. 8–4b). The major pathway for the release of neurotransmitters, however, appears to be *vesicular secretion*. At the synaptic terminals of neurons, many secretory vesicles can normally be found (see Chapter 1). When a neuron is stimulated, these transmitter-containing vesicles fuse with the plasma membrane to release their contents into the synaptic cleft (Fig. 8–4c). This pathway probably evolved from the requirement of many cells to release large peptides and proteins that would not normally be able to cross the lipid membrane, and also from the need to insert proteins such as ion channels and other integral membrane proteins into the plasma membrane. Thus it is not surprising that many insights into the release of neurotransmitters have come from work with non-neuronal cells. In the remainder of this chapter, we shall discuss the general mechanisms for secretion of proteins and other transmitters in neurons and exocrine cells. In the next chapter we shall consider in more detail the physiology of transmitter release at three well-studied chemical synapses.

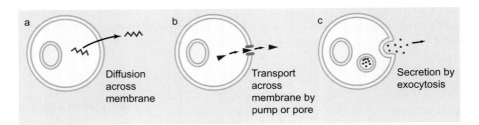

Figure 8–4. Three pathways (*a–c*) by which cells release substances into the external medium.

Constitutive and Regulated Secretion of Protein

In all eukaryotic cells, proteins that are destined for release into the extracellular space are synthesized in the *rough endoplasmic reticulum* (Fig. 8–5). Newly synthesized proteins enter the *Golgi apparatus*, a tightly packed stack of intracellular membranes in which the proteins undergo a variety of post-translational modifications. The proteins are then transferred to vesicles that bud off the Golgi apparatus and move to the cell surface. Here the vesicles release their contents into the extracellular space by fusion with the plasma membrane, a process known as *exocytosis*. In some cases, the arrival of the vesicles at the cell surface is followed immediately by exocytosis. This form of secretion has been termed *constitutive secretion*. The insertion of integral membrane proteins, including ion channels and receptor molecules, into appropriate regions of the plasma membrane also occurs by constitutive secretion (Fig. 8–5). Recall from Chapters 1 and 2 that the organelles involved in these processes are not restricted to neurons, and that the neuronal cell body has many features in common with other cells.

Neurotransmitters and hormones, in contrast, are not released immediately following their synthesis. They are packaged into vesicles, and the vesicles are transported along microtubules to the release sites, such as those at axon terminals. They are then stored at these sites until an appropriate stimulus is received by the cell. In neurons, this stimulus is usually a depolarization that causes calcium to enter the cell through voltage-dependent calcium channels. A large amount of neurotransmitter can then be released by the cell when the vesicles fuse with the plasma membrane. Such secretion, which is regulated acutely by external stimulation, has been termed *regulated secretion* (Fig. 8–5).

Even within these general classes of constitutive and regulated secretion there must be a large number of variants. For example, different proteins are inserted at different locations in the plasma membrane of a cell by constitutive pathways. In addition, neurons contain at least two kinds of secretory vesicles that participate in regulated secretion. *Large dense-core vesicles* contain primarily peptide neurotransmitters; *small synaptic vesicles* contain nonpeptide transmitters and the enzymes required for their synthesis. This diversity of intracellular traffic implies that newly synthesized proteins must somehow be assigned to the correct vesicle pathway. At least part of the information for the correct assignment may be found in the sequence of the protein itself. We shall now list some of the signals that may exist in a newly synthesized protein that determine whether part or all of the protein will be secreted, and by which secretory pathway. The focus in this section will be on peptide neurotransmitters.

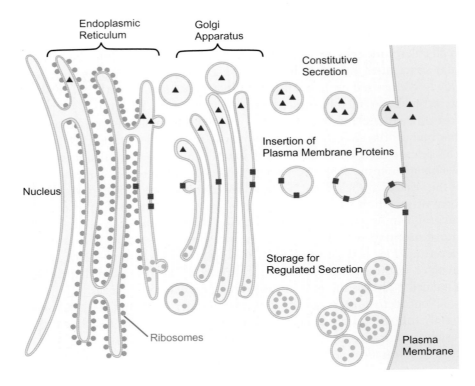

Figure 8–5. Constitutive and regulated secretion. The upper two pathways show the movement of proteins from ribosomes to the plasma membrane for constitutive release into the external medium or insertion into the plasma membrane. The lower pathway shows the buildup of protein-containing vesicles for regulated secretion.

Signals on Secreted Proteins

The signal sequence. The first part of a protein to be synthesized is the amino terminus. Proteins that are to enter one of the secretory pathways contain a stretch of hydrophobic amino acids at the amino terminus known as the *signal sequence* (Fig. 8–6). Although the exact signal sequence differs in different proteins, this hydrophobic stretch of amino acids is recognized by cytoplasmic factors (the *signal recognition particle*) and by components of the membrane of the rough endoplasmic reticulum. These factors then assist the protein to cross this membrane. Without crossing into the lumen of the endoplasmic reticulum, the protein could not eventually be packaged into membrane vesicles. Proteins that lack a signal sequence do not cross the membrane and thus become cytoplasmic proteins. The signal sequence is usually cleaved from the remainder of the protein by proteolytic enzymes shortly after it has crossed the membrane.

Sorting signals. Sorting of newly synthesized proteins must take place, so that different proteins are assigned to the appropriate vesicles and thus to the appropriate secretory pathway. In large part, this sorting appears

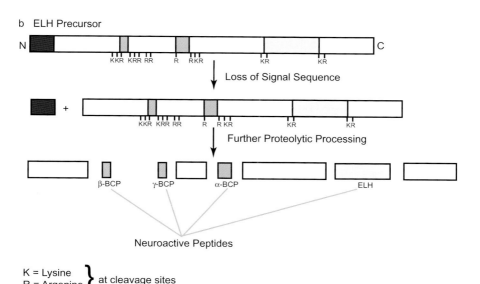

Figure 8–6. Processing of neuropeptide precursors. *a*: Formation of the hypothalamic peptide vasopressin. *b*: Formation of the *Aplysia* peptides α-, β-, and γ-BCP and ELH.

to occur within the Golgi apparatus. As we saw in Chapter 7, some proteins, such as certain ion channels that are destined for specific plasma membrane compartments, may be targeted there by specific auxiliary sub-units (see Table 7–2). In other cases, however, the sorting signal may be intrinsic to the protein sequence itself. For example, DNA that encodes the protein precursor for insulin (*proinsulin*) can be introduced into a pituitary cell line that does not normally make insulin. When this is done, not only is the DNA faithfully transcribed and translated into proinsulin, but this protein is also packaged into vesicles of the regulated secretion pathway, and insulin can be secreted from the cell in response to an appropriate stimulus.

The sorting of peptide transmitters and hormones into vesicles frequently is associated with a condensation of the protein into a dense aggregate. Such aggregates presumably serve to increase the concentration of peptide in a vesicle. Because they appear electron opaque when viewed in an electron microscope, peptide-containing vesicles are frequently described as *dense-core vesicles* or *granules* (Fig. 8–7).

Proteolytic processing. Most neuroactive peptides are not synthesized in the form in which they are eventually secreted. Instead, they are synthesized as part of a larger, inactive precursor protein, or *prohormone* (Fig. 8–6). A good example of this is the precursor of insulin, proinsulin, introduced in the previous section. Two others are the precursors for *vasopressin*, which is synthesized in certain neurons in the hypothalamus, and for *egg-laying hormone* (ELH), which is synthesized in *Aplysia* bag cell neurons. Proteolytic cleavage of the precursors to smaller fragments, including the active peptides, occurs in the secretory vesicles and also in the Golgi. In some cases, several different neuroactive peptides may be generated from a single precursor—for example, the production of the transmitters, α, β, and γ bag cell peptides (BCPs), together with ELH, from the ELH precursor (Fig. 8–6b). A sequence of two basic amino acids, either lysine or arginine, is frequently the site for proteolytic cleavage within the precursor (Fig. 8–6). Cleavage sites also exist, however, that are not marked by two basic amino acids.

Post-translational modifications. A newly synthesized protein and the peptide fragments generated from such a protein can be modified further by the action of enzymes in the Golgi or the secretory vesicles themselves. Table 8–1 lists the more common covalent modifications, some of which may be essential for the biological activity of a peptide. For example, one common modification is the amidation of the carboxyl terminal. This is

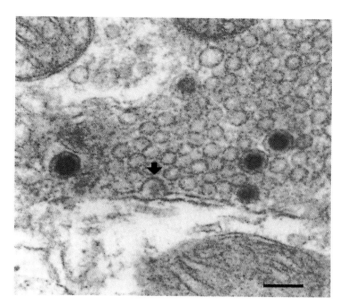

Figure 8–7. Dense-core granules and clear vesicles. An electron micrograph of part of the terminal of a rat neuron. Several larger dense-core granules coexist with smaller, clear synaptic vesicles. The arrow marks a vesicle that is undergoing exocytosis. The scale bar represents 100 nm. (Courtesy of Dr. Asa Thureson Klein.)

Table 8–1 Some Covalent Modifications of Peptide Transmitters and Hormones

Modification	Possible Function	Example
Conversion of N-terminal glutamate to pyroglutamate	Increase stability to proteases	Neurotensin
Amidation of C-terminal amino acid	Increase stability to proteases	Substance P
Glycosylation (usually addition of sugar residues to asparagine)	Targeting to appropriate location?	Thyrotropin—also commonly found in integral membrane proteins
Sulfation	Unknown	Cholecystokinin

carried out by the enzyme *peptidyl glycine (alpha)-amidating mono-oxygenase*, which is located within secretory granules. The enzyme acts on peptides that have a glycine residue at their carboxyl terminal by removing the glycine and amidating the penultimate amino acid residue. Figure 8–8 illustrates the amidation of the invertebrate neuropeptide FMRFamide and of vasopressin. It has been found that the biological activity of neuropeptides such as vasopressin is greatly reduced if the

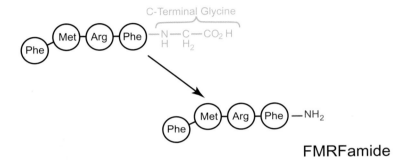

FMRFamide

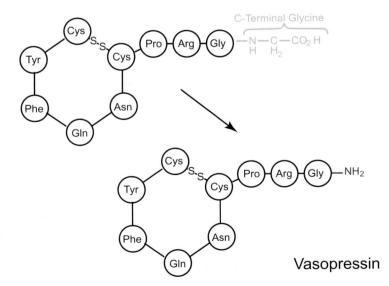

Vasopressin

Figure 8–8. C-terminal amidation. Formation of the neuropeptides FMRF-amide and vasopressin by removal of a C-terminal glycine.

peptide is not amidated. This and other modifications may also influence the stability of the peptide once it has been released. Some post-translational modifications of newly synthesized proteins may participate in the sorting of the protein into the appropriate vesicles. A precedent for this is found in proteins of the lysosomes. Newly synthesized lysosomal proteins contain asparagine residues to which mannose phosphate groups are added. These mannose phosphate groups are then recognized by receptors in the Golgi, causing these sugar-linked proteins to be targeted selectively to lysosomes.

"Classical" Neurotransmitters Use a Specialized Secretory Pathway

In contrast to peptide transmitters, small neurotransmitter molecules—including acetylcholine; the amino acid transmitters such as γ-aminobutyric acid (GABA), glycine, and glutamate; and the amines such as dopamine, norepinephrine, histamine, and serotonin—are not synthesized only at the soma (see Chapter 10). Instead it is the enzymes, which catalyze their synthesis, that are synthesized at the soma, then transported to the release sites at axon terminals. Depletion of neurotransmitter by electrical activity can therefore be followed by rapid resynthesis within the terminal. These small neurotransmitters are frequently termed *classical* neurotransmitters, because their function in neurons was recognized well before that of the neuropeptides.

Vesicles that contain these classical neurotransmitters are small and very homogeneous in size (typically 50 nm in diameter). These small synaptic vesicles are associated with the very rapid release of transmitter that occurs at specialized synaptic junctions (see Chapter 9). Because they do not generally contain a dense aggregate of neuropeptides, they are not electron dense and therefore appear as clear vesicles in electron micrographs (Fig. 8–7). Unlike the peptide-containing granules, which are formed at the soma and then transported to the synaptic endings, small synaptic vesicles are assembled within the nerve ending. The individual proteins in the membranes of small synaptic vesicles are first synthesized at the soma and then transported along axons to the terminals (see Fig. 8–5). On their arrival at nerve endings they are inserted into the plasma membrane by the constitutive secretion pathway and are then retrieved by endocytosis to form functional small synaptic vesicles (Fig. 8–9).

Both large dense-core vesicles and small synaptic vesicles release their neurotransmitter contents by exocytosis, resulting in the incorporation of

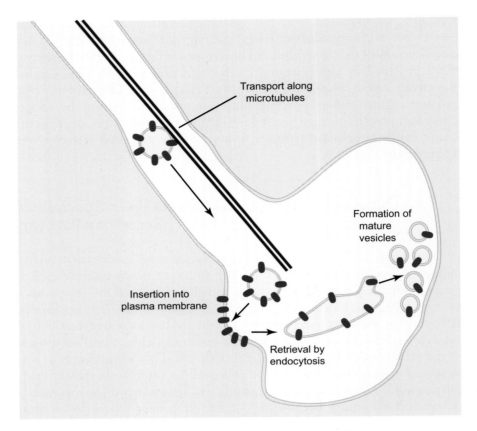

Figure 8–9. Formation of small synaptic vesicles. Vesicle proteins are transported down the axon to the terminals, where they are inserted into the plasma membrane before being retrieved by endocytosis and then packaged into mature synaptic vesicles.

vesicle membrane proteins into the plasma membrane. There is a fundamental difference, however, between these two types of vesicles. A dense-core vesicle is a "one-shot" apparatus; it can release its peptide neurotransmitter only once, after which the membrane proteins of the vesicle must be degraded or returned to the soma where new dense-core vesicles filled with peptide neurotransmitters are formed. In contrast, small synaptic vesicles can be recycled many times at the terminal, by repeated exocytosis followed by retrieval of vesicle proteins and refilling with neurotransmitter.

Exocytosis of Neurotransmitter-Containing Vesicles

Regulated secretion in many non-neural cells occurs through exocytosis. For example, in a chromaffin cell of the adrenal medulla, stimulation by a transmitter causes the secretory granules to fuse with the plasma membrane, releasing their contents to the extracellular space. The vesicle membrane that has been added to the plasma membrane is then rapidly resorbed by endocytosis. In many such cells, the process of exocytosis can be observed by light microscopy of living cells.

Release of neurotransmitter from small synaptic vesicles at synaptic junctions also occurs through exocytosis of transmitter-containing vesicles, but visualization of this process has been harder to achieve, largely because of the technical difficulties of working with small synaptic terminals. Direct visualization of exocytosis at a synapse was first achieved in the 1970s, by using electron microscopy combined with the freeze fracture technique. This work used the *neuromuscular junction*, the specialized synapse between motor neurons and skeletal muscle.

The vertebrate neuromuscular junction. Figure 8–10 is a drawing of the frog neuromuscular synapse, which uses acetylcholine as its transmitter. Large numbers of synaptic vesicles are associated with specialized areas of the presynaptic membrane that have been termed *active zones*. These are located close to structures called *dense bars*, which are at the cytoplasmic side of the presynaptic membrane. The dense bars may serve to align the synaptic vesicles at the sites of neurotransmitter release. Behind the active zones is a high density of *mitochondria* that provide the energy for secretion, reuptake of vesicles, and, as we shall see in Chapter 9, also regulate calcium levels in the terminal. On the muscle cell, clusters of *receptor molecules*, to which the released acetylcholine binds, are located in the areas of membrane closest to the active zones of the presynaptic terminals.

Morphological evidence for exocytosis during synaptic transmission. Earlier in this chapter, we saw the use of freeze fracture to detect arrays of gap junction particles. When applied to presynaptic membranes at the neuromuscular junction of a frog, this technique reveals deformations of the plasma membrane, due to the presence of the dense bars under the presynaptic membrane. Aligned in two rows on either side of the dense bar are strings of intramembranous particles that correspond to the calcium channel proteins (Fig. 8–11). The electron micrographs of Figure 8–11 were made using freeze fracture coupled with a technique

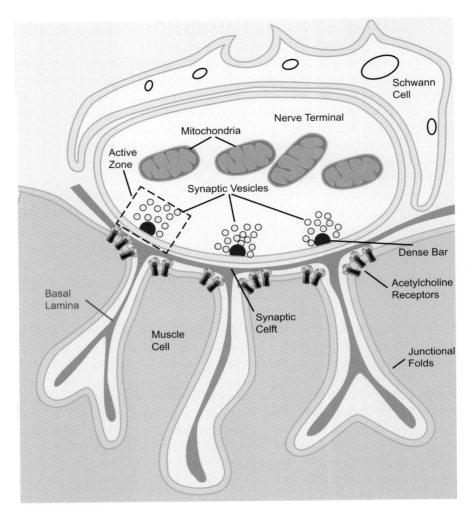

Figure 8–10. Diagram of a vertebrate neuromuscular junction.

that allows the terminals to be frozen very rapidly, and at precise times, following the stimulation of the motor nerve. In a terminal at rest, or in one that has been stimulated but that has not yet begun to release transmitter, the dense bar and the two rows of particles can be detected. At the very time that acetylcholine release occurs, however, further deformations of the membrane can be seen. These newly formed "pits" are thought to represent fusion of synaptic vesicles with the plasma membrane. The presence of these pits strongly suggests that release of transmitter occurs through exocytosis. ❽

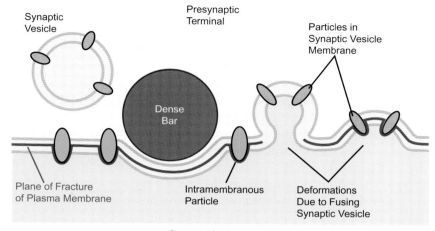

a

Synaptic
Vesicle

Presynaptic
Terminal

Particles in
Synaptic Vesicle
Membrane

Dense
Bar

Plane of Fracture
of Plasma Membrane

Intramembranous
Particle

Deformations
Due to Fusing
Synaptic Vesicle

Synaptic Cleft

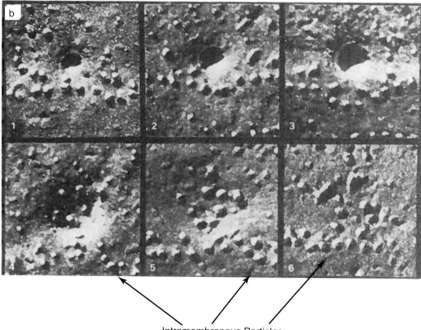

Intramembranous Particles

Figure 8–11. Freeze fracture evidence for exocytosis at the neuromuscular junction. *a*: Deformations of the plane of the plasma membrane are made by intramembranous particles, underlying structures, and fusing vesicles. *b*: Images of freeze fracture replicas of the presynaptic membrane, made by Tom Reese and colleagues, showing progressive stages (1–6) in the exocytosis of a single vesicle. (Courtesy of Tom Reese.)

Fluorescent dyes and proteins can be used to observe neurotransmitter exocytosis. Fluorescent styryl dyes have proven to be a particularly useful tool in studies of exocytosis of synaptic vesicles (Fig. 8–12). The first of these to be widely used is termed *FM1-43*. This dye is not fluorescent in solution but becomes so when it binds to cellular membranes, which it is able to do very rapidly and reversibly. Thus, when a synaptic terminal is exposed to FM1-43, the external membrane becomes fluorescent. If the neuron is stimulated at this time, the membrane of small synaptic vesicles is temporarily added to the plasma membrane, where it encounters the FM1-43. When these vesicular membranes are taken up again by endocytosis, the synaptic vesicles within the terminal become fluorescent. If the external FM1-43 is then removed, the plasma membrane loses its fluorescence, but the internal synaptic vesicles remain labeled. Subsequent stimulation of the terminal in an FM1-43-free solution allows one to measure the exocytosis of individual synaptic vesicles as a loss of the fluorescence of internal vesicles (Fig. 8–12).

A complementary optical technique for observing exocytosis was developed by Gero Miesenböck in 1998. This relies on a genetically engineered hybrid protein called *Synapto-pHluorin*, which contains two parts. The first part of the hybrid protein is a synaptic vesicle protein such as synaptophysin (described later in this chapter) that is normally present on the membrane inside synaptic vesicles (Fig. 8–12g). The second part of Synapto-pHluorin is a form of the fluorescent protein GFP (Green Fluorescent Protein) that is only fluorescent when it is not in an acidic environment. Because the inside of synaptic vesicles is normally acidic, this means that if this hybrid protein is introduced into neurons, it fails to fluoresce inside the vesicles. On stimulation of neurons to release neurotransmitter, Synapto-pHluorin is exposed to the external medium and the sites of transmitter release can be seen as points of fluorescence at the presynaptic terminals (Fig. 8–12h). These fluorescent spots disappear when the exocytosed membrane is subsequently retrieved and acidified.

Stages of transmitter release. As described earlier, exocytosis is the process of fusing one membrane, the vesicle membrane, to a second membrane, the plasma membrane. Such fusion of membranes is a very general biological process; it is required for the movement of vesicles from the endoplasmic reticulum to the Golgi apparatus, for example (see Fig. 8–5). Exocytosis is a special case of such fusion in which the acceptor membrane is the plasma membrane. In the case of neurotransmitter release, three distinct stages have been distinguished (Fig. 8–13).

FM1-43 technique

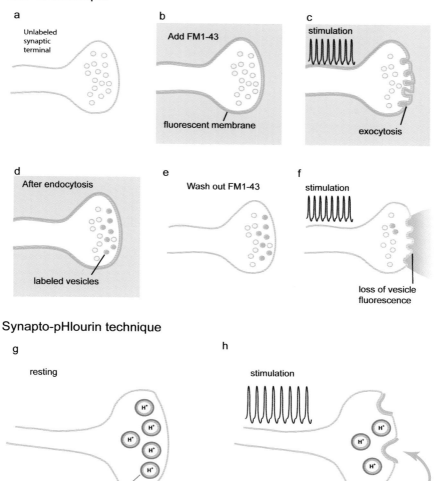

a
Unlabeled synaptic terminal

b
Add FM1-43

fluorescent membrane

c
stimulation

exocytosis

d
After endocytosis

labeled vesicles

e
Wash out FM1-43

f
stimulation

loss of vesicle fluorescence

Synapto-pHlourin technique

g
resting

Synapto-pHlourin (non-fluorescent)

h
stimulation

Synapto-pHlourin (fluorescent)

Figure 8–12. Measurements of endocytosis and exocytosis in synaptic terminals. Panels *a–f*: Use of the FM1-43 dye. The plasma membrane of a synaptic
terminal (*a*) becomes fluorescent upon exposure to the dye (*b*). Stimulation (*c*)
produces exocytosis and exposes the membrane of synaptic vesicles to the
external FM1-43. Endocytosis of the synaptic vesicle membranes (*d*) causes
them to become fluorescent, and they remain so after washout of the external
dye (*e*). The exocytosis of these labeled vesicles can then be detected by subsequent stimulation in a dye-free solution (*f*). Panels *g–h*: The Synapto-pHluorin
technique: When Synapto-pHluorin is introduced into neurons, it localizes to
the inner surface of the synaptic vesicle membrane, where it does not fluoresce
because of the high pH (*g*). Following endocytosis, Synapto-pHluorin is exposed
to the external medium, where it emits fluorescence.

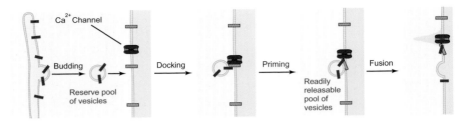

Figure 8–13. Steps in the movement of synaptic vesicles from internal membranes to fusion with the plasma membrane. See Animation 8–1. ⊙

Docking. Before the contents of a synaptic vesicle can be released, the membrane of the vesicle must first become tightly associated with the plasma membrane, a process termed *docking*. Before they are docked, vesicles are sometimes said to be in a reserve pool.

Priming. Simple association with the plasma membrane is not enough to ensure fusion. A priming reaction converts the vesicle to a form that can fuse when an action potential invades the terminal. The collection of primed vesicles at a presynaptic ending is frequently referred to as the *readily releasable pool* of vesicles.

Fusion. The active fusion of the vesicle membrane with that of the plasma membrane occurs when the calcium concentration is locally elevated to a high level by the opening of plasma membrane calcium channels during an action potential. The fusion of synaptic vesicles is followed by retrieval of the vesicular membrane by endocytosis and its recycling to form a new vesicle that has been refilled with transmitter. We shall consider this recycling process later in this chapter.

Molecules That Shape Neurotransmitter Release

Many discoveries have been made through the study of the proteins that are in, or attached to, the membrane of synaptic vesicles. Relatively pure synaptic vesicles from the nervous system may be prepared by subcellular fractionation. In this technique, nervous tissue is homogenized in a medium that allows subcellular organelles, such as mitochondria and synaptic vesicles, to remain intact. Because different organelles have different sizes, they can then be separated by centrifugation through layers of sucrose of different densities. In this way it is possible to obtain relatively pure preparations of vesicles from neuronal as well as other secretory tissues.

The availability of pure vesicles has enabled some of the key molecules in the process of exocytosis to be identified and the relationships between these molecules to be unraveled. ❻ It should be pointed out that the small synaptic vesicles, which appear clear in electron micrographs (Fig. 8–7) and contain small neurotransmitter molecules such as acetylcholine, glutamic acid, GABA, or glycine, have a somewhat different protein composition than that of the slightly larger dense-core vesicles that are predominantly peptide containing.

GTP-binding proteins regulate the interaction of vesicles with the plasma membrane. In all neurons, the exocytosis of vesicles containing classical neurotransmitters occurs at specialized active zones similar to those we saw for the neuromuscular junction. Attached to inside the plasma membrane at active zones is a very dense web of proteins that is physically linked to the membrane and to the cytoskeleton. This web is comprised of proteins with the musical names Piccolo and Bassoon as well as many other proteins. ❽

A set of proteins that bind GTP are known to regulate the budding and fusion process that controls the traffic of all membranous vesicles through cells (Fig. 8–14). These GTP-binding proteins are termed *rab* proteins ❽, and different members of this family associate with different membranes, such as those of the Golgi apparatus or the endoplasmic reticulum. The synaptic vesicle member of this protein family, which is termed *rab3*, binds to both synaptic vesicles and to proteins at the plasma membrane at the active zone, and it regulates both the docking and priming of vesicles.

The rab3 protein leads a dual life (Fig. 8–14). Much of the time it is anchored to the membrane of synaptic vesicles through a lipid chain, termed a *geranylgeranyl group*, which is attached to the rab3 protein. This attachment to the membrane only occurs when the protein is bound to GTP. Rab3 also binds two other proteins, *Munc13* and *RIM* (**R**ab3-**I**nteracting **M**olecule), a protein that is linked to the plasma membrane. This complex of three proteins, Rab3/Munc13/RIM, both serves to dock the vesicles to the active zone and participates in priming them for subsequent release.

When vesicle fusion occurs, the bound GTP is hydrolyzed to GDP and rab3 dissociates from the vesicle membrane to begin its second life as a soluble protein in the cytoplasm of the nerve ending. To become solubilized, the GDP-bound form of rab3 binds a soluble protein termed *GDI* (for **G**DP-**D**issociation **I**nhibitor). The GDP on the rab3 protein is

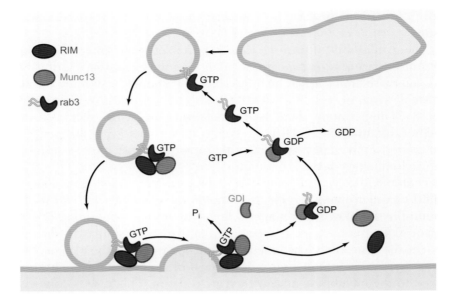

Figure 8–14. Cycle of GTP-dependent association and dissociation of rab3 from synaptic membranes.

subsequently exchanged for GTP in the cytoplasm, allowing rab3 with its lipid modification again to bind vesicles that have not yet docked at the plasma membrane. Thus each cycle of budding and fusion at the synaptic terminals is associated with one cycle of GTP hydrolysis (Fig. 8–14).

Exocytosis and the SNARE Complex

Three molecules central to the exocytosis of synaptic vesicles are *synaptobrevin, syntaxin,* and *SNAP-25.* Of these, only synaptobrevin is a synaptic vesicle protein. It has a very asymmetrical distribution across the membrane of the vesicle. Only a few amino acids, at the carboxyl-terminal end of the protein, are found inside the vesicle, and most of the mass of the protein is in the cytoplasm. Syntaxin has a very similar structure, but it is located in the plasma membrane of the synapse, with its bulk also in the cytoplasm. In contrast, SNAP-25 (named for SyNaptosomal Associated Protein of size 25 kDa) is not a true integral membrane protein but is firmly anchored to the plasma membrane by palmityl chains. These palmityl groups are lipid chains that are attached at one end to the protein at a cluster of cysteine residues and insert firmly into the lipid environment of the plasma membrane. These three proteins together form a complex that links the synaptic vesicle to the plasma membrane and whose structure has been determined by X-ray crystallography (Fig. 8–15). Even though this complex forms a close association between the vesicle and plasma membrane, its formation does not constitute either docking or priming. Instead, it appears to function in the fusion event itself, promoting the fusion of the two types of membrane; this is called the *fusion* or *SNARE complex*.

Prior to docking, the SNARE complex is not assembled. Instead, each of the components of the complex is bound to other proteins (Fig. 8–16a). For example, the synaptic vesicle SNARE protein synaptobrevin may be bound to other synaptic vesicle proteins, such as *synaptophysin*. Each synaptophysin protein spans the vesicle membrane four times and forms a homo-oligomer complex with other synaptophysin molecules. Although synaptophysin is one of the most abundant proteins in the membrane of synaptic vesicles, it appears not to be required for neurotransmitter release, and its real function is not yet understood. Similarly, the plasma membrane SNARE proteins have alternative partners when their vesicular mates are not around. Within nerve endings, *munc-18* is a soluble protein that, like synaptophysin, is not part of the fusion complex but is able to bind syntaxin.

After docking brings the synaptic and vesicular membranes into close proximity, the SNARE components leave some of their alternative partners and form the fusion complex. This complex consists of four long bundles of helical regions of the three proteins (Fig. 8–15). When the three proteins first come together, they assemble into a pattern that is energetically very unfavorable, termed the *trans*-state SNARE complex (Fig. 8–16b). It is the subsequent relaxation of the shape of the three-protein complex into an energetically more favorable state, termed the *cis* state, in which the proteins are "zippered" together, that pushes the lipids of the vesicle into the plasma membrane (Fig. 8–16e).

In the absence of other proteins, SNARE complexes always spontaneously collapse from the *trans* state to the *cis* state. In a nerve terminal at rest, however, this transition is prevented by two other proteins, the cytoplasmic *complexin* molecule and the synaptic vesicle protein *synaptotagmin* (Fig. 8–16b). These two proteins ensure that the SNARE complex does not zipper to its *cis* state until calcium enters the terminal through the plasma membrane. In particular, complexin binds to the assembled SNARE complex and helps to keep it in the *trans* state (Fig. 8–16c).

The rapid fusion of vesicles with the membrane that follows an action potential in the terminal is triggered by calcium entry through voltage-dependent calcium channels (Fig. 8–16d). The speed at which this

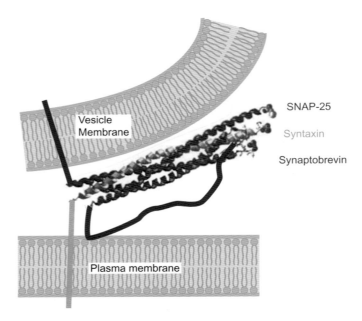

SNAP-25

Syntaxin

Synaptobrevin

Figure 8–15. Structure of the SNARE complex. The interactions of synapto-brevin, syntaxin, and SNAP-25 have been determined by X-ray crystallography. Locations of the vesicle and plasma membranes and the transmembrane regions of the proteins have been drawn in for clarity (Sutton et al., 1998).

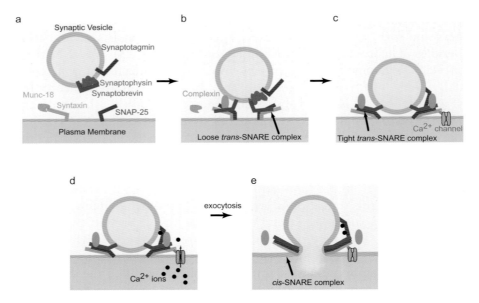

Figure 8–16. Sequence of the interactions of the SNARE complex with synap-tic vesicle and plasma membrane proteins before (*a–d*) and after (*e*) exocytosis. See Animation 8–1.⬤

happens is helped by the fact that the syntaxin molecule within the SNARE complex binds calcium channels directly and is therefore at the site of greatest calcium elevation (Fig. 8–16c). The proteins of the SNARE complex itself, however, are not sensitive to calcium. Instead, the major calcium-sensor is synaptotagmin, which has two regions in the cytoplasmic part of the molecule, called C2 domains, which closely resemble regions in other proteins that are known to bind both calcium and phospholipids. When calcium levels are raised to the level that occurs in the synaptic terminal during an action potential (Fig. 8–16d), synaptotagmin is thought to displace complexin from the SNARE complex, allowing fusion to proceed (Fig. 8–16e).

Toxins have identified the SNARE proteins involved in exocytosis. As we have seen and shall see repeatedly in this book, important insights into the mechanisms of neuronal function have been gained through the use of toxins that poison the nervous system. In the case of neurotransmitter release, evidence that synaptobrevin and SNAP-25 are essential for exocytosis was obtained in studies using three deadly bacterial neurotoxins: tetanus toxin and botulinum toxins A and B. Poisoning by these agents, known as *clostridial neurotoxins* after the bacteria that make them, causes paralysis by blocking neurotransmitter release. Tetanus toxin produces spastic paralysis, a characteristic rigidity of the body, by preventing transmitter release in the central nervous system, whereas botulinum B toxin causes flaccid paralysis by preventing acetylcholine release at the neuromuscular junction. The clostridial neurotoxins are produced by the bacteria as single proteins that are then cleaved into two different subunits, known as the *light chain* and the *heavy chain*. These subunits have different functions and are held together by disulphide bonds (Fig. 8–17a). The heavy chain binds to the external membrane of the neuron and facilitates the entry of the light chain into the cytoplasm of synaptic terminals. Thus, the different heavy chains in tetanus toxin and the botulinum toxins determine which particular neurons they will affect.

It is the light chains of the toxins, which bind zinc ions, that actually do the damage. Once these light chains are released from the heavy chains and enter the cytoplasm, they act as proteases that rapidly destroy selected proteins within the synaptic endings. For example, botulinum toxin B and tetanus toxin destroy only synaptobrevin, whereas botulinum toxin A selectively cleaves SNAP-25 (Fig. 8–17b). Another toxin, botulinum toxin C1, destroys syntaxin. Because each of these toxins is very specific in its action, and all of them impair neurotransmitter release, they have provided direct evidence that synaptobrevin, syntaxin, and SNAP-25 are critical components of the exocytotic machinery.

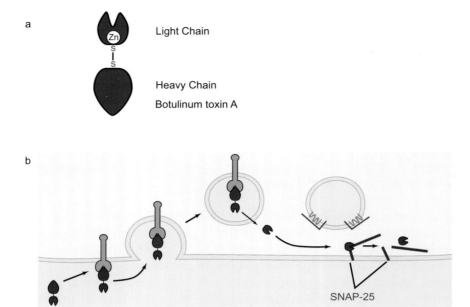

a: Light Chain

Zn

Heavy Chain

Botulinum toxin A

b

SNAP-25

Figure 8–17. Blockage of neurotransmission by clostridial neurotoxins. *a*: Structure of botulinum toxin A. *b*: Uptake of botulinum toxin A into the nerve terminal, followed by release of the light chain and proteolysis of SNAP-25. See Animation 8–1.⬤

Certain other toxins also regulate transmitter release. For example, α-*laterotoxin*, a component of the venom of the black widow spider, produces massive exocytosis of small synaptic vesicles. α-Laterotoxin binds a protein termed *neurexin* in the plasma membrane of the presynaptic terminal. Neurexins are receptors that normally bind postsynaptic proteins termed *neuroligins*, forming a bridge across the synaptic gap, and may participate in the development of synapses (see Chapter 17). It appears, however, that the toxin uses neurexin only as a convenient binding site. The mechanism of action of the toxin, at least in part, is to form a large nonselective channel in the plasma membrane, which produces massive calcium influx.

The SNARE complex is disassembled by NSF. The interaction between synaptobrevin, syntaxin, and SNAP-25 is a very tight one. For transmitter release to proceed normally, the SNARE complex must also become disassembled. This may occur at some point after exocytosis when the synaptic vesicle membrane is recovered and recycled. Some studies have

suggested, however, that such ungluing occurs in the preparation of vesicles for release, perhaps during the priming step.

Another set of proteins is responsible for this ungluing process
(Fig. 8–18). These include a homotrimeric protein, made up of three 76
kDa subunits, termed *NSF* (named for one of its chemical properties,
N-ethylmaleimide **S**ensitive **F**actor), and **S**oluble **NSF A**ccessory **P**roteins
(SNAP). By a true perversity of scientific nomenclature, these SNAPs are
entirely unrelated to SNAP-25. Three different SNAPs, α-, β-, and γ-SNAP,

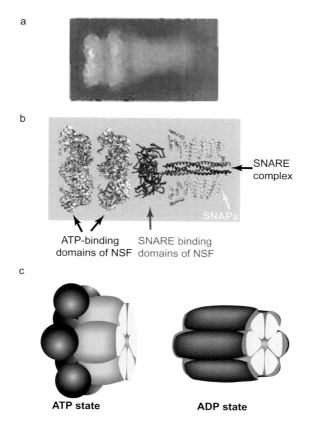

Figure 8–18. Ungluing the SNARE complex by NSF and the SNAPs. *a*: Electron microscopic image of the SNARE complex together with its interacting
proteins the SNAPs and NSF to form a large multi-protein complex termed the
20S particle (Hohl et al., 1998). *b*: A model based on X-ray crystal structures,
showing how NSF, SNAPs, and the SNARE complex come together to form the
20S particle (Zhao et al., 2007). *c*: Diagram of the change in the structure of
the NSF complex produced by ATP (Hanson et al., 1997). This change may
cause the unraveling of the interactions between the plasma membrane and
vesicular SNARE proteins.

are known to exist, and a functional ungluing complex requires the presence of γ-SNAP together with either α- or β-SNAP. Indeed, the SNARE proteins were first named for their association with these proteins (SNAP-REceptors). Electron microscopic and structural studies have shown that the SNAPs wrap around the elongated SNARE complex and that several NSF molecules assemble at one end of this complex. The conformation of the NSF complex is altered by ATP, and this change in shape may serve to unravel the SNARE complex (Fig. 8–18).

Recycling of Small Synaptic Vesicles

Once a vesicle has released its contents by fusing with the plasma membrane, the membrane of the vesicle, together with the synaptic vesicle proteins that have entered the plasma membrane, must be retrieved. This process occurs for both the small synaptic vesicles containing classical neurotransmitters and for the larger neuropeptide-containing dense-core granules. In the case of small synaptic vesicles, the retrieved membrane can be recycled to form new small vesicles filled with neurotransmitter. In contrast, the membrane proteins of the neuropeptide-containing granules must be destroyed or returned to the soma.

The recovery of small synaptic vesicles from the plasma membrane involves several new rounds of budding and fusion and introduces a new set of protein components (Fig. 8–19 ❸). A major pathway for recovery involves coating the membrane to be removed in a layer of protein termed *clathrin*. The structure of this coat resembles a cage made of chicken wire. Each link in the cage is a "three-legged" protein complex consisting of three heavy chains and three light chains of the clathrin molecule. This process is termed *clathrin-mediated endocytosis*.

The first step in the formation of the clathrin cage is the identification of which specific proteins in the plasma membrane have to be retrieved. This is accomplished by *adaptor proteins*, which bind to the target proteins that are destined for endocytosis and then recruit clathrin to the membrane. Binding of adaptor proteins to their targets only occurs in the presence of the phospholipid *phosphatidylinositol-4,5-bisphosphate* (abbreviated PIP_2), which is found only in the plasma membrane. This ensures that clathrin-mediated endocytosis begins at the plasma membrane and that clathrin coats do not form over internal organelles such as the vesicles themselves. For example, adaptor proteins termed *AP2* and *Stonin-2* bind directly to synaptotagmin when it is in the plasma membrane, while the adaptor protein *AP180* binds one form of synaptobrevin.

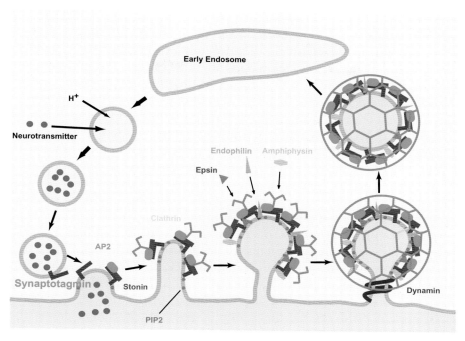

Figure 8–19. Recycling of synaptic vesicle membranes. After exocytosis, vesicle proteins such as synaptotagmin interact with PIP_2, AP2, and Stonin, and the infolding membranes begin to be coated with clathrin. Following the recruitment of additional molecules, including epsin, endophilin, and amphyphysin, the clathrin-coated vesicles are pinched off from the plasma membrane by the action of the motor protein dynamin. The membranes are then transported to the early endosome and reformed into new vesicles, which are refilled with neurotransmitter.

AP2 also activates the kinase that makes PIP_2, allowing more adaptor proteins to enter the surrounding membrane.

The binding of adaptor proteins not only recruits clathrin to the membrane but also brings in a second set of proteins required for the complete uptake of the membrane. These include protein termed *epsin, endophilin,* and *amphiphysin,* which bind lipids in a way that promotes the deformation of the lipid membrane into a sphere. Finally, another protein, *dynamin,* is required for the budding of the vesicle membrane from the plasma membrane. Like kinesin and related proteins we considered in Chapter 2, dynamin is a motor protein and is capable of generating movement in the presence of GTP. This force-generating protein, however, does not belong to the same protein family as the kinesins. As membrane that contains the synaptic vesicle proteins begins to bud off from the plasma membrane, dynamin forms a collar of protein around the neck of the newly budding vesicle (Fig. 8–19). This collar presumably allows the bud to pinch off from the remainder of the membrane. In cells that have defective dynamin proteins, vesicle proteins fail to be retrieved from the plasma membrane. An example of this occurs in the *Drosophila* strain *shibire*. These flies are defective in that their dynamin proteins fail to function at elevated temperatures; raising the temperature produces rapid and reversible paralysis because synaptic vesicles cannot be recycled.

After the coated vesicle buds off the plasma membrane, it moves to another membranous structure known as the *early endosome* (Fig. 8–19). This is the structure from which new synaptic vesicles are formed. After fusion of the coated vesicles with the early endosome, the retrieved synaptic vesicle proteins are ready to be reorganized into new small synaptic vesicles. To form mature synaptic vesicles, the vesicles that bud from the early endosomes must be replenished with neurotransmitter. Before this happens, the interior of the vesicle is first acidified by the activity of a proton pump. Such a proton pump uses ATP in the cytoplasm to pump protons across the membrane into the lumen of the vesicle. The acidic environment produced by this pump is essential, in that it drives the uptake and storage of neurotransmitter by specific transport proteins present in the membrane of the vesicles.

Detection of Exocytosis and Endocytosis by Capacitance Measurements

There are still many unanswered questions about the dynamics of exocytosis and endocytosis, even in those non-neuronal cells in which the process can be visualized readily in the microscope. For example, must the membrane of a vesicle undergo complete fusion into the plasma membrane to release transmitter? Alternatively, can a vesicle fuse with the plasma membrane transiently, release its neurotransmitter through a pore that spans the two membranes, then reseal without losing its integrity? The latter form of hypothetical release mechanism has been termed a *kiss-and-run mechanism* (Fig. 8–20a). A variant of the patch clamp technique, devised by Erwin Neher and colleagues, is providing answers to these sorts of questions. The technique is based on the fact that a cell's capacitance, which can be measured using the whole-cell configuration of the patch clamp, is directly proportional to the surface area of the cell membrane. To provide a continuous measure of cell capacitance, a high-frequency (~800 Hz) sinusoidal voltage command is applied to the cell. The current that flows across the membrane is measured with an amplifier known as a *lock-in amplifier*. This determines the magnitude of the currents that are in phase with the voltage oscillation and those that are 90° out of phase. The latter can be related directly to the capacitance of the cell. When a vesicle fuses with the plasma membrane, there is a

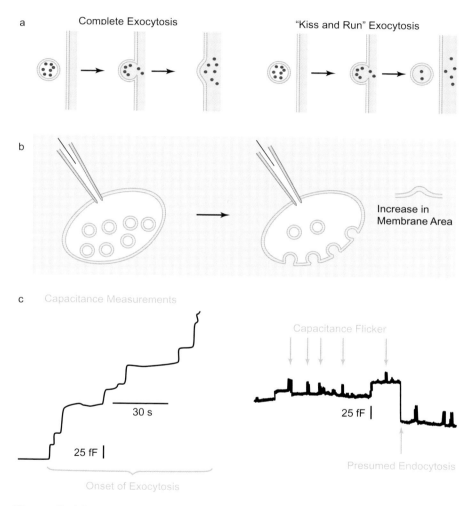

Figure 8–20. Exocytosis and capacitance measurements. *a*: Diagram contrasting complete exocytosis with the "kiss-and-run" mode of exocytosis. *b*: Exocytosis leads to an increase in the area of the plasma membrane. *c*: Capacitance measurements in a mast cell demonstrating stepwise increases (*left*) in capacitance at the onset of exocytosis, capacitance flicker, and stepwise decreases (*right*) in capacitance. Experiments of this sort have been carried out by Erwin Neher, Wolf Almers, and their colleagues.

small increase in the size of the surface membrane (Fig. 8–20b). This is registered as an increase in cell membrane capacitance, allowing the whole-cell patch clamp apparatus to detect directly the exocytosis of single secretory vesicles.

Figure 8–20c shows measurements of capacitance in a mast cell from the peritoneum of a mouse. When stimulated, mast cells release massive amounts of histamine from their secretory granules; their activity is responsible for many of the symptoms of an allergic attack. In whole-cell recordings, introduction of a nonhydrolyzable analog of GTP into the patch pipette is sufficient to trigger this release, suggesting that a GTP-binding protein (see Chapter 12) may contribute to the onset of secretion. A stepwise increase in capacitance is recorded as each vesicle fuses with the plasma membrane (Fig. 8–20c, left). Stepwise decreases in membrane capacitance can also be observed (Fig. 8–20c, right); these represent membrane retrieval by endocytosis.

For this technique to work, the secretory vesicles must be large, as is the case in chromaffin cells and mast cells, so that the amount of added membrane will be detectable as an increase in capacitance. Although it has not yet been possible to apply the technique to the exocytosis of synaptic vesicles in most nerve terminals, the approach has provided several insights into the process of exocytosis. An early step in exocytosis in mast cells is the formation of a pore between the inside of the vesicle and the external medium. The initial conductance of the pore is about 230 pS. With time, the conductance increases as the pore dilates and the vesicle fuses with the plasma membrane. The initial stages of pore formation, however, appear to be reversible. This is observed as the occurrence of flicker, in which the capacitance fluctuates between two levels (Fig. 8–20c). This indicates that vesicles can fuse with the plasma membrane and dilate partially, but then return to a "closed" state without undergoing full exocytosis. Capacitance measurements, coupled with techniques such as the FM1-43 approach described earlier, support the conclusion that the kiss-and-run mode of exocytosis occurs for both large dense-core vesicles and small synaptic vesicles.

Testing the Roles of Synaptic Proteins

In this chapter we have introduced some of the molecules that are important players in the game of exocytosis at synaptic endings. Most of the information on synaptic proteins presented in this chapter has been gained through biochemical studies of the interactions among these proteins. We have seen that naturally occurring toxins have played a critical role in identifying which proteins are essential to the exocytotic process. There are other experimental approaches to test the function of specific synaptic proteins. These include techniques such as the injection of proteins into synaptic endings, or the production of mutant animals that lack a specific protein. To understand these approaches, however, we must first understand in more detail the physiology of synaptic transmission. This will be covered in Chapter 9, which will also introduce some additional molecules that regulate transmitter release.

Nonvesicular Release of Neurotransmitters

Exocytosis is the major form of transmitter release at most synaptic junctions. Nonetheless, there are several examples of the release of transmitter by neurons by mechanisms other than calcium-dependent exocytosis. The first examples were found in the retina, where some of the synapses made by the *photoreceptors*, as well as synapses made by a retinal cell termed the *horizontal cell*, lack synaptic vesicles and do not require extracellular calcium to release their neurotransmitter. Subsequent work has demonstrated that this can also occur in central neurons. For the neurotransmitter GABA, nonvesicular release across the plasma membrane can occur through the GABA transporter protein, termed *GAT1*. This normally operates to remove GABA from the extracellular space, but under certain physiological conditions, the process can reverse its direction, allowing GABA to flow from the cytoplasm into the extracellular space.

Summary

Two ways that neurons communicate with one another are by direct electrical coupling and by the secretion of neurotransmitters. Electrical coupling arises from the existence of proteins, known as *connexins*, that form pores linking the cytoplasm of adjacent cells. Ions (as well as small molecules) can carry signals from one cell to another through these pores. *Neurosecretion* is a more complex process in which different categories of molecules are sorted into vesicles in the cytoplasm. A variety of chemical processes within these vesicles ensure that they contain biologically active transmitters or hormones. SNARE complex proteins, together with other proteins such as rab3, complexin, and synaptotagmin, then cooperate to allow synaptic vesicles that contain neurotransmitter to release their components into the external medium following calcium entry into nerve terminals. Such exocytosis of synaptic vesicles can be monitored with imaging techniques using fluorescent dyes or proteins such as synapto-pHluorin, or by capacitance measurements. A second set of molecules, which includes AP2, clathrin, amphiphysin, and dynamin, have the task of retrieving the membrane of synaptic vesicles back from the plasma membrane through the process of endocytosis.

Synaptic Release
of Neurotransmitters

I n the previous chapter we saw that secretion of proteins occurs through the exocytosis of membranous vesicles. This is a process that has been elaborated throughout evolution in endocrine, exocrine, and neuronal cells. It allows cells to send specific chemical signals that diffuse to, and act on, recipient cells. In many cases, the properties of neurons are much like those of endocrine cells, whose business is chemical communication. Some neurons release peptides directly into the blood, just as an endocrine cell does. Other neurons release their transmitters locally into the extracellular space, where the transmitter diffuses slowly over some distance and influences many other neurons. Many neurons, however, differ from these other cell types in that they have been under evolutionary pressure to develop very rapid chemical communication with specific target neurons and muscle cells. To this end, they have developed long axons that bring the source of the messenger substance right up to the membrane of their target cell. In addition, neurons often use small chemical transmitters that can be synthesized directly at the terminal rather than being transported from the soma, as is the case with peptide transmitters. The characteristics of the release of such transmitters at synaptic terminals differ from those of many other secretory cells. In this chapter we give an account of transmitter release at three thoroughly studied synaptic junctions: the vertebrate *neuromuscular junction*, the *giant synapse* of the squid, and the large *calyx of Held* synapse in the auditory brainstem of mammals.

Transmitter Release Is Quantized

We have seen that in many cells, secretion occurs through the exocytosis of packets of peptide or hormone that are stored inside vesicles. As expected for a process that involves the exocytosis of synaptic vesicles, the release of neurotransmitter at most chemical synapses also occurs in small packets, or *quanta*. This was first demonstrated by Bernard Katz and colleagues through electrophysiological studies at the neuromuscular junction, which we introduced in the previous chapter (see Figure 8–10). They placed electrodes in the postsynaptic muscle cell and measured the extent of depolarization of the muscle in response to synaptic stimulation. This depolarization provides a direct measure of the amount of synaptic transmitter released under various experimental conditions.

EPPs and MEPPs. When the presynaptic nerve is stimulated, an action potential travels along the axon to the presynaptic terminal. The depolarization of the terminal causes the release of acetylcholine. This in turn acts on the receptors in the postsynaptic membrane to depolarize the muscle by mechanisms that will be discussed in Chapter 11. This depolarization is termed an *end-plate potential* (EPP) (Fig. 9–1). Under normal conditions, the end-plate potential is many tens of millivolts in size, sufficient to trigger an action potential and the subsequent contraction of the muscle. However, even in the absence of nerve stimulation, small depolarizations can be recorded with an electrode in the muscle (Fig. 9–1). These spontaneously occurring depolarizations are only about 0.5 mV in amplitude, but in most other respects they are very similar to the larger end-plate potential evoked by nerve stimulation. In particular, the time course of the small depolarizations matches that of the end-plate potential. In addition, the spontaneously occurring potentials, like the end-plate potential, can be blocked by antagonists of the acetylcholine receptor such as curare (see Chapter 11). The potentials can also be prolonged by agents that prevent the hydrolysis of acetylcholine by the enzyme acetylcholinesterase in the synaptic cleft. Finally, the frequency of occurrence of these small depolarizations increases on depolarization of the presynaptic terminal. These and other findings indicate that these small potentials are due to the spontaneous release, at random intervals, of a small amount of acetylcholine from the presynaptic terminal. The small potentials were therefore named *miniature end-plate potentials* (MEPPs).

EPPs are made up of multiple MEPPs. Katz then showed that the stimulus-evoked synaptic potential, that is, the end-plate potential, was

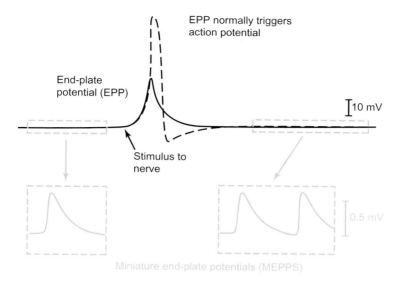

EPP normally triggers
action potential

End-plate
potential (EPP)

10 mV

Stimulus to
nerve

0.5 mV

Miniature end-plate potentials (MEPPS)

Figure 9–1. End-plate potentials (EPPs) and miniature EPPs (MEPPs). Membrane potential recordings in a muscle cell. An EPP evoked by a nerve stimulus normally triggers a postsynaptic action potential. When its amplitude is decreased or the action potential is blocked, however, its time course matches that of MEPPs.

caused by the simultaneous occurrence of a large number of individual potentials, each of which appeared identical to a MEPP. To analyze the synaptic potentials in detail, the neuromuscular junction was bathed in a medium that contained lower than normal levels of calcium ions and a higher concentration of magnesium ions. As we shall see, such a medium reduces the amount of transmitter that is released on stimulation. In these particular experiments, release was reduced to such an extent that the average amplitude of the postsynaptic potential following nerve stimulation was a few millivolts, only a few times larger than a spontaneously occurring MEPP. Under these conditions, the evoked end-plate potential did not have a fixed amplitude with each stimulus to the nerve. Instead, some stimuli failed to generate an end-plate potential, some generated end-plate potentials that were equal in amplitude and duration to a single MEPP, while others generated end-plate potentials that were equal in amplitude to two or more individual MEPPs (Fig. 9–2a). When a count is made of the number of times that end-plate potentials of different amplitudes occur in such an experiment, a histogram such as that in Figure 9–2b is generated. Although there is some variability in the size of

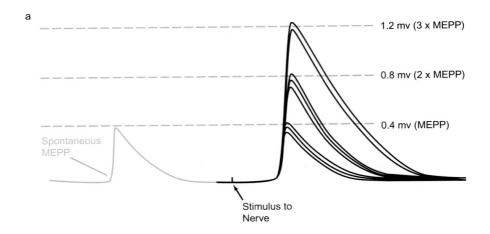

a

1.2 mv (3 x MEPP)

0.8 mv (2 x MEPP)

0.4 mv (MEPP)

Spontaneous
MEPP

Stimulus to
Nerve

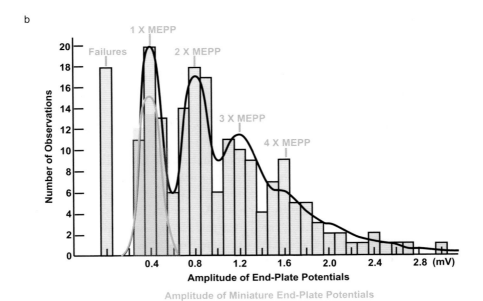

b

1 X MEPP

Failures

2 X MEPP

3 X MEPP

4 X MEPP

Number of Observations

Amplitude of End-Plate Potentials

Amplitude of Miniature End-Plate Potentials

Figure 9–2. End-plate potentials (EPPs) are made of multiple miniature EPPs (MEPPs). *a*: Under conditions of low transmitter release, nerve-evoked EPPs have amplitudes that correspond to a unit number of MEPPs. *b*: Histogram showing the relation between the size of EPPs and MEPPs in an experiment by Boyd and Martin (1956) on the cat neuromuscular junction.

individual MEPPs, the peaks in the histogram of evoked responses clearly correspond to the amplitudes of integral numbers of MEPPs.

A simple calculation shows that a single MEPP arises from the simultaneous action of a large number of acetylcholine molecules in the synaptic cleft. We now know that the conductance of the channel that is opened by acetylcholine is about 40 pS, and that the current flowing through the open channel is in the range of 5 pA near the muscle resting potential. The resistance of the muscle membrane is such that the opening of a single acetylcholine-gated channel in the postsynaptic membrane can produce a depolarization of less than 1 μV. Because much of the acetylcholine released into the synaptic cleft is hydrolyzed before it is able to interact with the receptor, and two molecules of acetylcholine are required to open a single acetylcholine-gated channel (see Chapter 11), it can be estimated that approximately 5000 acetylcholine molecules are released synchronously into the synaptic cleft to generate a single MEPP.

We shall see later that many factors can alter the strength of synaptic transmission. The fact that transmitter release occurs in quanta can sometimes make it possible to determine whether an alteration in the strength of synaptic transmission results from a change in the amount of transmitter that is released (rather than from a change in the sensitivity of the postsynaptic cell to the transmitter). In particular, if a change in number of quanta released on stimulation is detected, this immediately implies that some alteration has occurred in the presynaptic terminal. This quantal analysis is therefore a highly useful technique for investigating the properties of those relatively few synapses where such measurements can be made.

In the 1950s, Katz and colleagues suggested that the packets or quanta of acetylcholine released from the presynaptic terminals at the neuromuscular junction correspond to the content of acetylcholine within single synaptic vesicles. They suggested further that, even at rest, vesicles occasionally fuse with the presynaptic membrane, resulting in exocytosis of the contents of a single vesicle and the generation of a MEPP in the postsynaptic muscle cell. Depolarization by a presynaptic action potential greatly increases the probability of exocytosis of vesicles, leading to the simultaneous release of the contents of many vesicles. A normal end-plate potential would result from the exocytosis of several hundred synaptic vesicles. As we saw in the previous chapter, direct visualization of exocytosis at the neuromuscular junction was achieved only 20 years later, by using electron microscopy combined with the freeze fracture technique (see Fig. 8–11). ❽

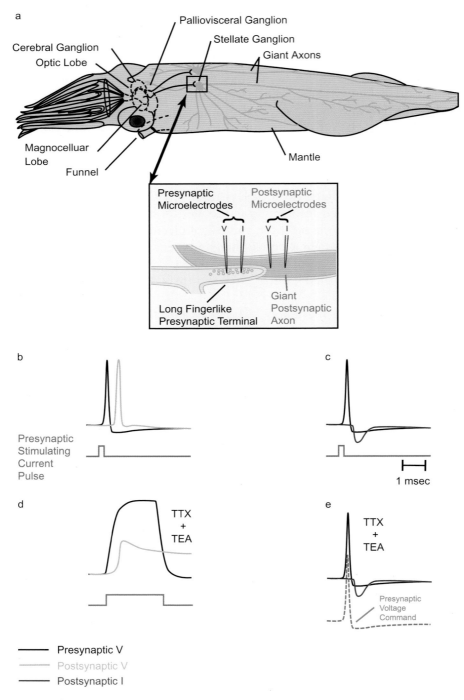

a

Cerebral Ganglion
Optic Lobe

Pallioviseral Ganglion

Stellate Ganglion

Giant Axons

Magnocelluar
Lobe

Funnel

Mantle

Presynaptic
Microelectrodes

Postsynaptic
Microelectrodes

V   I   V   I

Long Fingerlike
Presynaptic Terminal

Giant
Postsynaptic
Axon

b

Presynaptic
Stimulating
Current
Pulse

c

1 msec

d

TTX
+
TEA

e

TTX
+
TEA

Presynaptic
Voltage
Command

——— Presynaptic V
——— Postsynaptic V
——— Postsynaptic I

Figure 9–3.

Synaptic Transmitter Release Is Dependent on Calcium

It has been known since the work of Sidney Ringer in the nineteenth century that calcium ions in the extracellular medium are required for the normal function of the neuromuscular junction and other synapses. At most chemical synapses, eliminating calcium from the external medium prevents the release of transmitter evoked by nerve stimulation. In contrast, an increase in the external concentration of calcium frequently enhances the amount of neurotransmitter that is released.

The squid giant synapse. A major drawback to the investigation of neurotransmitter release is the small size of most vertebrate presynaptic terminals, including those of the neuromuscular junction. This precludes normal electrophysiological recordings from the terminals. A synapse at which this is not a problem is the giant synapse in the stellate ganglion of the squid (Fig. 9–3). This synapse is used in escape behavior; when stimulated, this synaptic pathway triggers the ejection of water through the mantle of the squid, propelling the animal rapidly away from a source of danger. Several cell bodies of the average molluscan neuron and many of the average mammalian neuron can be enclosed comfortably in the presynaptic terminal of the giant synapse (the terminal is about 50 µm in diameter and 700 µm in length). Accordingly, two or more independent microelectrodes can be placed in the presynaptic terminal as well as in the postsynaptic cell (Fig. 9–3a). As in the case of the neuromuscular junction, the amplitude of the postsynaptic voltage change is taken as a measure of the amount of transmitter released.

Work with the squid giant synapse confirmed and extended the findings of Katz on the neuromuscular junction and definitively established a role for calcium and for voltage-dependent calcium channels in the release of transmitter following a presynaptic action potential. For example, it is possible to evoke transmitter release at this synapse by direct injection of

---

Figure 9–3. The squid giant synapse. *a*: Position of the giant synapse within the stellate ganglion. *b–e*: Presynaptic and postsynaptic responses recorded in experiments by Bernard Katz and Ricardo Miledi, and by Rodolfo Llinas and coworkers. *b*: The normal response. *c*: The postsynaptic axon is voltage clamped and the postsynaptic current is recorded. *d, e*: Presynaptic sodium and potassium currents have been blocked by tetradotoxin (TTX) and tetraethylammonium (TEA). *d*: The prolonged postsynaptic potential is evoked by a long presynaptic depolarization. *e*: The presynaptic voltage is driven to follow that of a normal action potential, resulting in a normal postsynaptic current.

calcium ions into the presynaptic terminal. As in the somata of many neurons, voltage-dependent calcium currents can be recorded in the presynaptic terminal of this synapse (together with the much larger, voltage-dependent sodium and potassium currents that shape the action potential). Stimulation of an action potential in the terminal usually liberates sufficient transmitter to depolarize the postsynaptic axon to threshold and to trigger a postsynaptic action potential (Fig. 9–3b). If, however, the voltage of the postsynaptic cell is clamped near its resting potential, then the stimulation of the synapse generates an inward current in the postsynaptic cell (Fig. 9–3c). This inward current represents the current flowing through the transmitter-activated ion channels in the postsynaptic membrane.

It is possible to block the sodium and potassium channels in the presynaptic membrane, leaving calcium flux as the only voltage-dependent ion current. This is accomplished by applying the sodium channel blocker tetrodotoxin (TTX) to the external medium, and by injecting the potassium channel blocking agent tetraethylammonium (TEA) into the terminal (see Chapters 4 and 5). In this condition, postsynaptic responses can still be recorded when the presynaptic terminal is depolarized (Fig. 9–3d). Of course, in the presence of TTX and TEA, normal action potentials no longer occur in the presynaptic terminal. However, when the voltage in the terminal is driven by the voltage clamp to follow the normal shape of an action potential, the postsynaptic current exactly matches the normal postsynaptic response (Fig. 9–3e). This indicates that the normal presynaptic sodium and potassium fluxes are not required for release of neurotransmitter, but that calcium entry alone is sufficient.

Transmitter release occurs very rapidly following calcium entry. When the presynaptic terminal of the squid is depolarized to allow calcium channels to open, release generally increases after a delay of a few milliseconds. This, however, is not a good measure of the rate at which the release process itself is activated by calcium, because the presynaptic calcium current activates over a period of milliseconds. A more direct measurement of the response time of release can be obtained during measurements of calcium *tail currents*.

If in a voltage-clamp experiment the presynaptic terminal is stepped from its resting potential to a very positive potential close to the equilibrium potential for calcium ions, then calcium channels open in response to the depolarization. Nevertheless, at this potential calcium ions do not enter the terminal, because there is little or no driving force for calcium entry (Fig. 9–4a). Thus no transmitter release occurs with such large depolarizations. The voltage-clamp circuitry can then be used to step the membrane potential very rapidly back to the resting potential. Immediately after this step, the calcium channels remain open for a finite time at the resting potential, before they eventually close. Because the driving force for calcium entry is large at the resting potential, calcium ions enter through the still open calcium channels, resulting in a rapid but transient calcium tail current, measured after the end of the depolarizing pulse. This tail current is accompanied by transmitter release, measured as a postsynaptic depolarization. The delay between the onset of the sharp tail of calcium current in the presynaptic terminal and the onset of the

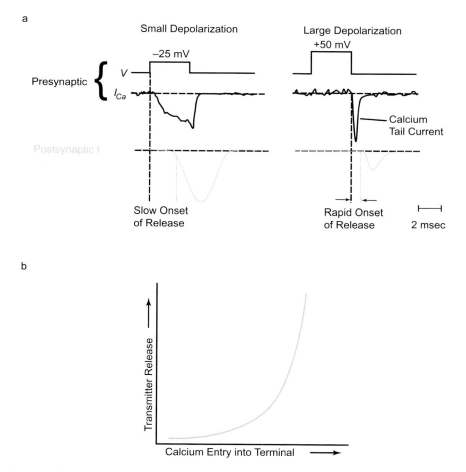

Figure 9–4. Calcium dependence of release. *a*: The postsynaptic response occurs very rapidly after a calcium tail current (after Augustine et al., 1985). I_{Ca}, calcium current; V, voltage. *b*: Steep dependence of release on the amount of calcium entry into the squid terminal.

postsynaptic response has been found to be as little as 200 μsec (Fig. 9–4a). This means that the release process at this synapse is so rapid that the actions of calcium are unlikely to involve complex, multistep biochemical reactions. As described in Chapter 8, the impressive speed at which the vesicles fuse is believed to result from rapid collapse of the SNARE complex from its *trans* state to the *cis* state following the binding of calcium-sensor to synaptotagmin (see Fig. 8–16).

Transmitter release requires binding of several calcium ions. As described earlier, transmitter release depends on the presence of extracellular calcium. At low concentrations of extracellular calcium, release of transmitter at the frog neuromuscular junction depends on the fourth power of the external calcium concentration. Experiments at the squid giant synapse, in which calcium currents and increases in calcium concentration can be measured directly, indicate that it is the release process itself that depends on a high power of intracellular calcium (Fig. 9–4b). Clear proof of this has come from using yet another specialized synapse, the *calyx of Held* in the brainstem of mammals.

The calyx of Held is found in the *medial nucleus of the trapezoid body* (MNTB); we shall learn its biological role in how the brain localizes sounds, in Chapter 18. What interests us here is that the calyces of Held are the largest presynaptic terminals in the central nervous system. A single synaptic ending (calyx) onto a postsynaptic MNTB neuron is shaped much like a baseball glove and completely envelops its cell body (Fig. 9–5a). As a result, it is relatively straightforward to patch clamp the presynaptic terminal—just the way that one records from the cell body of a neuron—and, perhaps more importantly, one can exchange the contents of the patch pipette with those of the cytoplasm of the presynaptic terminal (Fig. 9–5b).

By loading the cytoplasm of the terminal with different solutions, as well as by using techniques that allow calcium concentration to jump rapidly to fixed levels, it has been possible to determine how the rate of vesicle release depends on internal calcium. Figure 9-5c shows that release is sensitive to calcium over a wide range of concentrations. These can be divided into three general domains that correspond to *spontaneous release, asynchronous release*, and *synchronous release*.

Calcium levels that correspond to spontaneous release (<~200 nM) are associated with the random release of single vesicles such as those that give rise to MEPPs at the neuromuscular junction. When these are recorded in a neuron, rather than a muscle, and the presynaptic neuron is an excitatory one, they are termed mEPSPs, for miniature Excitatory PostSynaptic Potentials. At internal calcium concentrations around 1 μM,

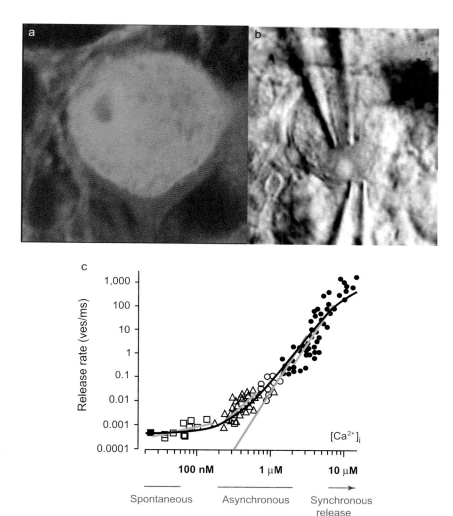

Figure 9–5. The calyx of Held. *a*: Image of the presynaptic terminal (red) and postsynaptic MNTB neuron (green). *b*: Image of patch pipettes placed simultaneously on the presynaptic terminal and the postsynaptic cell body (courtesy of Felix Felmy and Benedikt Grothe). *c*: plot of the rate of release of vesicles from the calyx as a function of the amount of calcium in the presynaptic terminal (Kochubey and Schneggenburger, 2011).

there is an enhancement of release that occurs randomly in time, hence the term *asynchronous release*. This corresponds to an increase in rate of spontaneous vesicle release that is observed following one or more action potentials, but that is not locked precisely in time to the action potentials.

High calcium levels around 10 µM or greater correspond to rapid synchronous release of many vesicles, such as occurs during an action potential. In this range, calcium dependence is very steep, depending of the fourth power of calcium concentration. The simplest interpretation of such findings is that some reaction in the terminal, perhaps the interaction of synaptotagmin with the SNARE complex, requires the binding of several calcium ions before release of transmitter occurs.

In the next section, we shall examine what physical processes make vesicle release accurately timed to the action potential during synchronous release. ❷

Nanodomains of calcium entry at vesicle release sites. To understand how neurotransmitter release can occur so rapidly in response to a presynaptic action potential, we must first consider the spatial pattern of changes in calcium within the terminal. Because the calcium channels that allow calcium to enter the terminal are discrete membrane proteins, one would not expect calcium concentrations to rise uniformly throughout the terminal following a depolarization. Instead, calcium levels would be expected to be initially much higher directly under the membrane where the calcium channels are located, resembling a "volcano" of calcium centered on the mouth of the channel (Fig. 9–6). Such a local area of high calcium is defined as a *nanodomain* if its diameter is about 100 nm or less. Immediately after the opening of a single calcium channel, calcium concentrations could reach millimolar levels at the inner mouth of the channel itself. As calcium diffuses from the channel, its concentration would be expected to drop. ❷

Experiments to test these expectations and to provide direct measurement of calcium levels in the terminal have required the use of substances

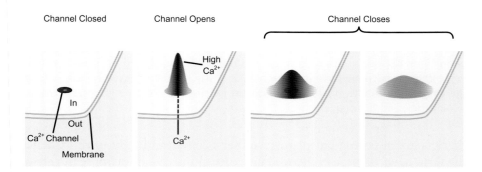

Figure 9–6. High levels of calcium occur at the intracellular mouth of calcium channels. "Volcanos" of calcium form inside a cell near the mouth of an open channel and disperse over time. See Animation 9–1. ◉

that could be introduced into the terminal to measure changes in calcium concentration. The first of these to be used was a protein known as *aequorin*, which emits light when it binds calcium. This protein was used to provide direct evidence that the calcium concentration in the terminal is raised during transmission. Subsequently, other calcium-indicator dyes have been used for this purpose, including the dye *Arsenazo III* and the now widely used calcium indicator *fura-2*. In its structure, fura-2 resembles the common chelator of calcium ions, EGTA (Fig. 9–7a). It is, however, a fluorescent compound, whose excitation spectrum changes when it binds calcium (Fig. 9–7b).

Using a fluorescence microscope coupled to a computer, digital images can be made of changes in calcium concentration occurring during transmitter release. This technique was first applied to the squid giant synapse; investigators demonstrated that trains of action potentials produce steep gradients of calcium across the terminal and that the gradient dissipates after the end of stimulation (Fig. 9–7c). More recently, using another calcium indicator called *Calcium Green-1*, it has been possible to observe the individual "volcanoes" of calcium directly, in response to stimulation of the presynaptic nerve at the frog neuromuscular junction (Fig. 9–7d). The fact that the positions of these sites of calcium influx vary from stimulus to stimulus indicates that they can result from the opening of a single calcium channel or, at most, a cluster of a very small number of channels (Fig. 9–7e).

Calcium channels are clustered near release sites. The finding that elevations of calcium are so spatially inhomogeneous during action potentials suggests that proteins located close to the mouth of the channel will be exposed to higher concentrations of calcium ions than those seen further from the channel. In the last chapter we saw that the SNARE complex is able to bind calcium channels (see Fig. 8–16). As a result, synaptotagmin, the calcium sensor for synchronous transmitter release, as well as other vesicular proteins, would be expected to encounter the high calcium levels within the calcium nanodomain immediately following the opening of a calcium channel.

The first direct evidence that the opening of a single calcium channel can trigger the exocytosis of a synaptic vesicle came from experiments by Elise Stanley. She used a cell-attached patch pipette to record the opening of single calcium channels on the transmitter-releasing face of a calyx-type presynaptic terminal that had been pulled away from its postsynaptic partner, while simultaneously monitoring release of the neurotransmitter using an optical technique. A variety of other types of experiments have now confirmed that, for synapses at which synchronous transmitter release is tightly linked to presynaptic action potentials, the calcium channels and vesicles are colocalized in a nanodomain, allowing transmitter

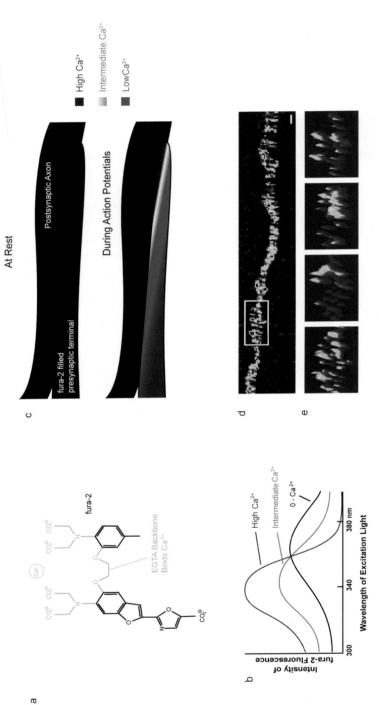

Figure 9–7. Measurement of intracellular calcium. **a**: Structure of the calcium indicator dye fura-2, which was synthesized by Roger Tsien. **b**: The intensity of fura-2 fluorescence varies with calcium concentration (Grynkiewicz et al., 1985). **c**: Diagram of how calcium levels in the squid presynaptic terminal, measured as ratios of fura-2 fluorescence, change during stimulation (Smith et al., 1993). **d**: An image of calcium elevations in presynaptic terminals of the frog neuromuscular junction following a presynaptic action potential (Wachman et al., 2004). **e**: Enlargement of the area in the white box in **d** for four different stimulus pulses. This shows that the activation of "volcanoes" of calcium entry, where calcium channels are located, occurs at a different set of sites with each stimulus.

release to occur with minimal elevation of the bulk cytoplasmic calcium concentration.

It should be pointed out that, as in all things biological, there many different types of synapses in the brain, and not all of these require nano-domain coupling between channels and the vesicles. For example, at the calyx of Held in the auditory system, such tight coupling only occurs after the onset of hearing. At earlier times, the synapses are fully functional, but release requires the simultaneous opening of many tens of channels to elevate bulk calcium sufficiently to trigger exocytosis.

Homosynaptic Plasticity: Facilitation, Potentiation, and Depression of Transmitter Release

As a series of action potentials invades a nerve terminal, the amount of neurotransmitter released with each action potential does not always remain constant. Depending on the synapse that is studied and the frequency at which it is stimulated, a train of action potentials in the presynaptic terminal may produce either a progressive increase or a progressive decrease in the amount of release. This property, which allows the amount of transmitter release to change as a result of previous activity in the terminal, has been termed *homosynaptic plasticity*. Three major forms of homosynaptic plasticity are *depression, facilitation*, and *potentiation*. These may be contrasted with a change in release induced by the action of other cells. The latter phenomenon has been termed *heterosynaptic plasticity*, and examples will be provided in Chapter 19. We will now describe in turn each of the forms of homosynaptic plasticity.

Synaptic depression. This describes a progressive decrease in the amount of transmitter released during a train of action potentials (Fig. 9–8a). This phenomenon is often encountered when a synapse is stimulated at a high rate and, in many cases, results from the depletion of the readily releasable pool of neurotransmitter at active zones. Although most synapses have an abundance of synaptic vesicles near the active zone, only a small proportion of these may actually be docked and primed, and comprise the readily releasable pool. Thus there are fewer and fewer available vesicles with each incoming action potential. The proportion of synaptic vesicles released by a single action potential (also termed the *release probability*) differs widely for synapses in different parts of the nervous system. In general, the amount of depression is greatest when this initial release probability is high.

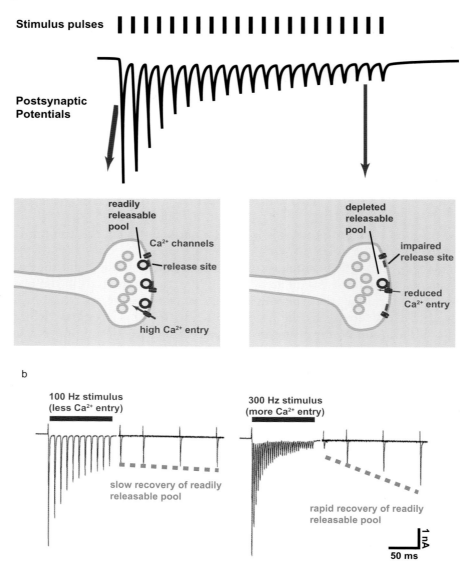

a

Stimulus pulses

Postsynaptic Potentials

readily releasable pool

Ca²⁺ channels

release site

high Ca²⁺ entry

depleted releasable pool

impaired release site

reduced Ca²⁺ entry

b

100 Hz stimulus (less Ca²⁺ entry)

300 Hz stimulus (more Ca²⁺ entry)

slow recovery of readily releasable pool

rapid recovery of readily releasable pool

1 nA

50 ms

Figure 9–8. Synaptic depression. *a:* A progressive decrease in amount of neurotransmitter released during repetitive stimulation may result from a decrease in the size of the readily releasable pool of neurotransmitter, decreases in calcium channel activity, or a change in the ability of release sites to sustain exocytosis. *b:* Recovery of the readily releasable pool of neurotransmitter at the calyx of Held occurs more rapidly following a high-frequency train (300 Hz) than a lower-frequency train (Wang and Kaczmarek, 1998).

Depletion of releasable vesicles is not the only way that synaptic depression can occur. At some synapses, the calcium channels undergo inactivation with repeated stimulation, reducing the amount of calcium that enters with each action potential. At others, it appears that the release sites at which vesicles dock become less able to support repeated release (Fig. 9–8a).

Recovery of the readily releasable pool during and after a train of stimuli is a tightly regulated process and can be greatly enhanced by biochemical events set in motion by calcium entering during stimulation. Thus, at the calyx of Held, high rates of stimulation (>200 Hz), which produce a greater overall influx of calcium into the terminal than low rates (<100 Hz), can accelerate the rate of recovery of the readily releasable pool (Fig. 9–8b).

Facilitation. This describes a progressive increase in the amount of transmitter released by successive action potentials during a brief stimulus train lasting up to a few seconds (Fig. 9–9). A leading hypothesis for the

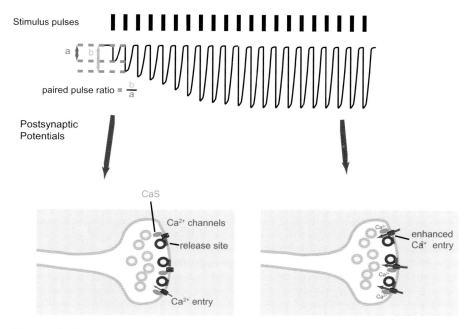

Figure 9–9. Synaptic facilitation. A progressive increase in transmitter release during repetitive stimulation may be caused by an enhancement of calcium channel activity and can be measured as the paired pulse ratio for two successive stimuli.

cause of facilitation is that the calcium concentration at the release sites does not have time to return to basal levels between the action potentials. Thus, at the occurrence of each action potential after the first, a small amount of residual calcium remains near the release site. The level of residual calcium is, by itself, insufficient to cause a significant amount of transmitter release during the interval between action potentials. At least one of the targets of this residual calcium can be the calcium channels themselves. $Ca_v2.1$ calcium channels bind a *neuronal calcium sensor protein* termed *CaS*. During repetitive stimulation, residual calcium acting on CaS can progressively enhance the activity of the channel, leading to enhancement of calcium entry with each action potential (Fig. 9–9). Recovery from facilitation is rapid, occurring within a few hundred milliseconds after the end of stimulation.

A common way to measure facilitation is the *paired pulse ratio*. When two stimuli are given closely spaced in time, the paired pulse ratio is simply the ratio of the size of the second postsynaptic current to that of the first. Because facilitation is primarily a presynaptic process, any change in the paired pulse ratio, for example, produced by application of a drug, is frequently taken as evidence that the drug acts on the presynaptic terminal rather than on the postsynaptic cell.

Potentiation. This describes an increase in transmitter release by an action potential *following* repetitive stimulation of a synapse. Unlike facilitation, potentiation is long-lasting and is slow in onset, usually requiring seconds to develop. For example, following a rapid train of action potentials, termed a *tetanus*, the amount of release in response to a single action potential may be enhanced over that prior to the tetanus for up to several minutes. This phenomenon can be observed at a wide variety of synapses, including the neuromuscular junction, and has been termed *post-tetanic potentiation* (PTP) (Fig. 9–10a). At least at some synapses potentiation, like facilitation, is due to an elevation of resting calcium levels over those in the terminal prior to stimulation. The long-lasting nature of this residual calcium results from the activity of mitochondria. In most synapses there is a high density of these organelles behind the active zones (see Fig. 8–10). In addition to providing energy for transmission in the form of ATP, mitochondria are able to take up large amounts of calcium. During the tetanus, calcium that enters through calcium channels spreads throughout the terminal and is rapidly accumulated into the matrix of the mitochondria (Fig. 9–10b). After the end of the tetanus, this accumulated calcium is slowly released back into the synaptic cytoplasm, providing residual calcium that can potentiate transmitter release by action potentials for several minutes. At many central synapses, potentiation may

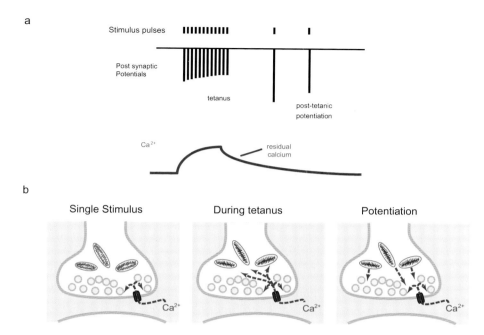

Figure 9–10. Post-tetanic potentiation. *a*: Enhancement of transmitter release for several minutes following a rapid train of presynaptic stimuli (tetanus). *b*: Calcium accumulation by synaptic mitochondria during a tetanus. Initially, a single action potential elevates calcium locally near the mouth of the channel (*left*). During the tetanus, calcium rises in the cytoplasm and is taken up by the mitochondria (*center*). After the tetanus, calcium released from the mitochondria adds to that entering through calcium channels (*right*) (Tang and Zucker, 1997).

endure for periods ranging from tens of minutes to hours or days, depending on the duration and intensity of the stimulus train. This has been termed *long-term potentiation* (LTP). The mechanisms of LTP appear to be distinct from those of PTP, and differ from synapse to synapse. The phenomenon of LTP has been studied as a model for the establishment of memories by the nervous system; we will discuss LTP further in this context in Chapter 19.

Facilitation, potentiation, and depression may all occur at a single type of synapse, resulting in a complex time course of changes in neurotransmitter release following the onset of a train of action potentials. ❽ In general, synapses with a high probability of release tend to suffer depression with repetitive stimulation, whereas low-release probability synapses enjoy facilitation and post-tetanic potentiation.

Testing the Roles of Synaptic Proteins in Neurotransmission

We have now described some of the dynamic features of transmitter release at synaptic terminals. In the previous chapter, we covered the biochemistry of some of the proteins of synaptic vesicles and their partners on the plasma membrane. An ongoing task for neurobiologists is to discover how each of these proteins contributes to neurotransmission and, perhaps even more importantly, to determine how the modulation of these proteins may alter characteristics of release such as potentiation and depression.

One approach being widely adopted to test the importance of a particular protein in the process of synaptic release is the genetic manipulation of animals such as mice or fruit flies to produce a *knockout* of a selected protein. Using a technique known as *homologous recombination*, it is possible to replace the normal gene for a selected mouse protein with a nonfunctional mutant gene. Such mutant animals can then be bred to produce strains of mice that lack the functional protein. As might be surmised, if an important protein is selected, the mutant animals may fail to develop normally, and experimental tricks often have to be employed to keep embryos alive long enough to isolate cells to study.

Nevertheless, results have been generated from this approach, and some of them have been surprising. For example, the elimination of synaptotagmin I, one of the two major forms of this presumed calcium sensor, does not eliminate synaptically evoked transmitter release but instead significantly alters the timing of the release following invasion of an action potential. Specifically, it eliminates the very rapid synchronous release that is evoked on calcium entry during a presynaptic action potential and converts it into asynchronous release (Fig. 9–11a). Interestingly, the total amount of release evoked by a stimulus is unchanged by the loss of synaptotagmin I (Fig. 9–11b). This result suggests that synaptotagmin I is the calcium sensor for the rapid synchronous release whose mechanism we discussed earlier in the chapter, but that other calcium-sensitive proteins may be able to sustain asynchronous release. This is consistent with the finding that asynchronous release can be triggered with lower levels of calcium (see Fig. 9–5c). The identity of this other calcium sensor is, however, not yet known. ❽

Findings from such knockout experiments will eventually have to be incorporated into our understanding of how processes such as docking, priming, fusion, and reuptake contribute to the rate and amount of transmitter release from a synapse during physiological patterns of activity, and how biochemical modulation of these processes adapts a neuron to

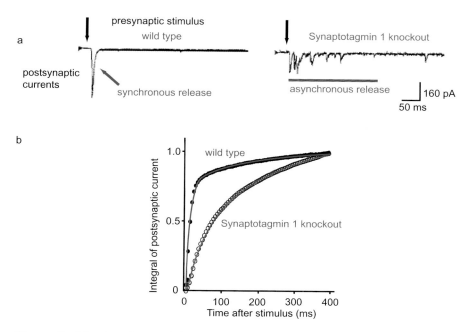

Figure 9–11. Knockout of the synaptotagmin I gene greatly slows the release of neurotransmitter. *a*: Postsynaptic recordings by Tom Sudhof and his colleagues using hippocampal neurons in cell culture showed that loss of synaptotagmin I eliminates synchronous release and that asynchronous release is enhanced (Geppert, Goda, et al., 1994). *b*: Recordings by Tei-ichi Nishiki and George Augustine showed that the integral of the postsynaptic current, which reflects the overall amount of transmitter release, is not altered by loss of synaptotagmin I, but its time course is delayed (Nishiki and Augustine, 2004).

increase or decrease its neurotransmitter output in response to changes in the environment. We shall now provide an example of the combined use of knockout and injection experiments in testing the role of another synaptic vesicle protein, *synapsin I*. ❽

Synapsins, Calcium/Calmodulin-Dependent Protein Kinases, and Neurotransmitter Release

There are three known synapsin proteins: synapsins I, II, and III. Synapsin III is present in cells before synapses are formed, and its function is not known. Synapsins I and II are proteins that associate with the cytoplasmic surface of synaptic vesicles. In electron microscopy experiments they have been found to be associated primarily with the clusters of vesicles located away from active zones, those that are sometimes referred to as the *reserve pool* (see Fig. 8–13). In addition to binding to the surface of synaptic vesicles, synapsins are able to bind the cytoskeletal protein actin. The properties of synapsins can be substantially modified by the action of several enzymes termed *protein kinases*. Before we discuss the effect of these modifications, we must cover some basic biochemistry of protein kinases.

Protein kinases. Members of this large family of enzymes transfer the terminal phosphate group from ATP to a variety of proteins. The addition of phosphate to a protein alters the electrical charge on the protein (phosphate groups are negatively charged) and may also change the three-dimensional shape, or *conformation*, of the protein. This in turn may induce a change in the biological activity of the phosphorylated protein (for a more detailed discussion, see Chapter 12). The activity of protein kinases can be detected in a homogenate of the nervous system by incubating the homogenate with ATP that has been radiolabeled in the terminal (γ) phosphate. The transfer of the radioactive phosphate group to specific proteins can then be measured readily. If calcium ions are also added to the homogenate, the incorporation of phosphate into certain proteins is much enhanced. Further enhancement is also usually observed with the addition of *calmodulin*, a small (16.7 kDa) calcium-binding protein that is found in all eukaryotic cells. The calcium-dependent phosphorylation of proteins is carried out by several different protein kinases, one of which is termed the *calcium/calmodulin-dependent protein kinase type II* (abbreviated Ca^{2+}/Cam kinase II). This enzyme is one of the myriad of cellular activities regulated by calcium and calmodulin.

Ca^{2+}/Cam kinase II phosphorylates synapsin I. Ca^{2+}/Cam kinase II is found in most cells. It is particularly abundant in the nervous system, however, where it accounts for 0.5%–1.0% of total protein, an extraordinarily high concentration for an enzyme. It is a large, multisubunit complex consisting of two different subunits, α and β, with molecular weights of 50,000 and 60,000, respectively. The active enzyme appears to contain 12 subunits (Fig. 9–12a), and the relative ratio of α to β subunits in the active enzyme varies in different types of cells. For example, in some cells, the enzyme is composed entirely of α subunits, whereas in other cells the larger β subunit predominates.

    Following an elevation of the calcium concentration inside a cell, the calcium ions bind to calmodulin in the cytoplasm. The calcium/calmodulin complex, in turn, binds to the individual subunits of the Ca^{2+}/Cam kinase II.

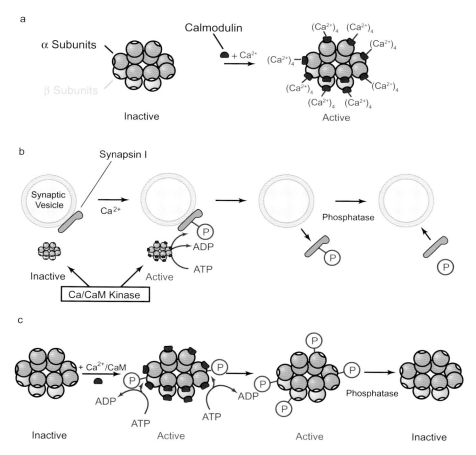

Figure 9–12. Calcium/calmodulin-dependent protein phosphorylation. *a*: Activation of Ca²⁺/Cam kinase II. *b*: Phosphorylation of synapsin I. *c*: Auto-phosphorylation of Ca²⁺/Cam kinase II.

When the enzyme is bound in this way, it becomes active and is able to transfer phosphate groups from ATP to a wide variety of proteins. One interesting feature of Ca²⁺/Cam kinase II is that it is also able to phospho-rylate itself. This is termed *autophosphorylation*. When this happens, the enzyme remains active but becomes independent of calcium and calmod-ulin until the phosphate groups on the enzyme are removed by other enzymes, termed *phosphoprotein phosphatases*. The autophosphorylation may thus allow Ca²⁺/ Cam kinase II to remain active for a considerable period of time, even after the calcium concentration in the cell has returned to its basal level (Fig. 9–12c).

Because Ca²⁺/Cam kinase II is so abundant, it is likely to regulate many different functions of cells. It is, for example, the major protein found in the region of a neuron immediately under the neurotransmitter receptors at the postsynaptic junction (see Chapter 19). Nevertheless, a major part of the Ca²⁺/Cam kinase II within the nerve terminal is found on the surface of synaptic vesicles, where it is known to bind synapsin I. Indeed, synapsin I is one of the best substrate proteins for this enzyme. As mentioned earlier, synapsin normally binds the surface of synaptic vesicles as well as to actin. When it undergoes phosphorylation by Ca²⁺/Cam kinase II, however, this association is greatly weakened, and the phosphorylated synapsin I may be released from the surface of the vesicles (Fig. 9–12b).

Role of synapsin I and Ca2⁺/Cam kinase II in transmitter release. Evidence that Ca²⁺/Cam kinase II plays a role in regulating neurotransmitter release has been provided by injecting this enzyme into the giant presynaptic terminal of the squid (Fig. 9–13). This causes an increase in neurotransmitter

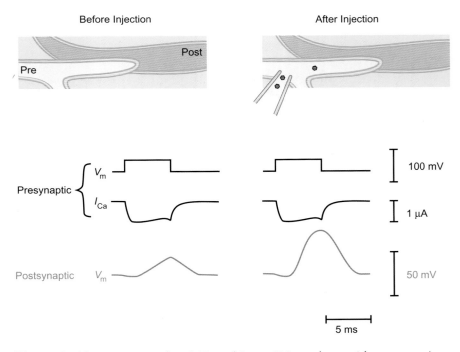

Figure 9–13. Injection of Ca²⁺/Cam kinase II into the squid presynaptic terminal. In a collaborative experiment between the laboratories of Paul Greengard and Rodolfo Llinas, injection of this enzyme caused increased transmitter release, measured as an increase in the postsynaptic depolarization (Llinas et al., 1985). I_{Ca}, calcium current; V_m, membrane voltage.

release. However, there is no change in the amplitude of the presynaptic calcium current, which suggests that the injection of the enzyme increases the efficiency of transmitter release. We have already seen that the rapid release of neurotransmitter at synaptic junctions occurs too rapidly to be mediated by a complex enzyme reaction. Ca^{2+}/Cam kinase II is therefore unlikely to be involved in triggering release directly, but instead may determine the amount of transmitter available for release by an action potential.

It is possible that this action of Ca^{2+}/Cam kinase II at the squid synapse occurs through the phosphorylation of synapsin I or a related molecule. A variety of experimental approaches, including genetic knockouts in mice, protein injections into squid and lamprey synapses, and electron microscopy, suggest that unphosphorylated synapsin I acts as a glue that holds together the reserve pool of synaptic vesicles behind the active zone. Phosphorylation of synapsin I may free some of these vesicles, allowing them to approach the plasma membrane and dock at release sites. For example, injection of unphosphorylated synapsin I into the presynaptic terminal decreases release, whereas injection of synapsin I that has previously been phosphorylated by Ca^{2+}/Cam kinase II has no apparent effect on release. This would be expected if excess unphosphorylated protein kept the reserve pool glued together and unavailable for release. Genetic knockout of the synapsin gene in mice or injection of anti-synapsin antibodies into synaptic endings disrupts the tight clusters of reserve vesicles (Fig. 9–14). These manipulations do not, however, prevent normal

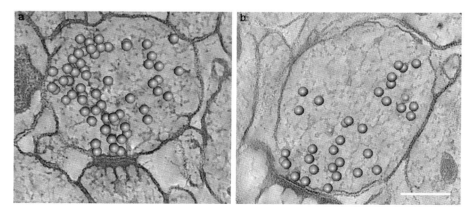

Figure 9–14. Genetic elimination of synapsins. *a*: Synaptic terminals on neurons in the hippocampus of normal mice are densely packed with small, clear synaptic vesicles, whose positions have been indicated by colored spheres. Vesicles docked at the active zone are colored blue. *b*: Electron microscopy shows that elimination of synapsins I, II, and III causes a dramatic loss of vesicles in synaptic terminals (Siksou et al., 2007).

synaptic transmission, but rather decrease release at high rates of synaptic stimulation, when the reserve pool is most likely to be needed.

Studies such as these suggest that phosphorylation of synapsin I by Ca^{2+}/Cam kinase II plays a role in ensuring continued neurotransmitter release during high rates of synaptic activity. There are still many questions remaining about the role of phosphorylation in the control of synapsins as well as other synaptic proteins. For example, the synapsin II protein resembles synapsin I in its structure but is not phosphorylated by Ca^{2+}/Cam kinase II. However, both of these proteins are phosphorylated by other protein kinases, many of which we will encounter in subsequent chapters. Each of the different kinases may alter the characteristics of neurotransmission in specific ways.

Summary

Several specialized synapses, including those in the squid stellate ganglion, the frog neuromuscular junction, and the calyx of Held, have been instrumental in advancing our understanding of the release of neurotransmitters from presynaptic terminals. Studies of rapid synaptic transmission have shown that neurotransmitters are released in packets, or quanta, which may correspond to the exocytosis of individual synaptic vesicles. Following an action potential, release is very closely linked in space and time to the entry of calcium though voltage-dependent channels. The amount of transmitter released by a single action potential is not fixed but can undergo depression, facilitation, or potentiation during and after repetitive stimulation of a synapse. We are still far from having a complete understanding of the mechanisms that induce neurotransmitter release and of the factors, such as synapsins, that modulate release. Biochemical and genetic experiments that characterize, modify, or eliminate components of synaptic vesicles and release sites, coupled with physiological experiments at specific synapses, are providing further insights into the release process.

Neurotransmitters and Neurohormones

We have seen in the previous two chapters that a neuron is in large part an elaborate and intricately regulated machine for the secretion of a variety of chemicals. Why do cells go to all this trouble? Although direct electrical connections between nerve cells also play an essential role (see Chapter 8), it is chemical signaling that mediates much of the intercellular communication among nerve cells within the central nervous system. In addition, the transfer of information into the nervous system from sensory organs and the output from the nervous system in the form of muscular contraction are mediated by extracellular chemical messengers. In this chapter we will discuss in a systematic way the different classes of neurotransmitters and neurohormones, some of which we have already met in other contexts. Subsequent chapters will deal with receptors on the target cell that recognize and bind these substances and with the transduction mechanisms involved in converting the extracellular chemical signal into an appropriate response (usually electrical) in the target cell.

What Is a Neurotransmitter and What Is a Neurohormone?

Which neuroactive substances are transmitters and which are hormones? Classically, synaptic *neurotransmitters* have been thought of as substances that are released locally into an anatomically well-defined synaptic cleft, and influence the activity of only one or a few adjacent cells (Fig. 10–1). The prototype neurotransmitter is *acetylcholine*, which was first demonstrated to be the chemical mediating nerve-to-muscle synaptic transmission in cardiac and skeletal muscle (see Chapter 1) and was subsequently found to be an important neuron-to-neuron transmitter as well. In contrast, *hormones* have been defined as substances that are released from the tissue in which they are synthesized, and travel via the blood to other (often remote) organs whose activities they influence (Fig. 10–1). Another criterion that has often been used to distinguish between transmitters and hormones is a temporal one: Transmitters have been thought to produce rapid-onset and rapidly reversible responses in the target cell, whereas the actions of hormones can be slower and much longer lasting.

Because hormone synthesis and secretion had been studied most thoroughly in organs such as the adrenal gland and the gonads, it came as something of a surprise when it was demonstrated many years ago

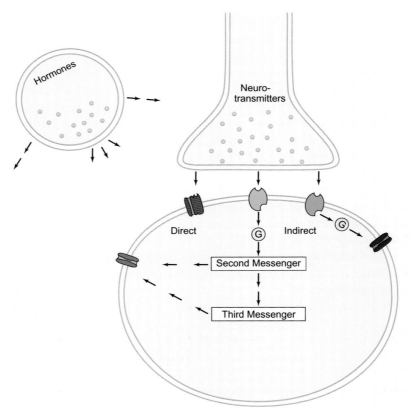

Figure 10–1. Intercellular communication. This chapter will emphasize the neurotransmitters and hormones that mediate much cell-to-cell communication in the nervous system. G, guanyl nucleotide binding protein.

that the nervous system also synthesizes and releases hormones—
neurohormones—that act on remote organs. The classical neurohormones are the closely related nine–amino acid peptides *oxytocin* and
vasopressin, which are (1) synthesized in the so-called *magnocellular
neurons* in the hypothalamus, (2) transported down the axons of these
neurons to the posterior pituitary where they are released, and (3) dis-
persed via the bloodstream to regulate smooth muscle contraction and
water balance, respectively.

More recently it has become evident that oxytocin, vasopressin, and
other neuroactive peptides can have local transmitter-like actions as well.
In addition to being secreted into the blood, they may be released at synapses instead of—or even together with—more classical transmitter substances. Furthermore, we know now that the actions of some classical
neurotransmitters may be slow in onset and long in duration, and indeed
they may be released into the bloodstream to act as true hormones,
whereas many peptide "hormones" may produce rapid transmitter-like
effects on target cells. In other words, the classical distinctions between
neurotransmitters and neurohormones are becoming obsolete. One person's transmitter is another person's hormone, and it may be more useful
to classify neuroactive substances on the basis of other criteria—for example, the nature and mechanism of the response they evoke in the target cell
(see Chapters 11 and 12).

Acetylcholine

The first chemical to be implicated as a neurotransmitter was acetylcholine, the structure of which is shown in Figure 10–2. It had been known
since the early part of the twentieth century that acetylcholine could influence the physiological properties of nerve and muscle cells. In the 1920s
it was demonstrated that acetylcholine is the transmitter at *neuromuscular synapses*, which are synapses between neurons and cardiac, smooth,
and skeletal muscle, as well as at a variety of neuron–neuron synapses in
the central and peripheral nervous systems. It is instructive to recall at this
time the first experiment to demonstrate unequivocally the chemical mediation of synaptic transmission by acetylcholine, the classic double heart
experiment published by Otto Loewi in 1921 (see Fig. 1–7), because it is
so beautiful an example of a *bioassay*, the use of a physiological response
to assay for a biologically active compound. The bioassay remains an
essential tool by which neuropharmacologists identify and quantitate neuroactive substances for which no sufficiently sensitive or specific chemical
assay is available.

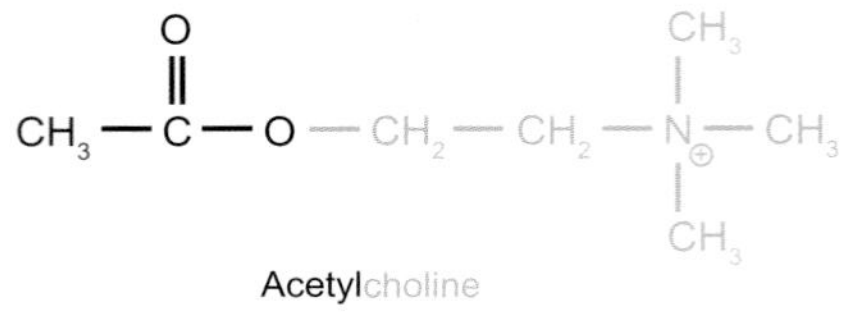

$$CH_3 - \overset{\overset{\textstyle O}{\|}}{C} - O - CH_2 - CH_2 - \overset{\overset{\textstyle CH_3}{|}}{\underset{\underset{\textstyle CH_3}{|}}{N_\oplus}} - CH_3$$

Acetylcholine

Figure 10–2. Chemical structure of acetylcholine.

Synthesis and release. Unlike the other neurotransmitters and neurohormones that we shall discuss later in this chapter, acetylcholine is not simply one member of a class of closely related compounds. Rather, it has some unique properties that place it in a class by itself. It is synthesized in nerve terminals from acetyl-CoA and choline, in a reaction catalyzed by the enzyme *choline acetyltransferase* (Fig. 10–3). Although acetyl-CoA and choline are common metabolites present in all cells, choline acetyltransferase (and hence acetylcholine) is not. In fact, the presence of choline acetyltransferase in a neuron is sufficient to define it as a cholinergic neuron. The acetylcholine is packaged in synaptic vesicles (see Chapter 8) and is released following the arrival of an action potential at the nerve terminal via vesicle exocytosis. However, not all the acetylcholine in nerve terminals is in vesicles. Some is in the cytoplasm, and evidence exists for the direct release of acetylcholine from these cytoplasmic stores at some synapses.

Degradation and resynthesis. Following its release into the synaptic cleft, acetylcholine can bind to at least two distinct classes of receptor molecule and produce different responses in different target cells (see Chapter 11). However, the neurotransmitter does not remain at a high concentration in the synaptic cleft for very long. The fate of the released acetylcholine is to be destroyed by a powerful hydrolytic enzyme, *acetylcholinesterase,* which produces acetate and choline (Fig. 10–3). This enzyme is clustered at high concentrations in the synaptic cleft. In addition, its catalytic rate (of the order of 10^4 to 10^5 substrate molecules hydrolyzed per second) ranks it among the most rapid enzymes known, ensuring that the concentration of acetylcholine in the cleft drops very quickly following its release. Much of the choline is then taken up again into the nerve terminal and utilized for the replenishment of the terminal's acetylcholine. Cholinergic nerve terminals contain a high-affinity sodium-dependent choline uptake system (Fig. 10–3) that provides a large proportion of the choline in the terminal and whose activity is probably rate limiting for acetylcholine synthesis. As we shall see later in the chapter, sodium-dependent uptake systems (often called *transporters*) specific for

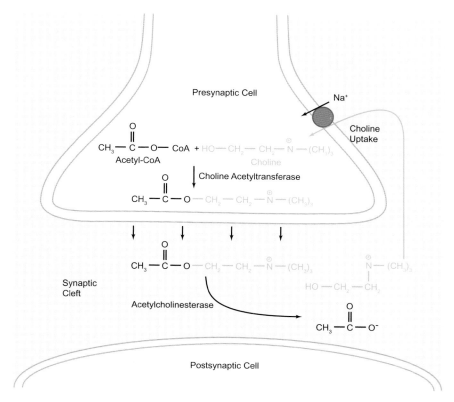

Figure 10–3. Synthesis, release, and degradation of acetylcholine. The transmitter is synthesized from choline and acetyl-CoA by the enzyme choline acetyltransferase. Following its release, it is broken down rapidly by the enzyme acetylcholinesterase to choline and acetate. The choline can be taken back up into the terminal via a sodium-coupled choline transport system.

particular neurotransmitters or for their precursors or metabolites are ubiquitous in nerve terminals and play an essential role in regulating the amount of neurotransmitter available for release.

The Amine Neurotransmitters

Epinephrine and norepinephrine. During the period when Loewi and others established that acetylcholine is the chemical transmitter in peripheral *parasympathetic* nerves such as the vagus nerve, other neuropharmacologists were investigating the nature of synaptic transmission in the *sympathetic* nervous system, another component of the peripheral *autonomic* nervous system. There is a classic set of criteria that must be fulfilled to conclude that a particular compound is indeed a physiological transmitter substance. It was demonstrated early that the compound *norepinephrine* (Fig. 10–4) fulfills all the criteria for the sympathetic neurotransmitter, in that

1. it is synthesized and stored in high concentration in sympathetic nerve terminals;
2. it is released on stimulation of sympathetic nerves; and
3. it mimics the action of the endogenous transmitter when it is applied exogenously to target organs.

In addition, pharmacological agents that block the response to sympathetic nerve stimulation also antagonize the actions of exogenously applied norepinephrine. Norepinephrine can bind to several different classes of *adrenergic receptor* on the postsynaptic cell; these receptors can be distinguished on the basis of their pharmacological properties and molecular structures, and the mechanisms they use to transduce the signal from the neurotransmitter into a response in the target cell (see Chapter 12).

For technical reasons, it was much more difficult to establish a neurotransmitter role for norepinephrine in the central nervous system, but now this role is also widely accepted. Most of the brain synapses that use norepinephrine arise from neurons whose cell bodies lie in the *locus ceruleus*, a cluster of only several thousand neurons located in the midbrain.

Adrenergic synapses use either norepinephrine (noradrenaline) or its N-methylated derivative *epinephrine* (adrenaline) as their neurotransmitter (Fig. 10–4). The fuzziness in the distinction between neurotransmitters and hormones arises even in this most classic case of adrenergic transmission. Although most adrenergic synapses use norepinephrine as their neurotransmitter, epinephrine is synthesized and stored at high concentrations in the *chromaffin cells* of the adrenal medulla and is released into the bloodstream to act as a hormone on many of the same target organs that receive sympathetic innervation. During embryonic development, the adrenal medulla arises from the same precursor cells as the sympathetic neurons. Adrenal chromaffin cells share many properties with sympathetic neurons and thus are widely used as a readily accessible model system for investigating adrenergic transmission (see Chapter 15).

Dopamine. Both norepinephrine and epinephrine belong to the general family of compounds known as *catecholamines*, organic molecules that contain an amine group as well as a catechol nucleus (a benzene ring with

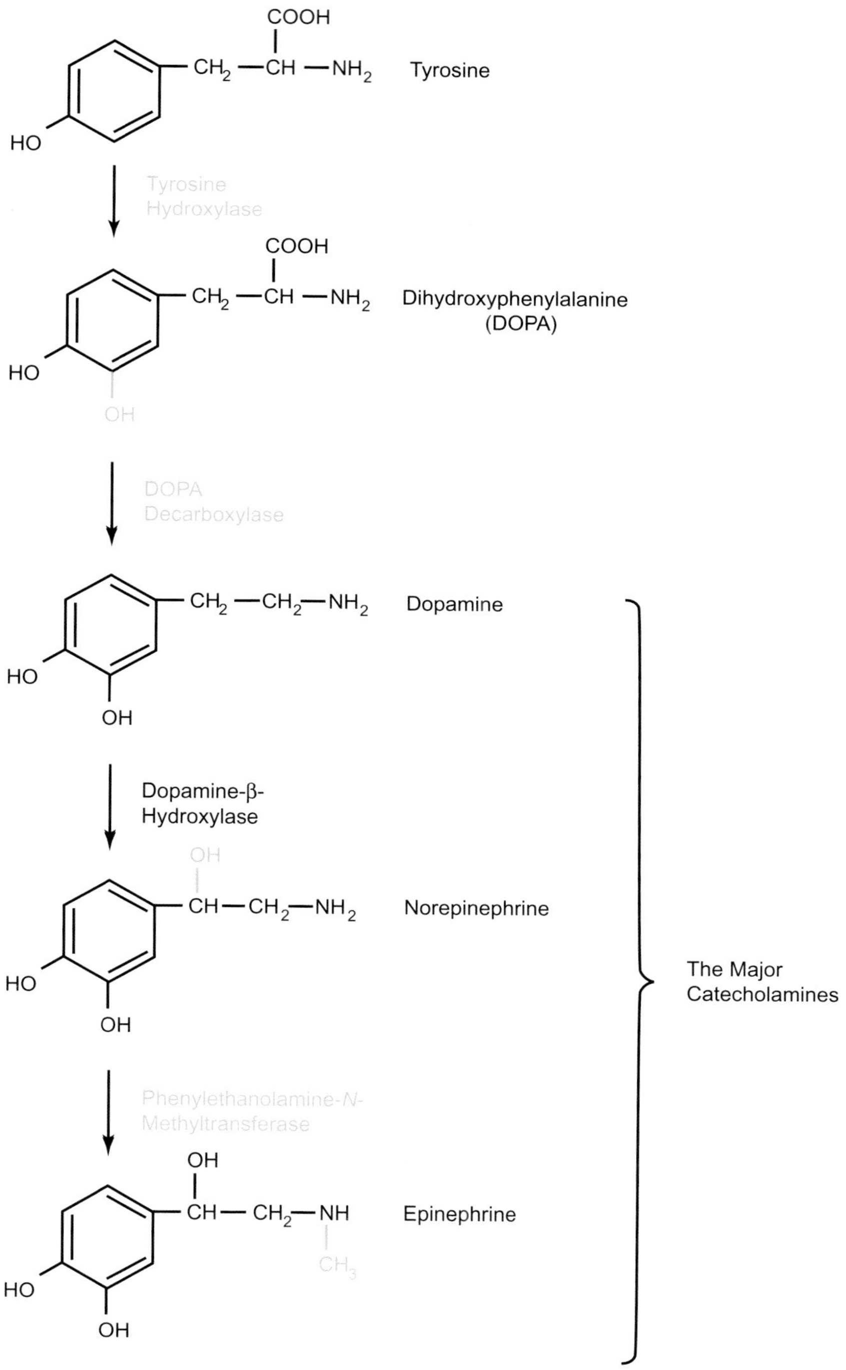

Figure 10–4. Biosynthesis of the catecholamines. The starting point for the synthesis of the catecholamines is the amino acid tyrosine. Hydroxylation of tyrosine to dihydroxyphenylalanine (DOPA) by the enzyme tyrosine hydroxylase is the rate-limiting step in the biosynthetic pathway.

two adjacent hydroxyl substitutions—see Fig. 10–4). Another important member of this family is *dopamine,* which is an intermediate in the biosynthesis of norepinephrine and epinephrine, and is a major central nervous system neurotransmitter in its own right. In contrast to the relatively few adrenergic neurons in mammalian brain, there are several distinct nuclei—large collections of neurons—that contain dopamine, and dopaminergic axons are distributed in complex patterns to many parts of the brain. Dopamine systems play an essential role in certain motor functions as well as in behavior, mood, and perception. As we shall discuss briefly later, a number of debilitating diseases, including Parkinson's disease, schizophrenia, and bipolar illness, can be attributed at least in part to a dysfunction in dopaminergic pathways.

Synthesis, storage, and release of the catecholamines. The biosynthetic pathway for the catecholamines is diagrammed in Figure 10–4. The first step is the hydroxylation of the common amino acid tyrosine to dihydroxyphenylalanine (DOPA) via the enzyme *tyrosine hydroxylase,* the expression of which is diagnostic of an adrenergic or dopaminergic neuron (analogous to choline acetyltransferase for cholinergic neuron types). Tyrosine hydroxylase is the rate-limiting enzyme for catecholamine biosynthesis and is subject to complex regulatory control; for example, its activity can be modulated by products in the biosynthetic pathway, and its synthesis is controlled by factors (such as nerve growth factor, see Chapter 15) that affect the growth and differentiation of sympathetic neurons.

The subsequent step in the pathway involves the removal of the carboxyl group from DOPA via the enzyme *DOPA decarboxylase* to produce dopamine, the first of the major catecholamine neurotransmitters. Dopamine can then be hydroxylated on the β carbon by dopamine β-hydroxylase to produce norepinephrine, which can in turn be methylated to epinephrine by a phenylethanolamine-*N*-methyltransferase (Fig. 10–4).

The catecholamines, like acetylcholine, are packaged into vesicles. The properties of catecholamine-containing vesicles have been studied most thoroughly in adrenal chromaffin cells (where the vesicles are called *chromaffin granules*) and sympathetic noradrenergic neurons, and it is assumed that the properties of the storage vesicles in central catecholaminergic neurons are similar. In addition to the neurotransmitter, catecholamine storage vesicles contain high concentrations of ATP (perhaps as high as 100 mM, about one-quarter the intravesicular concentration of the neurotransmitter) as well as a protein called *chromogranin,* whose function is only poorly understood but which may be involved in packaging and storage. Dopamine β-hydroxylase is also present in the granules, suggesting that at least one of the steps in the biosynthesis of norepinephrine occurs within the vesicles themselves.

The release of catecholamines from chromaffin cells occurs by exocytosis following the entry of calcium into the cell, presumably via voltage-dependent calcium channels, as is the case for cholinergic neurons. Because of the large size of chromaffin granules (in the range of 0.1 μm), exocytosis can be observed directly in the microscope and can be studied with techniques such as capacitance measurements, as described in Chapter 8. Exocytosis has not yet been directly established, however, for other catecholamine-releasing neurons. In fact, it seems somewhat puzzling that a neuron would dump the energetically expensive contents of its adrenergic granules into the extracellular space. There is evidence that the ATP may play some role in communicating with adjacent cells, but this is probably not the case for the chromogranin and dopamine β-hydroxylase, which would have to be replaced via new protein synthesis and axonal transport, processes that are slow and might place an apparently unnecessary demand on the cell's energy resources. It is possible that the proteins remain associated with the vesicle membrane in some way that allows their reuptake, perhaps via active transport or endocytosis. Alternatively, the suggestion has arisen, as in the case of acetylcholine, that nonexocytotic release of the transmitter occurs.

Uptake and metabolism of the catecholamines. The actions of catecholamines on their target cells are terminated much more slowly than those of acetylcholine. There is no rapidly acting extracellular enzyme analogous to acetylcholinesterase. Instead, catecholamines are removed from the synaptic cleft by reuptake into the presynaptic cell. The uptake is the sodium-dependent transport process we have come across in our discussion of choline uptake and will meet again often before this chapter ends. The high-affinity active transport mechanism moves catecholamines against a concentration gradient and causes them to accumulate inside the presynaptic cell at a much higher concentration than that in the extracellular space.

The two major enzymes involved in the catabolism of catecholamines are *monoamine oxidase* (MAO) and *catechol O-methyltransferase* (COMT). MAO catalyzes the metabolism of catecholamines to their corresponding aldehydes (Fig. 10–5), which can then be broken down further to products that leave the brain and are excreted. MAO is localized in the mitochondrial membrane of catecholaminergic terminals and helps to regulate the levels of catecholamines in these terminals. It is also present in other cell types in which its function is not understood. COMT catalyzes the methylation of one of the hydroxyl groups of the catechol nucleus (Fig. 10–5), again producing a product that is metabolized further and then excreted. Although COMT functions to metabolize catecholamines throughout the body, its precise localization and function in catecholaminergic transmission have yet to be determined. Inhibitors of MAO and COMT are important psychoactive drugs.

Catecholamine pharmacology and nervous system dysfunction. There is evidence that dysfunction in brain catecholamine pathways contributes to bipolar disorder and schizophrenia. The evidence is indirect, however, and is based largely on the fact that drugs that ameliorate the symptoms of these diseases interact with catecholamine systems. The classic finding is that a variety of MAO inhibitors, such as pargyline, which cause a rise in brain catecholamine levels, are clinically effective antidepressants. A separate class of clinically effective compounds, the *tricyclic antidepressants*, such as imipramine, appear to prolong catecholamine action (predominantly at noradrenergic synapses) by inhibiting the high-affinity reuptake system. Findings such as these have given rise to the *catecholamine theory of affective disorder*, which in essence states that decreased activity at certain central noradrenergic synapses causes behavioral depression. In addition, according to this theory, mania results from excess activity at these synapses. The latter hypothesis is supported by the fact that the stimulant drug amphetamine increases activity in noradrenergic pathways, most likely by inhibiting the reuptake of the neurotransmitter. All of these data point to the sodium-dependent, high-affinity neurotransmitter uptake systems as essential for the proper functioning of central nervous system synapses (discussed later in this chapter). More recently, serotonin (see later discussion) has also been implicated in bipolar disorder.

The *catecholamine theory of psychotic illness* focuses on dysfunction at dopaminergic synapses. A variety of antipsychotic drugs, the classic example being chlorpromazine, are effective blockers of postsynaptic dopamine receptors. It is particularly compelling that several chemically diverse groups of compounds (such as the butyrophenones, thioxanthenes, and phenothiazenes) all block dopamine receptors and ameliorate the symptoms of schizophrenia, giving rise to the hypothesis that excessive activity in dopaminergic pathways is responsible for schizophrenia. Again, the evidence is indirect, but this has not deterred pharmaceutical companies from devoting enormous resources to the search for novel dopamine receptor blockers that might serve as antipsychotic drugs.

The role of dopaminergic pathways in certain motor functions is more firmly established in that there are anatomical findings to correlate with

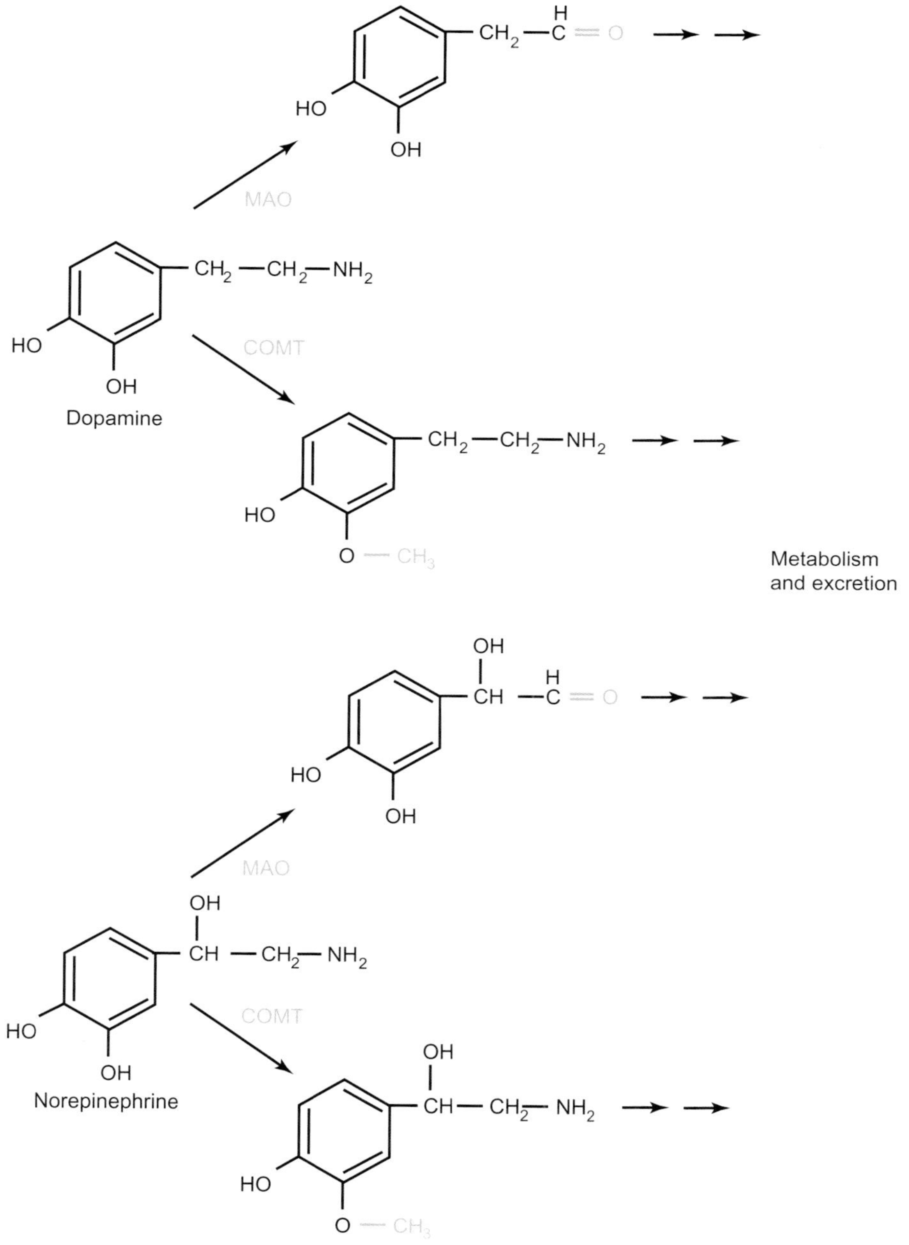

Figure 10–5. Metabolism of the catecholamines. Two enzymes, monoamine oxidase (MAO) and catechol-O-methyltransferase (COMT), catalyze the first steps in the degradation of the catecholamines.

the pharmacology. The *basal ganglia*, several *nuclei*, or collections of nerve cells deep in the brain, play an essential role in the control of body movements. The basal ganglia themselves receive inputs from the *substantia nigra*, a major dopamine-containing center in the brain, and influence the activity of the motor cortex via the thalamus (Fig. 10–6a).

Dopamine released by this pathway is essential for normal motor activity. The dopamine neurons that project from the substantia nigra to the *caudate nucleus*, one of the basal ganglia, can undergo pathological degeneration of unknown cause. As a result, the influence of the substantia nigra is removed (Fig. 10–6b), and this leads to characteristic motor dysfunction, including low-frequency tremor of the extremities at rest and

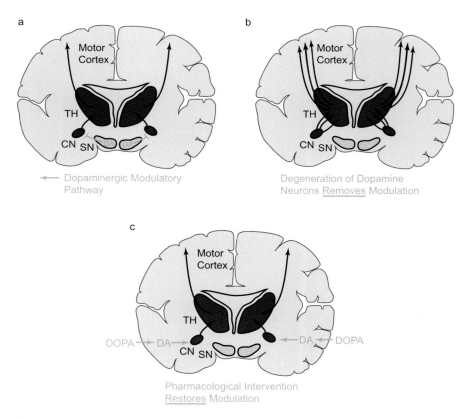

Figure 10–6. Dopamine pathways and motor functions (*a–c*). Dopamine (DA) neurons in the substantia nigra (SN) provide an important influence on motor functions. The DA neurons project to the caudate nucleus (CN), which in turn communicates with neurons in the thalamus (TH) that influence output from the motor cortex.

impairment of postural reflexes. This degenerative disorder is called *Parkinson's disease*, after the English physician James Parkinson who first described the symptoms in the nineteenth century. With the realization that Parkinson's disease involves a degeneration of dopaminergic neurons, Arvid Carlsson and colleagues reasoned that it might be possible to alleviate the symptoms by introducing dopamine into the brain. Because dopamine cannot readily enter the brain from the blood, the treatment of choice is to administer DOPA, which does enter the brain and is converted to dopamine via the action of DOPA decarboxylase (Fig. 10–4). When Parkinson's patients are administered DOPA, together with an MAO inhibitor (Fig. 10–6c), there is a dramatic, albeit temporary, alleviation of symptoms. This is a striking example of a disease of the nervous system that is attributable to the loss of a particular neurotransmitter and that responds to a rational treatment. Carlsson shared the 2000 Nobel Prize in Physiology or Medicine for his role in identifying dopamine as a neurotransmitter and introducing DOPA therapy for Parkinson's patients. Unfortunately, DOPA treatment does not reverse the course of the disease, which involves a progressive degeneration and deterioration over a period of several years, but simply relieves the symptoms for some brief interval. As we shall see in subsequent chapters, the degeneration seen in Parkinson's disease shares certain features with other neurodegenerative disorders, including Alzheimer's disease, and there has been exciting recent progress in defining some of the molecular events that lead to neurodegeneration in these disorders.

Serotonin. Serotonin, or *5-hydroxytryptamine* (5-HT), is another neuroactive compound that is neither exclusively a hormone nor exclusively a classical neurotransmitter. Large quantities of serotonin are found in the circulation, for example, in platelets, as well as in the central nervous systems of vertebrates and invertebrates. Like the catecholamines, serotonin is synthesized from one of the common amino acids, in this case, tryptophan, which is taken up into neurons via a specific sodium-dependent uptake system and is hydroxylated in a reaction catalyzed by *tryptophan hydroxylase* (Fig. 10–7). As in the case of catecholamine biosynthesis, this initial reaction is the rate-limiting step in the formation of serotonin. The resulting 5-hydroxytryptophan is then immediately decarboxylated to produce serotonin. Serotonin is packaged into secretory granules and is presumed to be released by a calcium-dependent exocytotic mechanism, then taken back up into the presynaptic terminal where it can be degraded by MAO (Fig. 10–7).

Serotonin is localized in discrete groups of neurons, largely in the brainstem, that send projections all over the brain. However, the precise

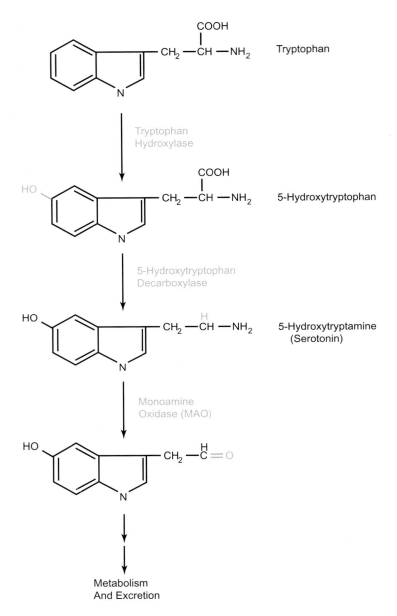

Figure 10–7. Synthesis and metabolism of serotonin (5-hydroxytryptamine, or 5-HT).

role of serotonergic pathways remains obscure. Serotonin has been implicated in the regulation of sleep and "vigilance," and it is widely believed that hallucinogenic drugs such as *lysergic acid diethylamide* (LSD) may produce their effects by interacting with serotonin pathways. The experimental evidence is complex and does not fall into a readily interpreted pattern, but it appears that LSD may antagonize some and mimic other actions of serotonin. The drug *Prozac* and its relatives such as *Zoloft*, all of which are selective blockers of serotonin reuptake, are highly effective and widely used antidepressants, and they have also been used successfully to treat a variety of less severe syndromes usually characterized as neuroses. The effectiveness of selective serotonin reuptake inhibitors (SSRIs) such as Prozac indicates that dysfunction in serotoninergic pathways must contribute to affective disorders; but again, both the anatomical and molecular details are only poorly understood. As we shall see in Chapters 18 and 19, in some invertebrates, for example, lobsters and the marine snail *Aplysia*, serotonin acts as a neurotransmitter/neurohormone to influence certain behaviors via mechanisms that have become reasonably well understood at the cellular and even molecular levels.

More amines. We have not yet exhausted the category of amine neurotransmitters. For example, there is evidence that *histamine* can affect neuronal activity in the mammalian (as well as in the invertebrate) central nervous system. Another important compound in many species is *octopamine*, which is closely related to dopamine and norepinephrine. The nucleotide *ATP* may be released at certain synapses together with catecholamines and influence neuronal properties, and the nucleoside *adenosine* can also regulate neuronal activity via so-called *purinergic* receptors. Although we will not discuss these other candidates further, we remind the reader that the list of amine neurotransmitters may be extended in the future.

Amino Acid Neurotransmitters

It should come as no surprise that it was at first difficult to demonstrate a neurotransmitter role for amino acids. Although such a role had been suspected for many years, how does one test the relevant criteria—for example, the presence of the compound in synaptic terminals—for compounds that occur in high concentration not only in neurons but in all cell types? Of the three major amino acid neurotransmitters in the mammalian nervous system—γ-aminobutyric acid (GABA), glycine, and glutamic acid (Fig. 10–8)—this problem has been particularly acute for the latter

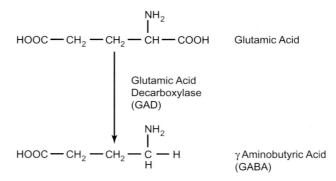

Figure 10–8. The major amino acid neurotransmitters.

two, which are ubiquitous constituents of proteins; GABA, in contrast, is present almost exclusively in brain and was recognized early on as a probable neurotransmitter. In any event, accumulating evidence has put the doubts to rest, and it is now widely accepted that the large majority of central nervous system synapses use amino acids as their neurotransmitters.

Glutamate and other excitatory amino acid transmitters. Glutamic and aspartic acid and various synthetic analogs of these amino acids produce excitatory (depolarizing) responses on neurons in virtually every part of the mammalian brain. Glutamate has long been established as an excitatory neurotransmitter at insect and crustacean neuromuscular junctions, and it is now evident that it is the major excitatory brain neurotransmitter. Much of the evidence has been based on the demonstration of several different classes of receptor for glutamate in brain (see Chapter 11), and it is clear that glutamate receptors play an essential role in long-term plastic changes at synapses, both during development and in the adult (see Chapters 17 and 19).

The details of glutamate synthesis and catabolism may be found in any standard textbook of biochemistry, so we will not dwell on them here. As is the case for most of the neurotransmitter candidates we have discussed, there are sodium-dependent high-affinity glutamate transport systems in neurons (as well as one in glia) that are involved in terminating the synaptic actions of glutamate (see below).

GABA and glycine: the inhibitory amino acid neurotransmitters. The involvement of GABA in neurotransmission has been much easier to determine. Again, it was first established as an inhibitory neurotransmitter at an invertebrate synapse, in crustacean muscle. Attention was focused on its role in mammalian brain when it was found that its concentration there is much higher than in any other tissue. GABA is synthesized from glutamic acid via a reaction catalyzed by the enzyme *glutamic acid decarboxylase* (GAD) (Fig. 10–8), the presence of which is considered to be positive identification of a GABAergic neuron.

The location of proteins within the brain (or other tissues) can be found using a technique called *immunohistochemistry*, an important technique that we have seen before and shall encounter again, particularly in the chapters dealing with neuronal development and plasticity (Part IV). An antibody that binds specifically to a protein is incubated with fixed or frozen sections of brain tissue. The sections are then incubated again with a fluorescent or colored reagent that binds to and reveals the location of the antibody, and the pattern of distribution of the protein to which the antibody binds can be observed in the microscope. Such immunohistochemical investigations, using specific anti-GAD antibodies, have revealed that GABAergic terminals are present throughout the brain. Thus GABA appears to be the major inhibitory neurotransmitter in the mammalian central nervous system.

Synaptic terminals contain a sodium-dependent high-affinity GABA uptake system, and the major route of catabolism is via a mitochondrial GABA-α-oxoglutarate transaminase that regenerates glutamic acid. GABA is a particularly interesting neurotransmitter because its actions are subject to modulation by a variety of pharmacological agents that have profound effects on brain function; in the next chapter we shall discuss the ways some of these agents interact with GABA receptors to modulate GABAergic transmission.

The role of glycine in neurotransmission has not been easy to determine because, like glutamate, there is too much of it. It appears, however, to be the most important inhibitory neurotransmitter in the spinal cord and lower brainstem (and probably in the retina as well). As in the case of glutamate, much recent progress has come from the identification and isolation of specific receptors that mediate the postsynaptic actions of glycine. Glycine can also interact with at least one class of brain glutamate receptor and modulate glutamatergic transmission. The actions of glycine in the spinal cord are terminated by a specific sodium-dependent transport system in presynaptic endings (discussed later in this chapter).

Neurotransmitter Transporters

We have emphasized throughout this chapter that a characteristic common to all of the small-molecule neurotransmitters is the existence of neuronal and glial plasma membrane uptake systems, which are responsible for halting neurotransmitter action and replenishing their supply in presynaptic terminals. In the case of acetylcholine at the neuromuscular junction, its action is terminated by enzymatic hydrolysis, as we have seen earlier in this chapter, but the choline that is produced is then taken up by a specific transport system as well (see Fig. 10–3). The individual uptake systems are mediated by distinct transporter proteins, each of which is specific in that it prefers to transport a particular neurotransmitter. However, the various transporters do have common features. In particular, all require sodium to be present in the external medium and use the energy provided by the sodium concentration gradient across the membrane to concentrate neurotransmitters within the nerve terminal. That is, the movement of the neurotransmitter molecule into the cell is coupled obligatorily to the movement of sodium ions across the membrane.

Families of neurotransmitter transporters. Through molecular cloning we have learned that, like the ion channels, neurotransmitter transporters can be grouped into several families that exhibit distinct features of structure and function. One family contains transporters present in the membranes of synaptic vesicles. They are responsible for concentrating neurotransmitters from the cytoplasm into the vesicle prior to exocytosis during synaptic transmission; we will not discuss them further here. A second group, the *solute carrier 6 (SLC6)* family of transporters, contains the plasma membrane transporters for GABA, glycine, and a number of amines, including norepinephrine, dopamine, and serotonin. The members of this family exhibit substantial sequence homology with one another and from hydrophobicity plots are predicted to have 12 membrane-spanning segments (Fig. 10–9a). The crystal structure of LeuT, a bacterial SLC6 family homolog from *Aquifex aeolicus*, confirms this membrane topology predicted from the protein sequence and provides insight into the protein movements that are associated with neurotransmitter transport. For all of the SLC6 transporters, chloride ion is an obligatory cosubstrate, together with sodium. The *stoichiometry* of the transport, the ratios of the three cotransported species, is 1-3 Na^+:1

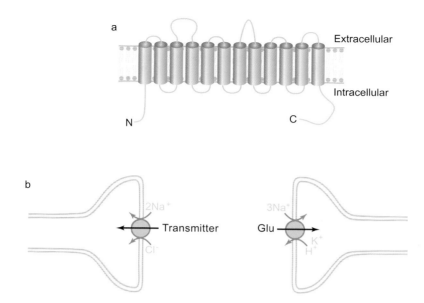

Figure 10–9. Neurotransmitter reuptake. *a*: Predicted structure of the sodium/chloride-dependent neurotransmitter transporters. *b*: Stoichiometries of the sodium/chloride-dependent (*left*) and the glutamate (Glu) (*right*) transporters.

Cl^-:1X, where X is the neurotransmitter (Fig. 10–9b). As a result, the transport may be *electrogenic*—that is, it will tend to change the membrane voltage, and indeed, it is possible to record transmembrane currents associated with the activity of some of these neurotransmitter transporters. The combination of the chloride and sodium gradients allows the neurotransmitter to be concentrated in the synaptic terminal to concentrations as much as 10,000-fold higher than those in the extracellular space. Mutational and structural analyses have uncovered discrete structural modules in transporters responsible for specific functions, reminiscent of the situation for ion channels described in earlier chapters.

   Interestingly, the transporters for glutamate fall into a distinct family of their own, the *SLC1* family of excitatory amino acid transporters, or EAATs. Their sequences exhibit little homology to those of the Na^+- and Cl^--dependent carriers, and they are predicted to have fewer membrane-spanning segments, a prediction confirmed by the determination of the crystal structure of Glt_{Ph}, an archaeal SLC1 family homolog from *Pyrococcus horikoshii*. In addition, the stoichiometry is different for the two transporter families. Transport of glutamate is still associated with the cotransport of sodium, but not chloride. Instead, there is countertransport of potassium out of the cell, probably together with protons (Fig. 10–9b), with a stoichiometry of 3 Na$^+$:1 K$^+$:1 H$^+$:1 glutamate. Thus, in addition to being electrogenic, the activity of the glutamate transporters may produce a change in intracellular pH.

Diversity of neurotransmitter transporters. Although the diversity of the neurotransmitter transporters does not rival that of the calcium or potassium channels, there exist genes encoding several distinct transporters for most of the small molecule neurotransmitters. This molecular heterogeneity was not unexpected, because it has been known for a long time that there are differences in the affinity of the transporter for the neurotransmitter and in transporter pharmacology in different brain regions. In addition, there are differences in the cellular distribution of the different transporters for a given neurotransmitter. Experiments using immunohistochemistry, as described earlier, as well as a companion technique known as *in situ hybridization*, through which a specific messenger RNA can be localized, have demonstrated that some transporters are expressed in neurons whereas others are restricted to glial cells. Recall from Chapter 2 that one of the important functions of glia is thought to be the regulation of synaptic transmission by the uptake and metabolism of neuronal neurotransmitters.

The Peptides

By the early 1970s, a consistent pattern was emerging of neurotransmitter candidates in the central nervous system. Acetylcholine, serotonin, and the catecholamines were well established as major brain neurotransmitters, and although the transmitters at many central synapses remained unidentified, suspicions were beginning to arise that amino acids might fill this gap. The recognition that *neuroactive peptides* are more than simply hormones released to carry out actions on targets outside the brain, that they play a crucial role within the central nervous system, has made it clear just how naive this simple picture was.

Neuropeptides share some of the characteristics of the small-molecule neurotransmitters. For example, like many classical transmitters, some neuropeptides are released to the circulation and act at a distance, while others are confined to a discrete synaptic cleft. As mentioned at the beginning of this chapter, it will not be profitable to focus on the largely semantic distinction between neurotransmitters and neurohormones. Nevertheless, the neuropeptides do differ in many respects from their

Table 10–1 Properties of Peptide Neurotransmitters

Synthesized as a larger precursor protein at the soma and transported to release sites; must be replenished by synthesis at soma

Slow postsynaptic effects

Actions terminated by extracellular proteases or by diffusion

Co-released with classical neurotransmitters

Can trigger complex coordinated behaviors

Actions do not require point-to-point synaptic connections

smaller counterparts. Table 10–1 lists some of the properties of peptide neurotransmitters that tend to distinguish them from the classical neurotransmitters. Because these are not all true at every location at which peptidergic transmission occurs, they are more appropriately termed *trends* rather than properties.

The roster of putative peptide neurotransmitters is large and diverse. We will limit ourselves here to describing just a few of the neuroactive peptides, with emphasis on the historic development of this field and some examples of the trends listed in Table 10–1.

Substance P. The story of the peptides actually began as far back as the 1930s, with the accidental discovery of substance P. While screening various tissues for acetylcholine, Ulf von Euler and John Gaddum found a compound in brain and intestine that caused a lowering of blood pressure and contraction of intestinal smooth muscle. Because its actions were not blocked by the cholinergic antagonist atropine, they concluded that this compound could not be acetylcholine, and named it substance P (for **pow**der). Although von Euler suggested as early as 1936 that substance P might be a protein, it was to be more than 30 years before its structure was determined. Substance P is an 11–amino acid peptide that is present, as determined by direct bioassay and immunohistochemical studies, throughout the mammalian brain. It is thought to be a synaptic transmitter in sensory pathways concerned with pain and touch. It is, in fact, found in neurons that innervate tooth pulp, where the only known sensory modality is pain. Substance P was the first, and remains among the best understood, of the *brain-gut peptides*, neuroactive substances found in both the central nervous system and the gastrointestinal tract. There are now numerous examples of this localization pattern, which reflects the fact that the gut is densely innervated.

Hypothalamic peptides. As mentioned previously, the hypothalamus is the site of synthesis of the two classical neurohormones, oxytocin and vasopressin. In addition to affecting peripheral tissues, oxytocin and vaso-pressin can alter the firing patterns of central neurons, and vasopressin in particular has been reported to have behavioral effects in rats and humans. The magnocellular neurons in which these peptides are synthesized proj-ect not only to the pituitary circulation but also to a variety of sites within the brain, and the immunohistochemical localizations of oxytocin and vasopressin are consistent with their involvement in synaptic transmission (Fig. 10–10).

Also synthesized in the hypothalamus are other biologically active peptides, the *releasing factors* or *releasing hormones.* The releasing hor-mones enter the portal circulation, which brings them to the anterior pituitary where they regulate the release of another group of pituitary peptide hormones. These in turn act on peripheral tissues. For example, the release of *thyroid-stimulating hormone* (thyrotropin or TSH), which enters the general circulation from the pituitary to act on the thyroid gland, is promoted by *thyrotropin-releasing hormone* (TRH) from the hypothalamus (Fig. 10–10). Much TRH is also found in other brain regions, where it may act as a local hormone or neurotransmitter. This is true for other releasing factors as well. ❽

Opioid peptides. The age of the peptides was ushered in with a vengeance in the mid-1970s with the description of the *enkephalins* and other opioid peptides. The story of the opioid peptides is particularly interesting because their discovery was not simply serendipitous but resulted from a rational search for endogenous compounds that might mimic the actions of morphine and other opiates.

Throughout recorded history it has been known that the juice of the poppy produces feelings of euphoria and is a highly effective painkiller. The active ingredient, morphine (named after Morpheus, the Greek god of dreams), was first isolated almost 200 years ago. By the early 1970s it was reasoned that the actions of morphine and other opiates must result from their binding to specific receptor sites in the brain, and it became possible to demonstrate the binding of radioactive opiates to brain mem-branes. The next assumption was that if the brain contains specific recep-tor sites, there must be endogenous ligands that bind to and activate these receptors.

Two compounds with such properties were indeed found. They were first isolated by their ability to produce *analgesia,* relief from pain, in bioassays for pain relievers. These compounds are pentapeptides of

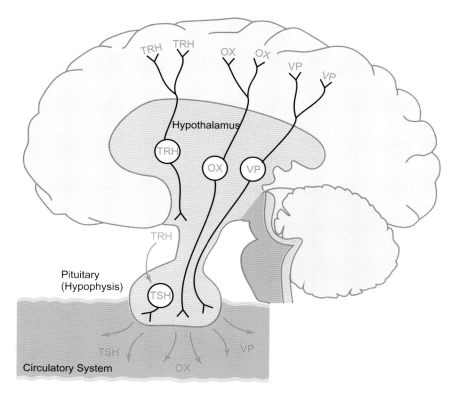

Figure 10–10. The hypothalamus is an important source of neuroactive peptides. Oxytocin (OX) and vasopressin (VP) neurons project to the posterior pituitary (also called the *neurohypophysis*), where the peptides are released to the general circulation. Releasing hormones, here exemplified by thyrotropin-releasing hormone (TRH), reach the anterior pituitary (*adenohypophysis*) via the portal circulation and stimulate release of hormones such as thyrotropin (TSH). The hypothalamic peptides can also act as brain neurotransmitters.

similar structure, which were named *Met-enkephalin* and *Leu-enkephalin*, because they differ only in their carboxyl-terminal amino acids:

Tyr-Gly-Gly-Phe-*Met*   *Met*-enkephalin

Tyr-Gly-Gly-Phe-*Leu*   *Leu*-enkephalin

Synthetic enkephalins were also soon shown to be potent opiates in the standard bioassays for analgesics.

It was noted that the entire structure of Met-enkephalin is contained in amino acid residues 61–65 of a larger pituitary hormone, the 91–amino acid β-lipotropin, which had been isolated and sequenced several years earlier but whose function was obscure. Within a year after the description of the enkephalins, it became evident that at least three additional longer peptides with opioid-like activity, the *endorphins*, are contained within the sequence of β-lipotropin. Immunohistochemical experiments have demonstrated that the enkephalins and endorphins are distributed widely in the brain, but do not appear to be colocalized within the same neurons.

The enkephalin/endorphin story emphasizes another important way the peptides differ from the more classical neurotransmitter candidates (see Table 10–1). The latter tend to be small molecules that can be synthesized and metabolized by enzymes present in nerve terminals, whereas the peptides must be synthesized by the protein biosynthetic machinery and undergo complex processing and packaging (see Chapter 8) before they can be released to exert their biological effects. Questions as to how peptide stores can be replenished rapidly to sustain release for long periods of time have not yet been resolved; this replenishment does not seem to be a problem for the classical neurotransmitters.

The finding that the endorphins and enkephalins are contained within a larger precursor was not entirely surprising, since it was known that many peptide hormones, for example, insulin and the original neurohormones oxytocin and vasopressin, are synthesized as larger prohormones and then processed proteolytically to produce the active hormone (see Chapter 8). In fact, β-lipotropin itself is contained within a much larger and more complex precursor. Molecular cloning approaches have enabled investigators to demonstrate that the actual precursor is a polypeptide that contains within its sequence the structure of the pituitary hormone corticotropin (ACTH) as well as of the opioid peptides and several other naturally occurring peptides of unknown function. This rather astonishing discovery has many ramifications. For example, in Chapter 18 we shall discuss the physiology of neurons that use several peptide neurotransmitters, each of which is cleaved from a different part of a single precursor protein (see Fig. 8–6b).

Some invertebrate peptides. Invertebrates have been a rich source of neuroactive peptides, some of which are also present in vertebrate nervous systems. Among the best studied of the invertebrate neuropeptides are the pentapeptide *proctolin*, first isolated from the cockroach hindgut; the snail tetrapeptide *FMRFamide*; and the 36–amino acid *Aplysia egg-laying hormone* (ELH). The genes encoding the precursors for some of these peptides show remarkable homologies to certain mammalian neuropeptides. Furthermore, the large size and ready identifiability of many invertebrate neurons are experimental advantages that have allowed the physiological roles of some of these peptides to be investigated in considerable detail. For example, the synthesis, processing, and release of ELH by the neurosecretory bag cell neurons of *Aplysia*, as well as its mechanism of action on target neurons, are well understood. Similarly, the actions of proctolin in modulating behavioral patterns in the lobster have been thoroughly analyzed at the cellular level. Some of these systems will be discussed in more detail in subsequent chapters.

Colocalization of peptides and classical neurotransmitters. The pharmacologist Henry Dale suggested many years ago that a given neuron synthesizes only a single neurotransmitter and releases only that transmitter at all its terminals. We now know that this is not true in immature neurons, where release of two classical neurotransmitters from the same terminal shapes the development of synaptic connections. Moreover, there is now convincing evidence that some mature neurons contain one or more neuropeptides *and* a classical neurotransmitter, packaged in different vesicles but often present in the same synaptic terminal.

It is not known precisely what advantages this offers a neuron. It has become clear, however, that the different transmitters need not be released at the same time. In several cases it has been found that only the classical transmitter is released by low-frequency stimulation, and co-release of the peptide requires short bursts of high-frequency stimulation. It is not evident how the selective release of one transmitter, and not another, comes about, but it may have to do with the very different patterns of calcium distribution expected in a nerve terminal invaded by low- and high-frequency action potentials. At low firing frequency there will be brief and highly localized changes in calcium immediately under the membrane and adjacent to the voltage-dependent calcium channels, whereas calcium levels elsewhere, for example, away from the membrane, will be little affected. At higher rates of stimulation, however, calcium levels will increase progressively with each action potential during a burst, and a substantial elevation will also occur away from the immediate vicinity of the calcium channels. Thus, if the peptide-containing vesicles undergo exocytosis only at high levels of calcium, or if they are preferentially located at a distance from the membrane, peptide release will occur only with higher frequencies of presynaptic stimulation. We can see, then, that the coexistence of different neurotransmitters in distinct vesicle populations within a single neuron allows that neuron to produce different effects on a postsynaptic target, depending on the precise pattern of stimulation.

Questions about the functional significance of such colocalization of neurotransmitters can be broadened to ask why, in fact, there are so many neuroactive substances. In principle, a nervous system could function with just two neurotransmitters—one to mediate excitation and the other for inhibition. Indeed, as we shall see in the next two chapters, since there are multiple receptors for most (if not all) neurotransmitters, even a *single* neurotransmitter might in principle be sufficient, and the excitatory or inhibitory nature of the response could be determined by the type of receptor that happens to be present on the target cell. We do not have a definitive answer to this dilemma, but again, a reasonable explanation is that multiple neuroactive substances provide both a wide range of times over which the response endures and a wide range in the character of the response. It is a mistake to think of neurons as simply on or off, active or inactive, excited or inhibited. Rather, as we have seen in Chapters 3–7, neurons contain a variety of different ion channels that allow their activity to be modulated in subtle ways, and it may be that a multiplicity of transmitters (as well as receptors and transporters) has evolved to participate in this fine-tuning of neuronal function.

Summary

A multitude of chemicals called *neurotransmitters* mediate intercellular communication in the nervous system. These include acetylcholine, the catecholamines, serotonin, glutamate, GABA, glycine, and a wide variety of neuropeptides. Although they exhibit great diversity in many of their properties, all are stored in vesicles in nerve terminals and are released to the extracellular space via a process requiring calcium ions. Their actions are terminated by reuptake into the presynaptic terminal or nearby glial cells by specific transporter proteins or by their destruction in the extracellular space. The role of neurotransmitters is to alter the properties, chemical, electrical or both, of some target cell. With the arrival on the scene of the neuropeptides, it has become evident that signaling in the nervous system occurs through the use of rich and varied forms of chemical currency, and that some neurons use more than one type of currency simultaneously.

11

Receptors and Transduction Mechanisms I: Receptors Coupled Directly to Ion Channels

A ll of the neuroactive substances we discussed in the previous chapter alter some property of a target neuron. This effect on the target cell is highly specific for the particular neurotransmitter or neurohormone. How does a neuron know just which of the many possible chemical signals it is being tickled by, and how does it decide on a response appropriate for that signal? Obviously, the answers to these questions are crucial for understanding intercellular communication in the nervous system. In this and the following chapter we will discuss the neurotransmitter and hormone receptors that recognize the signaling molecules, and the transduction mechanisms that convert the extracellular signal into a response, usually (but not always) electrical, in the target neuron. The focus of this chapter is on neurotransmitter receptors that are coupled *directly* to the ion channels whose activity they regulate (Fig. 11–1). In Chapter 12 we will discuss *indirect* receptor–channel coupling mechanisms (compare Fig. 12–1). We emphasize here, as we have previously, that these mechanisms are not restricted to nerve cells. Other cell types have receptors for intercellular signaling molecules and convert the signal into some biological response. Neurons are unusual in that their biological response is normally a change in voltage that ultimately alters transmitter release, allowing the signal to be passed along from one neuron to another in a multineuronal pathway.

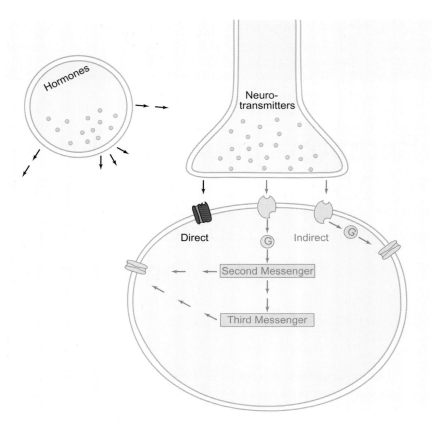

Figure 11–1. Intercellular communication. This chapter will discuss neurotransmitter receptors that are coupled directly to the ion channels that they regulate (purple).

Specificity of Responses

The pharmacological concept of specific receptor structures has been with us for more than a century. Pharmacologists recognized many years ago that drugs that produce specific effects must interact with specific sites on or in the cell. It became possible to build up a pharmacological profile of a receptor by identifying drugs that produce a particular response (receptor *agonists*) and other drugs that inhibit the response to agonists (receptor *antagonists*). Furthermore, the concentrations at which these various agonists and antagonists are effective can be measured, to provide a quantitative pharmacological profile. The earliest examples of such pharmacological characterization were studies with the first identified neurotransmitter, acetylcholine.

Although acetylcholine was found to be the transmitter substance at neuromuscular junctions, in sympathetic and parasympathetic ganglia of the peripheral autonomic nervous system, and in postganglionic parasympathetic neurons, the pharmacological profile was not identical for all responses to acetylcholine. Instead it was found that nicotine and related compounds can act as agonists at neuromuscular junctions in skeletal muscle, but not in cardiac muscle, and only at some cholinergic synapses in the autonomic nervous system (Fig. 11–2). Furthermore, responses at these *nicotinic* synapses, but not at the others, could be blocked by compounds such as hexamethonium and the plant alkaloid curare (Table 11–1). It is worth noting that this pharmacology has been understood for a very long time by South American tribes that use curare at the tips of their arrows to paralyze prey.

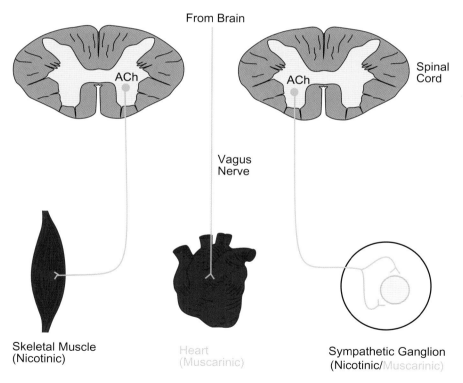

Figure 11–2. Acetylcholine (ACh) activates two different classes of receptors. Some actions of acetylcholine released from brain or spinal neurons can be mimicked by nicotine, whereas others can be mimicked by muscarine. The receptors that mediate these different classes of response are termed *nicotinic* and *muscarinic*, respectively.

Table 11–1 Two Pharmacologically Distinct Receptors for Acetylcholine

| Receptor Type | Some Examples of | | Responding Cell Type |
	Agonists	Antagonists	
Nicotinic	Acetylcholine	Hexamethonium	Skeletal muscle
	Nicotine	Curare	Sympathetic neurons
			Parasympathetic neurons
			Some brain neurons
Muscarinic	Acetylcholine	Atropine	Smooth muscle
	Muscarine	Scopolamine	Sympathetic neurons
	Oxotremorine	Quinuclidinylbenzylate	Gland cells
			Heart cells
			Many brain neurons

In contrast, the action of acetylcholine at other cholinergic synapses can be mimicked by muscarine but not by nicotine (Fig. 11–2) and blocked by atropine and quinuclidinylbenzylate but not curare (Table 11–1). It now appears that such *muscarinic* synapses account for a large proportion of the cholinergic synapses in the mammalian central nervous system. Although this breakdown of acetylcholine responses into two distinct groups seems to make life much more complicated for the harassed neuroscientist trying to understand transmitter actions, it is only the tip of the iceberg. Virtually every neurotransmitter is now known to interact with more than a single class of receptor; acetylcholine, for example, not only has two major classes of receptor but within the muscarinic and nicotinic classes multiple distinct subtypes exist. About the time the pharmacology of acetylcholine responses was being clarified, norepinephrine was also shown to interact with more than a single class of receptor, and the number of adrenergic receptor subtypes increased over the years as new, more specific pharmacological tools became available. Similarly, the amines, the excitatory and inhibitory amino acids, and the peptides all can mediate multiple responses in target cells (Table 11–2). As we shall see in the following discussion, the structural basis for these multiple responses has become well understood as methods for measuring receptor properties have become available and as the techniques of molecular cloning, genome sequencing, and protein structural analysis have been brought to bear on the receptor molecules.

Table 11–2 Examples of Neurotransmitters and Neuromodulators
That Interact with More Than One Class of Receptor

Neurotransmitter/Neuromodulator	Receptor Class
Acetylcholine	Nicotinic
	Muscarinic
Adenosine	A_1, A_2, A_3
Norepinephrine	α-Adrenergic
	β-Adrenergic
Histamine	H_1, H_2, H_3
Dopamine	Multiple subtypes
Serotonin	Multiple subtypes
GABA	$GABA_A$
	$GABA_B$
	$GABA_C$
Glutamate	NMDA
	KA/AMPA
	Metabotropic
Opioid peptides	μ, δ, κ

Receptor-Binding Assays: From Concept to Physical Entity

We have mentioned that the pharmacological concept of specific receptor sites for transmitters has been around since the early twentieth century. It is only more recently, however, that techniques have become available for the direct measurement of membrane receptors. The development of ligand-binding assays for this purpose has confirmed the diversity of receptors inferred from the pharmacology of neurotransmitter responses. Furthermore, binding assays have enormously expanded our understanding of receptor structure and function, and so it is important to understand this approach and the kinds of information it can provide.

The fundamentals of the ligand-binding assay are shown in Figure 11–3. The term *ligand* (from the Latin *ligare*, to tie or bind) is used to denote a molecule that will bind to the receptor under study. The ligand is labeled in some way, usually with radioactivity or with a fluorescent probe, and is incubated together with plasma membrane fragments from cells that contain the receptor (Fig. 11–3a). After an appropriate period of time to allow the ligand to bind to receptors on the plasma membrane fragments, the bound ligand is separated from the remaining free ligand (the ligand is added in large excess, so that even when all receptor sites are occupied there will still be some free ligand). The amount of radioactivity bound can be measured, and under appropriate conditions it provides a measure of the number of specific binding sites. Binding studies can also provide an estimate of affinity, a measure of how tightly the ligand binds to the receptor (Fig. 11–3b). ❷

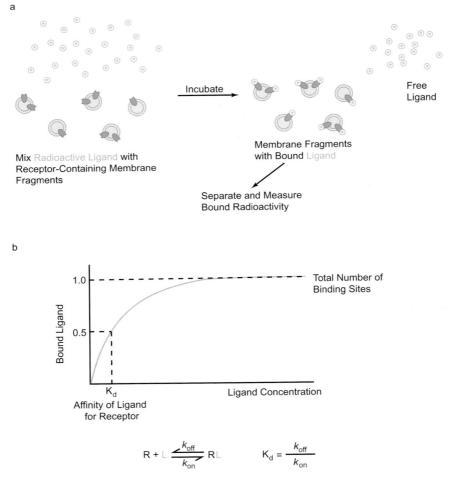

Figure 11–3. Ligand-binding assays for receptors. *a*: Radioactive ligands (blue) that bind to the membrane receptor under investigation (green) can be used to characterize some properties of the receptor. *b*: Under appropriate conditions, the ligand-binding assay can provide a measure of the number of receptors and of their affinity (K_d) for the ligand. k_{off}, dissociation rate constant. k_{on}, association rate constant.

Pharmacological Diversity of Receptors Reflects Structural Diversity

What is the molecular underpinning for pharmacologically distinct classes of receptor for a single neurotransmitter? Are there multiple receptor molecules, as we have seen for ion channels and neurotransmitter transporters, or can a single receptor molecule take on different pharmacological guises? These questions could be answered definitively only by isolating receptor molecules. It did not escape the attention of receptorologists that, in addition to providing an assay for receptors in their normal membrane-bound state, the binding of specific ligands might be used to measure numbers of receptor molecules during purification. Furthermore, appropriate ligands might be exploited not only for assay but also as reagents to be used in purification—as *affinity reagents*. Let us take as a classic example the first neurotransmitter receptor to be purified by methods based on these concepts, the nicotinic acetylcholine receptor and its associated ion channel. It was the first because it was the easiest; a high-affinity and highly selective ligand was available, and there was a rich source of receptor in the electric organs of certain eels and fish.

A toxin that binds to nicotinic acetylcholine receptors. Certain creatures appear to have been designed by evolution as gifts to neurobiologists. The story of the nicotinic acetylcholine receptor, like that of the ionic mechanism of the action potential, belongs to two groups of such animals, each with its unique specializations for survival. The first group is a family of snakes whose venoms contain toxins that bind with extremely high affinity and selectivity to the nicotinic acetylcholine receptor. These snakes paralyze and kill their prey by blocking neuromuscular transmission with the toxins. The most thoroughly characterized of these toxins is α-*bungarotoxin*, from the venom of the Taiwanese cobra *Bungarus multicinctus*. This toxin is a polypeptide of 74 amino acids, which binds to the nicotinic receptor with a dissociation constant in the range of 10^{-15} M (Fig. 11–4). Once it binds to the receptor, the half-time for its dissociation is many days! This makes it an ideal reagent for receptor assay and affinity purification.

A rich source of nicotinic acetylcholine receptors. The second set of creatures that has contributed so greatly to our understanding of the nicotinic receptor are various species of the genus *Torpedo*, the electric rays, and the genus *Electrophorus*, the electric eels. The electric organs in these

$$\text{NAChR} + \alpha\text{-BT} \; \underset{k_{on}}{\overset{k_{off}}{\rightleftharpoons}} \; \text{NAChR} - \alpha\text{-BT}$$

$$K_d = \frac{k_{off}}{k_{on}} = 10^{-15}$$

Figure 11–4. α-Bungarotoxin (α-BT) binds very tightly to the nicotinic acetylcholine receptor (NAChR). The affinity (K_d) of the toxin for the receptor is extremely high. k_{off}, dissociation rate constant. k_{on}, association rate constant.

animals consist of modified muscle cells, which are flattened and organized in parallel arrays known as *electroplaques*. These cells have lost their contractile apparatus and have neuromuscular junctions that cover virtually the entire surface of one of the flattened faces of the cell, but not the other. The geometry of the electroplaque is such that voltage differences across the many cells add in series, producing a very large voltage that can stun prey.

The advantage of using the electric organ as a source of nicotinic receptor becomes obvious when we examine the density of the receptor in the membrane of the *Torpedo* electroplaque cells. Most receptors and ion channels are relatively rare membrane proteins. Typically, one or a few receptor molecules are found per square micron of membrane surface area. In contrast, the *Torpedo* electroplaque membrane is packed tightly with nicotinic receptor, to a density as high as 10,000 to 30,000 per square micron, and indeed there appears to be little room for any additional protein in the membrane! Thus the rays have done a significant part of the purification for us, by inserting the protein and very little else into a membrane that can be isolated with relative ease.

Structure of the nicotinic acetylcholine receptor. With this rich source of material and the rapid, sensitive, and specific α-bungarotoxin assay, large amounts of homogeneous nicotinic receptor can be obtained. Biochemical analysis demonstrated that the receptor is a large complex consisting of five subunits (Fig. 11–5). There are two α subunits of molecular weight (MW) approximately 40,000, and one each of a β (MW 48,000), γ (MW 58,000), and δ (MW 64,000) subunit (Fig. 11–5a). Each α subunit contains a site for binding acetylcholine (and α-bungarotoxin) (Fig. 11–5b). This is consistent with physiological data, which indicate that the binding of two acetylcholine molecules is necessary for the opening of the nicotinic receptor's associated ion channel.

cDNA clones encoding the various subunits of the nicotinic receptor were obtained from amino acid sequence information, as described previously for the sodium channel. Mutagenesis and heterologous

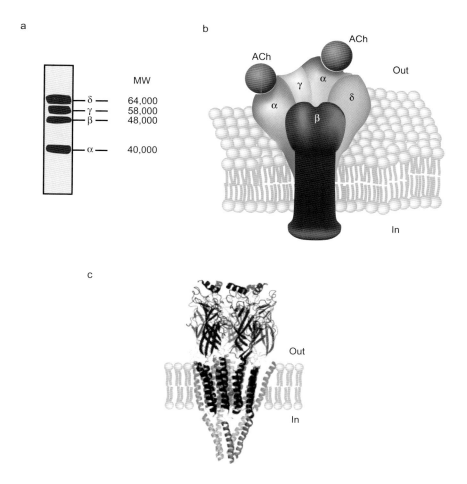

Figure 11–5. The nicotinic acetylcholine receptor from *Torpedo*. *a*: Polyacrylamide gel separation of the individual subunits of the purified nicotinic receptor. *b*: Cross-linking studies demonstrated that the functional receptor molecule contains two α subunits, each with an acetylcholine (ACh) binding site, and one each of the β, γ, and δ subunits. *c*: Three-dimensional structure of the nicotinic acetylcholine receptor derived from electron microscopy and X-ray crystallography. (Modified from Baenziger and Corringer, 2011. See also Unwin, 2005.)

expression experiments have provided a detailed picture of structure–function relationships in this receptor. The different subunits have substantial amino acid sequence homology with one another, which suggests that they may have evolved from a single ancestral protein. Each subunit has four hydrophobic domains that were predicted to be membrane-spanning sequences (Fig. 11–6). An important concept that we emphasize here is that, considerably later, when the muscarinic acetylcholine receptor from various sources was purified and cloned, its amino acid sequence was found to be completely unrelated to that of the nicotinic receptor. The muscarinic receptor has only a single kind of subunit, and the functional membrane receptor is an oligomer. The subunit molecular weight is 51,000, similar to that of the nicotinic receptor subunits, but a comparison of the amino acid sequences shows no relationship. From hydrophobicity measurements it was inferred that the muscarinic receptor has seven membrane-spanning domains, in contrast to the nicotinic receptor subunits, each of which crosses the membrane four times (Fig. 11–6).

Several different approaches have provided information about the three-dimensional structure of the nicotinic receptor (Fig. 11–5c). Extremely useful has been the surprising discovery that certain snails produce a soluble acetylcholine-binding protein that closely resembles the nicotinic receptor in amino acid sequence and subunit composition, and high-resolution crystal structures of this protein have been obtained. These crystal structures, as well as an earlier structural analysis by high-resolution electron microscopy, have confirmed the inferences about nicotinic receptor structure and function that were drawn originally from mutational analysis.

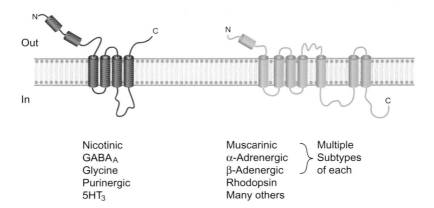

Nicotinic
GABA$_A$
Glycine
Purinergic
5HT$_3$

Muscarinic
α-Adrenergic
β-Adenergic
Rhodopsin
Many others

Multiple
Subtypes
of each

Figure 11–6. Topology of acetylcholine and other receptors. Hydrophobicity profiles predicted accurately that nicotinic acetylcholine receptor subunits, as well as those of the GABA$_A$, glycine, purinergic, and one subtype (5HT$_3$) of serotonin receptors, contain four membrane-spanning domains (pink). In contrast, muscarinic and related receptors were predicted to have seven membrane-spanning domains (green).

Common Structural Motifs in Receptors: Receptor "Superfamilies"

It would appear that the nicotinic and muscarinic acetylcholine receptors have virtually nothing in common other than the fact that they bind and are activated by acetylcholine. They are present in different kinds of nerve, muscle, and gland cells (Table 11–1); mediate very different physiological responses in terms of kinetics and mechanisms; and their amino acid sequences and inferred structures show no similarities (Fig. 11–6). Because these discrepancies have been a disappointment to those biologists who seek unifying concepts in nature, it was all the more satisfying that unifying concepts did indeed emerge. Just as there are commonalities among the voltage-dependent ion channels (Chapters 3–7), there are families of receptor molecules with common structural features. However, perhaps somewhat surprisingly, the family groupings are not based on the neurotransmitter that binds to the receptor; rather, they reflect the transduction mechanism, the molecular mechanism that the receptor uses to translate the extracellular signal into a physiological response in the target cell. We shall now discuss one large family, that of the *ligand-gated ion channels*, to which the nicotinic acetylcholine receptor belongs. We will describe the structural features that the family members share, as well as the transduction mechanism that they use to excite or inhibit a neuron. In Chapter 12 we shall describe other receptor families in which the transduction mechanisms are more complex. These include the family to which the muscarinic acetylcholine receptor belongs, the *G protein–coupled receptors*.

The Family of Directly Coupled Receptor–Ion Channel Complexes

The simplest way of transducing an extracellular signal into a change in excitability of the target neuron is by direct coupling of the neurotransmitter receptor to the ion channel whose activity it regulates. This is the mechanism used by the various ligand-gated ion channels. The ligand-binding site and the ion channel are part of the same molecule or macromolecular complex. Occupation of the receptor by the neurotransmitter leads to a conformational change that is passed along to the closely associated ion channel, and as a result channel properties are altered (Fig. 11–7). Note that in Figure 11–7 we show a channel that is closed in the resting state and opens as a result of transmitter action, allowing the ion (X^+) to flow across the membrane. Although this is the way most of the known ligand-gated ion channel systems appear to function, in principle it is possible that a transmitter might close some channel that is open in the resting state.

An important feature of this general mechanism is that the modulation of ion channel properties is dependent on continued occupation of the receptor by the transmitter. The conformational change induced by receptor occupancy is readily reversible, and as soon as the receptor is no longer occupied by transmitter, the channel returns to its normal resting state (reverse arrow in Fig. 11–7). Accordingly, it might be expected that the ligand-gated ion channel systems would mediate rapid-onset and rapidly reversible synaptic transmission, and this is indeed the case.

Neuronal nicotinic acetylcholine receptors. The nicotinic acetylcholine receptor complex is one of these ligand-gated ion channels, in that it contains not only the acetylcholine-binding site but also the ion channel

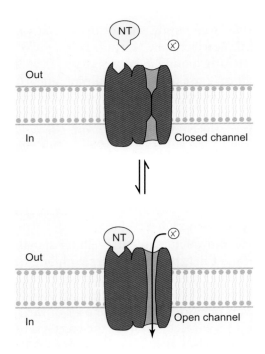

Figure 11–7. Direct receptor–channel coupling. In directly coupled receptor–channel systems, the neurotransmitter (NT) binding site and the ion channel are intimately associated in a single macromolecular complex. Contrast with Figures 12–2 and 12–6.

that is activated by acetylcholine binding. Most of the early data that support this conclusion came from work on the *Torpedo* and muscle nicotinic receptors. However, studies of a large family of neuronal nicotinic receptors place them in this same category of ligand-gated ion channels as well. ❷

The existence of nicotinic receptors in the brain should not come as a surprise in view of the well-established behavioral, cognitive, and addictive actions of nicotine. However, an understanding of neuronal nicotinic receptors was slower to develop because most of them are not sensitive to α-bungarotoxin, the tool that has proven so useful in characterizing *Torpedo* and muscle nicotinic receptors. Molecular cloning and genome sequencing have established that there exist genes for multiple kinds of nicotinic receptor subunits in mammals, including nine α subunits and four β subunits, most of which are expressed in brain but not muscle. Some functional neuronal nicotinic receptors are formed by a combination of α and β subunits, whereas several of the α subunits each are capable of forming functional homomeric nicotinic receptor/channels. Once again we see that there is the potential for enormous diversity generated both by multiple gene products and the formation of heteromultimeric channels, and this diversity is reflected in the pharmacological and biophysical properties of both native and heterologously expressed neuronal nicotinic receptors.

The GABA$_A$ receptor. Among the other ligand-gated ion channel systems that have been well characterized is the mammalian brain GABA$_A$ receptor. In fact, there are at least three classes of receptors for GABA: the GABA$_B$ receptor, which is not directly coupled to its ion channel; the GABA$_C$ receptor, which appears to be restricted to visual pathways; and the GABA$_A$ receptor, on which we will focus here. This receptor was first purified using specific affinity reagents. Its subunit structure resembles that of the nicotinic acetylcholine receptor. The inhibitory action of GABA results from activation of a chloride channel that is directly coupled to the GABA$_A$ receptor (Fig. 11–8a; contrast this with the nicotinic receptors, which are coupled to cation channels that allow sodium and potassium to flow). In the previous chapter we referred to the fact that the receptor for GABA has a rich pharmacology, in that it is the site of action of a number of clinically important drugs, including the benzodiazepines such as *Valium* and *Librium*, which are used widely as antianxiety and relaxant drugs, and the barbiturates, which are important anticonvulsants and sedatives. These agents produce their clinical consequences by enhancing the effect of GABA on the chloride channel that is coupled directly to the receptor (Fig. 11–8b). GABA$_C$ receptors are also coupled to chloride channels, but do not respond to barbiturates and have a pharmacological profile distinct from that of GABA$_A$ receptors.

The GABA$_A$ receptor was purified to homogeneity by isolating proteins that bind to benzodiazepines through the technique of affinity chromatography. This purifies a complex with most of the pharmacological binding sites of the GABA$_A$ receptor in native neuronal membranes. The purified complex contains α (MW 53,000) and β (MW 57,000) subunits with the stoichiometry α$_2$-β$_2$, as well as other subunits. Protein sequence information was used to clone cDNAs encoding the α and β subunits, and heterologous expression of these two subunits together produces GABA-activated chloride channels with most, but not all, of the properties of native GABA$_A$ receptors. Homology screening, using sequences from the cloned α and β subunits, resulted in the cloning of multiple subtypes of α and β subunits, as well as distinct γ, δ, and other subunits. The benzodiazepine binding site is only present in recombinant GABA$_A$ receptors when the γ subunit is expressed together with α and β, indicating that the native receptor must contain at least three different kinds of subunits.

The combination of multiple subunit genes (some 19), alternative splicing of messenger RNA, and the requirement for a heteromultimeric complex together gives rise to impressive diversity in the family of GABA$_A$ receptors. Particularly striking is the similarity of GABA$_A$ and nicotinic acetylcholine receptors in their overall structural features. The number and distribution of transmembrane domains are the same for all the subunits of the GABA$_A$ and nicotinic receptors, there is significant sequence homology in certain regions of the subunits, and the functional receptor/channel in both cases is a pentamer consisting of several different kinds of subunits.

The glycine and 5-HT$_3$ receptors. In some parts of the central nervous system, in particular in the brainstem and spinal cord, glycine, rather than GABA, is the major inhibitory neurotransmitter. At these sites, glycine functions just like GABA, by activating a receptor linked to a chloride channel (Fig. 11–8a). The convulsant drug strychnine blocks inhibition by selectively antagonizing glycine-mediated (but not GABA-mediated) inhibition, and strychnine has been used for assay and affinity purification of the glycine receptor. The receptor complex consists of α and β subunits, with molecular weights of about 48,000 and 58,000, respectively; cross-linking studies suggest an overall pentameric structure with a molecular weight of about 260,000. Both the α and β subunits share substantial sequence homology and predicted structure with those for the corresponding subunits of the nicotinic and GABA$_A$ receptors, particularly in

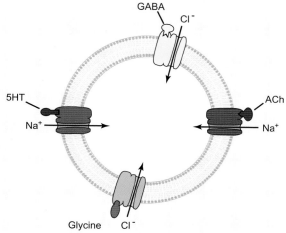

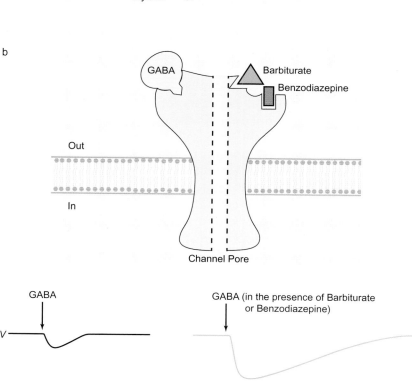

Figure 11–8. The GABA$_A$, glycine, and 5HT$_3$ receptors. *a*: The GABA$_A$ and glycine receptors are intimately coupled to channels selective for chloride ions. In contrast, the nicotinic acetylcholine (ACh) and 5HT$_3$ receptors are coupled to cation channels. *b*: The GABA$_A$ receptor–channel complex has binding sites for barbiturates and benzodiazepines, which can markedly enhance the hyperpolarizing response to GABA.

the putative membrane-spanning segments. Multiple genes and alternative splicing contribute to glycine receptor heterogeneity in the brain and spinal cord.

Serotonin (5-HT) is another neurotransmitter for which multiple classes of receptors exist. One of these classes, the 5-HT$_3$ receptor, falls into the family of ligand-gated ion channels by virtue of its rapid onset and rapidly reversible electrical response, and its overall structural homology with other family members. Like the nicotinic receptor/channels, 5-HT$_3$ receptors are directly coupled to a cation channel (Fig. 11–8a). A remarkable experiment has confirmed the close relationship among the various ligand-gated receptor/channels and revealed that discrete structural domains of these proteins can function as independent functional units. A chimeric receptor–channel complex, formed from the amino-terminal domain of the α7 subunit of the neuronal nicotinic receptor and the transmembrane region and carboxyl-terminal domain of the 5-HT$_3$ receptor, forms a functional receptor–channel complex that is activated by acetylcholine and other nicotinic agonists but has the channel properties of 5-HT$_3$ channels.

The glutamate receptors. The receptors for glutamate are of particular interest, in part because glutamate is the major excitatory neurotransmitter in the mammalian brain, but also because glutamate receptors are widely believed to play an important role in some kinds of learning and memory (see Chapter 19). At least four classes of glutamate receptor are known, based on amino acid sequence, agonist pharmacology, and transduction mechanism (Table 11–3). One of these classes transduces its signal via an intracellular second messenger cascade; this *metabotropic* glutamate receptor will be discussed in the next chapter. The remaining three classes consist of ligand-gated ion channels, or *ionotropic* receptors. They are named after the agonists that activate them most effectively: (1) kainic acid (KA), (2) α-amino-3-hydroxy-5-methyl-4 -isoxazole propionic acid (AMPA), and (3) N-methyl-D-aspartic acid (NMDA) (Fig. 11–9). The KA and AMPA receptors are closely related, whereas the NMDA receptors are distinct both functionally and structurally.

An essential difference lies in the properties of the different ion channels activated by the KA and AMPA receptors on the one hand and the NMDA receptors on the other (Fig. 11–10). The KA and AMPA receptors activate cation channels that allow sodium and potassium ions to flow. Near a neuron's resting potential the driving force for potassium is low and that for sodium is high, so activation of these channels leads to depolarization as a result of an inward sodium current (Fig. 11–10a). In

Table 11–3 Four Kinds of Glutamate Receptor

Receptor Subtype	Cloned Subunits	Agonists
AMPA	GluR1–GluR4	AMPA, quisqualate
KA	GluR5–GluR7	KA, quisqualate
	KA1–KA2	
NMDA	NMDAR1	NMDA
	NMDAR2A–NMDAR2D	
Metabotropic	mGluR1–mGluR6	trans-ACPD, quisqualate

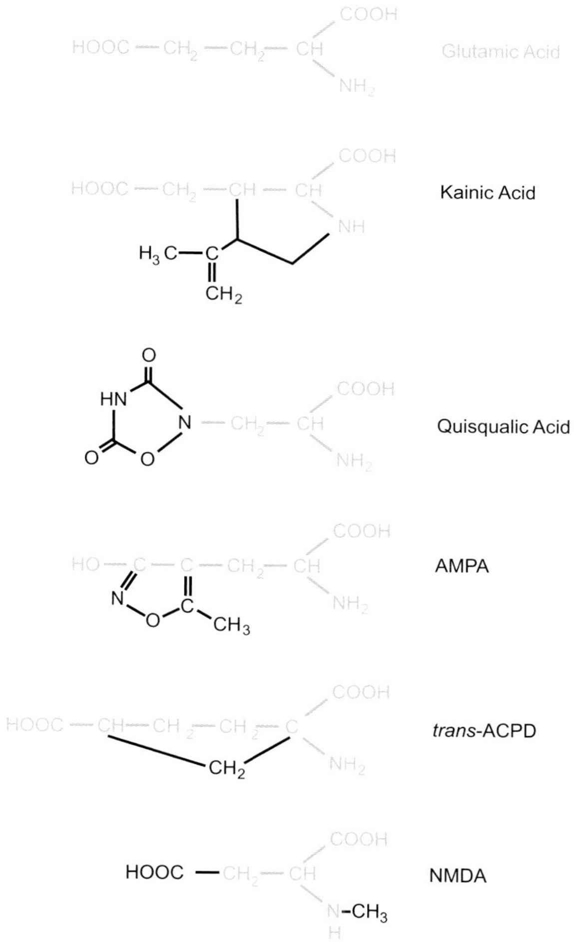

Figure 11–9. Ligands for glutamate receptors. The portion of the ligand molecule that resembles glutamate is shown in blue.

contrast, NMDA receptors activate cation channels that allow not only sodium and potassium but also calcium ions to flow (Fig. 11–10b). It might be thought that activation of these channels would also cause a depolarization that would simply add to that produced by the KA and AMPA receptor channels. Near the resting potential, however, this does not occur because of an interesting property of the NMDA receptor channels—they are blocked by extracellular magnesium ions in a voltage-dependent manner.

The way the NMDA receptor/channel works is also illustrated in Figure 11–10. When a neuron is near its resting potential, magnesium ions bind to the outside of the channel and effectively prevent the movement of other ions through the pore (Fig. 11–10a). When the cell is depolarized, however, the magnesium ions are driven out of the channel, allowing the other ions free access (Fig. 11–10b). The amount of current passing through NMDA receptor channels is therefore much greater at depolarized than at hyperpolarized membrane potentials. Thus, when the cell is depolarized in the presence of glutamate, calcium (as well as sodium) flows into the cell through the NMDA receptor channels (Fig. 11–10b).

The presence of several distinct receptor types coupled to different ion channels allows the target cell to respond differently to different intensities of synaptic stimulation. A moderate stimulus produces only a small depolarization, and no calcium entry, as a result of activation only of the KA/AMPA receptor channels. Although glutamate binds to the NMDA receptors, no current flows through the NMDA channels under these conditions. With more intense stimulation the depolarization becomes sufficient to relieve the magnesium block of the NMDA receptor channels, resulting in further depolarization and calcium entry.

Another important difference between the KA/AMPA and NMDA receptors is that the activity of the NMDA receptor is influenced by glycine, acting at a site distinct from that at which glutamate binds. In fact, it is thought that glycine is necessary for the activation of NMDA receptors, making it a *co-agonist* with glutamate. The extent to which regulated release of glycine occurs in higher brain regions (other than the brainstem and spinal cord) is not known, but clearly the involvement of glycine provides yet another means for fine-tuning the activity of NMDA receptors.

The presence of the three pharmacologically and functionally distinct ionotropic glutamate receptor subtypes also allows a postsynaptic cell to respond differently to presynaptically released glutamate, depending on whether the target cell is actively firing action potentials. A moderate synaptic stimulus to a silent cell will activate AMPA and KA channels, but will not produce sufficient depolarization to activate the NMDA channels. In contrast, the same stimulus to an active postsynaptic neuron,

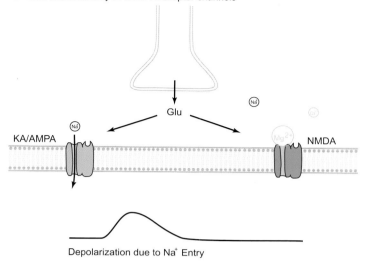

a Weak stimulus activates only KA/AMPA receptor channels

Glu

KA/AMPA

NMDA

Depolarization due to Na$^+$ Entry

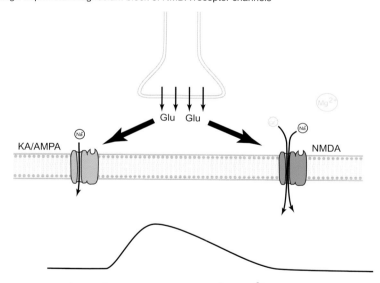

b Strong stimulus depolarizes sufficiently to relieve
voltage-dependent magnesium block of NMDA receptor channels

Glu   Glu

KA/AMPA

NMDA

Larger depolarization with both Na$^+$ and Ca^{2+} entry

Figure 11–10. Different kinds of responses mediated by the glutamate (Glu) receptor subtypes. *a*: The response to a weak presynaptic stimulus is activation of only the KA/AMPA receptor subtypes. *b*: In contrast, a stronger stimulus causes ions to flow through both the KA/AMPA and NMDA receptor channels. Voltage-dependent Mg^{2+} block of the NMDA receptor channel was first demonstrated by Philippe Ascher and colleagues (see Nowak et al., 1984) and by Mark Mayer and Gary Westbrook (see Mayer et al., 1984). See Animation 11–1. ◖

whose membrane is relatively depolarized by the action potentials, allows calcium to enter through the NMDA receptor/channels because their magnesium block has been relieved by the depolarization. Similarly, glycine, perhaps released by other active neurons, will contribute to the regulation of NMDA receptor activity and, hence, of calcium entry. As we shall see in Chapters 17 and 19, it is believed that the calcium that enters through the NMDA receptor channels at glutamatergic synapses may contribute to long-lasting changes in the properties of the postsynaptic neurons.

Molecular structures of the glutamate receptors. The glutamate receptors initially were difficult to clone. The lack of a rich source or suitable ligands made purification of the receptor protein difficult, and no mutations were available in appropriate organisms for positional cloning, as was the case for potassium channels (see Chapter 5). The glutamate receptors finally fell to the cloning strategy called *expression cloning*. Once these initial clones were available, further cloning by sequence homology uncovered an enormous number of related subunits (Table 11–3). &

Hydrophobicity plots predicted four membrane-spanning segments in the subunits of the ionotropic glutamate receptors, but these subunits are about twice as large as those of the nicotinic, $GABA_A$, glycine, and $5\text{-}HT_3$ receptors, and clearly form a separate family. This conclusion was confirmed by the three-dimensional structure, elucidated by X-ray crystallography, of a mammalian AMPA receptor formed from a single kind of subunit (Fig. 11–11). The receptor is a homotetramer, containing four subunits arranged around an overall axis of two-fold symmetry. Interestingly, most of the receptor is located on the extracellular side of the plasma membrane. Heterotetrameric receptors also exist, and the number of combinatorial possibilities is, as we have seen for other receptors and ion channels, enormous. It seems likely that KA and NMDA receptors share the general features of this structure as well.

Molecular architecture of the postsynaptic membrane. How are the various glutamate receptors distributed and organized in the postsynaptic membrane of excitatory synapses in the central nervous system? Clues have emerged from the identification of synaptic proteins that associate tightly with one or more glutamate receptor subtypes, often via highly specific protein–protein interaction domains. One such domain that is critical for the molecular organization of both the presynaptic terminal and the postsynaptic membrane is called the *PDZ domain*, because it was

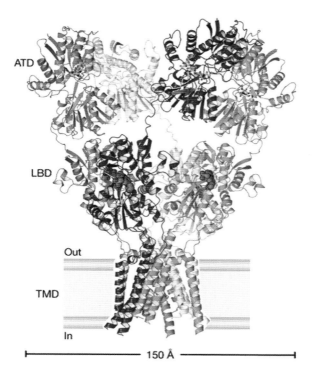

ATD

LBD

Out

TMD

In

├──────── 150 Å ────────┤

Figure 11–11. Crystal structure of a homomeric AMPA receptor (see Sobolevsky et al. 2009, and Wollmuth and Traynelis, 2009).

first described in three related proteins with strikingly uninformative names: PSD-95, DLG, and Z01. The domain is about 80–100 amino acids long and is defined by the presence of multiple repeats of the sequence Gly-Leu-Gly-Phe within this stretch; it binds with nanomolar affinity to short peptide sequences in target proteins. Many synaptic proteins contain within their sequences multiple PDZ domains as well as other functional domains (Fig. 11–12), each of which may interact with a different target protein to help build a macromolecular complex.

How does one identify the proteins in such a complex? The *yeast two-hybrid screen*, a powerful technique that takes advantage of yeast genetics to identify proteins that interact with other proteins, has proven extremely useful in this context. This approach has been used extensively with glutamate receptor channels and other synaptic proteins, and has provided evidence for a multicomponent protein scaffold (some components of which possess colorful names such as PICK, CASK, GRIP, GRASP, SHANK, and HOMER) that organizes the receptors and links them to the cytoskeleton and to various intracellular signaling pathways

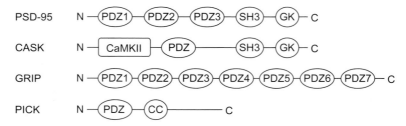

Figure 11–12. Molecular organization of some PDZ domain–containing proteins. Some proteins that participate in the organization of the postsynaptic membrane (see Figure 11–13) may contain several distinct PDZ domains with different amino acid sequences.

as well as to each other (Fig. 11–13; compare with the pictures of the postsynaptic membranes in Figs. 17–8 and 17–18). It is evident from studies of this sort that signaling at the synapse must involve the tightly coordinated activities of an astonishingly large complement of molecular components. ❧

RNA editing determines glutamate receptor/channel properties. An unusual and striking finding for several of the AMPA receptor/channel subunits was that the sequence of their cDNA differs from that of their genomic DNA in the region of the pore domain. This is not due to an error in transcribing the genomic DNA into messenger RNA (from

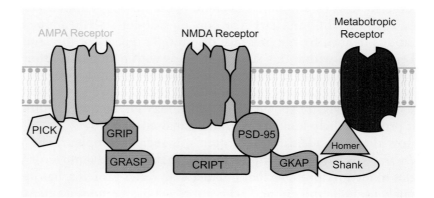

Figure 11–13. Molecular organization of the postsynaptic membrane at glutamatergic synapses. Protein–protein interactions, many of them identified by the yeast two-hybrid screen, link different kinds of glutamate receptors to one another and anchor them to the cytoskeleton. Compare with the organization of the neuromuscular junction illustrated in Figure 17–18.

which the cDNA is derived). Rather, the DNA is transcribed faithfully, and the sequence difference arises from *RNA editing*—specifically in this case, the enzymatic deamidation of a specific adenosine to inosine in the messenger RNA. The consequence of this editing is that a codon (CAG) that formerly coded for a glutamine residue in the pore domain is changed to one (CIG) that codes for arginine, and this results in a dramatic change in the conduction properties of the channel. RNA editing has also been discovered in the messenger RNAs for some kinds of voltage-gated ion channels. Regulated RNA editing provides yet another way of generating diversity in the voltage-dependent and ligand-gated ion channels.

Summary

The question of how cells respond to signals from their environment is relevant to all aspects of cell biology. Thus it will come as no surprise that many of the signaling molecules discussed in Chapter 10 are not restricted to the nervous system but can act on many cell types. The common feature that allows neurons and other cells to respond to these extracellular signals is the presence of specific receptors in the plasma membrane that may be very similar in their structure and functional properties in different kinds of cells. It has been known for a long time from pharmacological studies that there may be several different kinds of receptors for individual neurotransmitters. This has important implications for the way information is processed in neuronal networks. This heterogeneity has been confirmed as the structures of many receptors have been elucidated. A gratifying picture has emerged from these structural studies: Many receptors can be grouped into families based on structural, functional, and regulatory homologies that are far more extensive than had been appreciated previously.

Perhaps surprisingly, the family groupings reflect common receptor transduction mechanisms rather than common ligand-binding sites. One such family is that of the ligand-gated ion channels. The members of this family were first linked on the basis of a functional criterion—direct coupling between the receptor and the ion channel whose activity it regulates. Biochemical and molecular studies, as well as genome sequencing and structural determination, provide a basis for dividing these receptors into two subfamilies. One subfamily comprises the tetrameric KA/AMPA and NMDA receptors for the transmitter glutamate, while the other encompasses the pentameric ligand-gated ion channels for other classical transmitters, including acetylcholine, GABA, glycine and serotonin. The sequence homologies and remarkably similar predicted arrangement of transmembrane segments suggest that the various subunits of the different receptors within a subfamily may have evolved from a single ancestral subunit. Evolution has allowed the ligand specificity of the receptor site and the ion selectivity of the channel pore to diverge. However, the essential overall structural design of the ligand-gated receptor–channel complex has been preserved. In the next chapter we shall consider the molecular details of other, more intricate, signal transduction pathways.

12

Receptors and Transduction Mechanisms II: Indirectly Coupled Receptor/Ion Channel Systems

The recognition of extracellular signals by specific receptors on the target cell is not the final step in intercellular communication. Cells must also possess mechanisms to *transduce* the extracellular signal, to convert it into some biological response that is characteristic of the particular target cell. In neurons, the biological response is often the modulation of the properties of one or more membrane ion channels. We have seen that in the ligand-gated ion channel family, the tasks of recognition and transduction reside in a single protein complex (Fig. 11–1). Other families of receptors alter neuronal excitability through more complicated changes in the biochemistry of the cell (Fig. 12–1).

G Protein–Coupled Receptor/Ion Channel Systems

Most neurotransmitter receptors are not coupled directly to the ion channel whose activity they regulate. Among the indirectly coupled receptor/channel systems is a large family coupled via *guanyl nucleotide-binding proteins*, or *G proteins*. From an examination of Table 12–1 it is apparent at first glance that the various G protein–coupled receptors (GPCRs) have little in common. This receptor family includes several peptide receptors, the muscarinic acetylcholine receptor subtypes, and receptors for most of the major classes

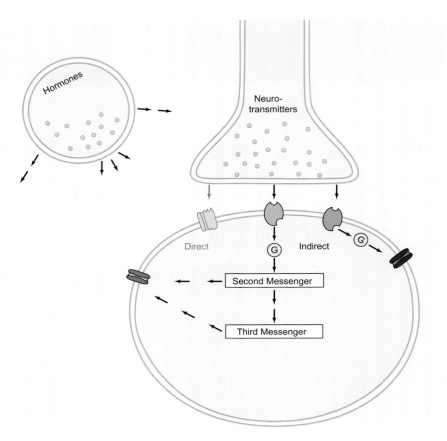

Figure 12–1. Intercellular communication. This chapter will deal with transduction mechanisms that are not mediated via direct receptor channel coupling, but rather involve either protein–protein interactions in the plane of the plasma membrane or intracellular messengers (color). Guanyl nucleotide-binding proteins (G) play an essential role in these transduction mechanisms.

of biogenic amines. It even contains, in addition to other sensory receptors, the visual pigment rhodopsin, which does not respond to a signaling molecule but can be thought of as a "receptor" for light (see Chapter 13). Note that cell surface receptors in yeast and in the slime mold *Dictyostelium*, cell types far removed from neurons, are included in the list, emphasizing the commonality of mechanisms in neurons and other kinds of cells.

Why would we possibly want to gather these diverse receptors in the same list? As in the example of the directly coupled receptor systems described in the last chapter, many of these are linked on the basis of a functional criterion—in this case, transduction of the extracellular signal via a G protein. As we will see later in this chapter, many receptors use

Table 12–1 A Few of the Many Receptors That Use
G Proteins to Transduce Extracellular Signals

Substance K receptor

Yeast mating factor receptor

Slime mold cyclic AMP receptor

Luteinizing hormone-releasing hormone receptor

Muscarinic acetylcholine receptors

Adrenergic receptors

Purinergic receptor

Metabotropic glutamate receptors

Endocannabinoid receptors

Rhodopsin

Olfactory receptors

G proteins to activate intracellular enzymes that produce *second messengers*. These are diffusible molecules that may influence a variety of cellular constituents, including ion channels. It is now realized that G proteins also interact directly with some ion channels to regulate their activities (Fig. 12–2), without the mediation of a second messenger. Such modulation of ion channel properties, via protein–protein interaction in the plane of the membrane, may be slower in onset and somewhat longer lasting than that mediated by the intramolecular conformational changes in the directly coupled systems (compare with Fig. 11–7).

Receptors coupled to G proteins belong to a family. Molecular cloning and genome sequencing have made it clear that this grouping of receptors in Table 12–1 is indeed appropriate. The first members of the family to be sequenced were the opsin visual pigments. The most striking feature in the sequence was the presence of seven stretches, each of 20–28 hydrophobic amino acids, which were interpreted to represent membrane-spanning domains. This, of course, is very different from the four membrane-spanning regions characteristic of the directly coupled receptors (see Fig. 11–6). When one of the mammalian β-adrenergic receptors was subsequently cloned and sequenced, it was also found to have seven hydrophobic membrane-spanning domains and substantial amino acid homology with bovine opsin. This is also the case for the α-adrenergic receptors, although they differ markedly from the β receptors in their binding site

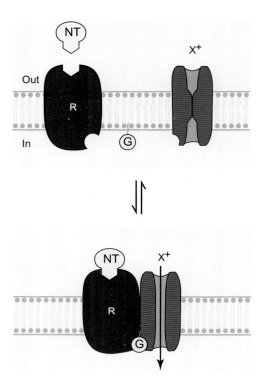

Figure 12–2. G protein–mediated receptor–channel coupling. Not all neuro-transmitter receptors are intimately associated with an ion channel as in Figure 11–8. In the case shown here, binding of neurotransmitter (NT) to its receptor (R) activates a G protein (G), which then interacts with the ion channel, causing it to open.

pharmacology. We have already pointed out in Chapter 11 that the various muscarinic acetylcholine receptor subtypes also exhibit these structural features (see Fig. 11–6).

This pattern is very consistent. Receptors whose transduction mechanism is not understood from physiological or biochemical experiments can be assigned with confidence to the GPCR family solely on the basis of their amino acid sequences. A good example of this is the receptor for substance K, one of several related peptides called the *tachykinins*. Among the biological actions of these peptides are important effects on sensory processing. Before its amino acid sequence was determined, the transduction mechanism for the actions of substance K was not known. However, the sequence showed the presence of seven membrane-spanning domains. This allowed the prediction that the receptor is coupled to a G protein. Similarly, the metabotropic glutamate receptors, which mediate slow responses to glutamate and are not directly coupled to ion channels like their ionotropic cousins described in Chapter 11, clearly fall into the family of GPCRs (Table 12–1) on the basis of their molecular structures.

Genome sequencing has revealed that an astonishing fraction of human genes, in the range of 4% of the total, encode GPCRs. Of these thousand or so genes, well over half are devoted to the very large family of olfactory receptors (see Chapter 13); nevertheless, there are several hundred GPCRs that bind neurotransmitters and hormones. Indeed, the ligands for some of them—the so-called *orphan* receptors—have not yet been identified, but it is evident from the amino acid sequence that, whatever the ligand, its physiological action will be mediated by a G protein.

How do G proteins work? To discuss G protein modulation of ion channels, we must first consider the structure of G proteins and some aspects of the molecular mechanism by which they act. G proteins are heterotrimers consisting of one each of α, β, and γ subunits (Fig. 12–3), and a large number of structurally and functionally distinct G proteins are known to exist. Molecular cloning and genome sequencing have been used to define a number of subfamilies of α subunits, and multiple β and γ subunits also exist. Although the β and γ subunits can be dissociated from one another under denaturing conditions in a test tube, they remain together as a functional βγ complex under physiological conditions. The βγ complex was originally thought to subserve only the important but relatively humdrum function of anchoring the G protein to the plasma membrane, but as we shall see later in this chapter, they do far more than that. The membrane anchoring is achieved via a post-translational modification that attaches an isoprenoid lipid group covalently to a cysteine residue at or near the carboxyl terminal of the γ subunit. The lipid moiety is extremely hydrophobic and interacts tightly with the plasma membrane (Fig. 12–3).

All of the known heterotrimeric G proteins operate via the same general mechanism illustrated in Figure 12–3. In its inactive state, the G protein has GDP bound to a specific guanyl nucleotide-binding site on the α subunit. As it moves in the plane of the membrane, the G protein occasionally bumps into an appropriate agonist-occupied membrane receptor (or, in the case of the photoreceptor G protein transducin, a light-activated rhodopsin molecule). This receptor–G protein interaction allows GTP to

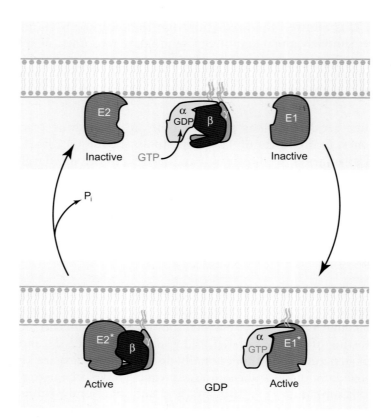

Figure 12–3. Mechanism of action of G proteins. G proteins are heterotrimers consisting of one each of an α, β, and γ subunit. They are anchored to the plasma membrane by a lipid group that is attached covalently to the γ subunit. In the presence of GTP, the α subunit dissociates from βγ, and both α and βγ can influence the activities of other proteins (E₁ and E₂). A GTPase activity intrinsic to the α subunit completes the cycle (see review by Oldham and Hamm, 2008).

displace GDP from the guanyl nucleotide-binding site. The displacement is accompanied by the dissociation of the α subunit from the βγ complex, and both α and βγ can then interact with and activate effector systems. The effector system might be an ion channel as in Figure 12–2, or some enzyme (denoted by E₁ and E₂ in Fig. 12–3). The α subunit carries an intrinsic GTPase activity, and so after some time, the bound GTP is hydrolyzed to GDP. Following this hydrolysis the α subunit recombines with βγ, allowing the cycle to begin again (Fig. 12–3).

A variety of biochemical probes have been instrumental in proving this sequence of events and in identifying and analyzing biological responses mediated by G proteins. These include several GTP analogs such as GTPγS and Gpp(NH)p (guanylylimidodiphosphate), which displace GDP from the guanyl nucleotide-binding site but are not hydrolyzed by the GTPase. These compounds act as essentially irreversible activators of the G proteins. In contrast, the GDP analog GDPβS, which binds very strongly to the binding site, maintains the G protein in the inactive GDP state, thus inhibiting G protein–mediated responses. Two bacterial toxins have also been extremely useful. *Cholera toxin* irreversibly activates some G proteins, by causing ADP-ribose to be covalently attached to an arginine residue on the α subunit. Pertussis toxin irreversibly inhibits other G proteins, by catalyzing the same reaction with their α subunits.

Some ion channels are modulated directly by G proteins. As we shall see later in this chapter, second messengers whose synthesis is controlled by G proteins influence the activity of many ion channels. Such indirect G protein involvement in ion channel modulation appears to be a rather ubiquitous phenomenon. Some ion channels, however, may interact directly with G proteins. That is, the effector system (E_1 and E_2 in Fig. 12–3) is not an enzyme but is the ion channel protein itself (Fig. 12–2). We shall illustrate this mechanism by considering the inwardly rectifying potassium channel in the heart, whose activity is increased by acetylcholine acting at muscarinic receptors. This K_{ACh} channel is responsible for the slowing of the heart on stimulation of the vagus nerve; thus this story is of historical interest (see the Otto Loewi experiment described in Chapter 1), as well as being the first and most thoroughly characterized example of this kind of channel modulation.

A variety of evidence indicates that there is no second messenger involvement in the muscarinic activation of K_{ACh}. However, a role for a G protein was suggested by the finding that potassium current can be activated only when GTP and GTP analogs are added to the electrode in the whole-cell recording configuration (Fig. 12–4a). Furthermore, the channel can be activated in detached membrane patches by GTP applied to the cytoplasmic membrane surface (Fig. 12–4b). In addition, pretreatment with pertussis toxin prevents the activation of the potassium current by muscarinic agonists.

Purified G proteins that have been activated by prior treatment with GTP analogs can be applied directly to the cytoplasmic membrane surface of detached patches containing K_{ACh} channels (Fig. 12–5). This treatment

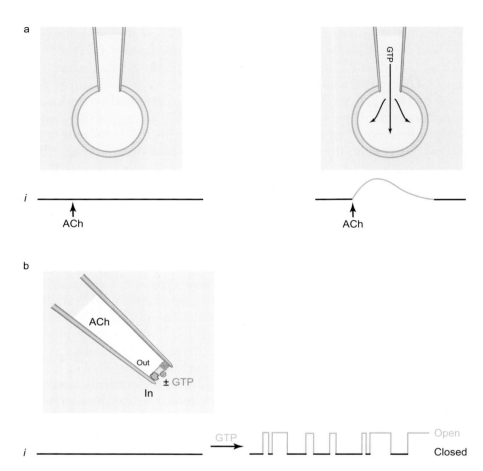

Figure 12–4. A G protein is involved in activation of a cardiac potassium channel by acetylcholine. Acetylcholine (ACh) binds to a muscarinic receptor and activates a potassium channel in cardiac muscle cells. This response, whether measured by whole-cell recording (*a*) or in a detached membrane patch (*b*), requires GTP (see Breitwieser and Szabo, 1985; Pfaffinger et al., 1985).

activates the channel, confirming a direct interaction of the G protein with the channel or with some unknown regulatory protein that is very closely associated with the channel. In an important extension of these experiments, isolated α subunits or $\beta\gamma$ complexes were applied to the inside of the patch, and it was found that $\beta\gamma$ can modulate channel activity. This was the first indication that $\beta\gamma$ (as well as α) can directly influence effector systems.

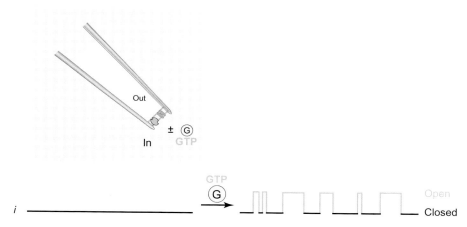

Figure 12–5. Purified G proteins (G) can activate the cardiac muscarinic potassium channel. In detached membrane patches in which the endogenous G proteins have been previously inactivated by treatment with pertussis toxin, addition of an activated G protein can produce gating of the potassium channel (see Logothetis et al., 1987).

Inwardly rectifying potassium channels can be regulated by G proteins. The inwardly rectifying potassium channels like the K_{ACh} channel in the heart remained uncloned during the extensive homology screenings that followed the initial characterization of *Shaker*. Like the glutamate receptor/channels described in Chapter 11, they finally fell to an expression cloning approach, and subsequent homology screens and genome sequencing have uncovered a large family of inward rectifiers that clearly is distinct from the other families of voltage-dependent potassium channels. The structure of the inward rectifier potassium channels is interesting (see Fig. 7–9). They contain two putative membrane-spanning segments, flanking a pore domain that is very similar in sequence to that of the voltage-dependent potassium channels. Interestingly, another of the inwardly rectifying potassium channels, whose activity is inhibited by intracellular ATP (a K_{ATP} channel), appears to be modulated by G protein α subunits rather than by βγ. In view of the sequence similarities between K_{ACh} and K_{ATP} channels, it appears that only subtle structural differences are sufficient to alter the specificity for G protein subunits. A crystal structure of an inwardly rectifying potassium channel in complex with βγ has revealed at atomic resolution the intramolecular protein rearrangements involved in channel gating.

Second Messenger–Coupled Receptor/Ion Channel Systems

The third category of receptor–channel coupling that we shall discuss is channel modulation via intracellular second messengers. In this case, the receptor and channel are not part of a single macromolecular complex (Fig. 11–7), nor do they interact directly in the plane of the membrane (Fig. 12–2). Rather, the receptor is coupled, via a G protein, to an enzyme, the activity of which is regulated by the occupation of the receptor by a neurotransmitter (Fig. 12–6). When the receptor is occupied by transmitter (the so-called first messenger), the enzyme is activated and can catalyze the formation of an intracellular second messenger (Fig. 12–6). The second messenger can then set in motion a sequence of events that ultimately influences the properties of, or *modulates*, an ion channel. Note that the channel properties do not immediately revert to the normal resting state when the transmitter dissociates from the receptor, as is the case for the directly coupled systems. Instead, the channel modulation persists (in the example in Figure 12–6, the channel remains open) until the actions of the second messenger are reversed. This may involve a sequence of molecular events, requiring seconds, minutes, or even hours. Thus second messenger coupling provides a mechanism for ion channel modulation that may be delayed in onset. In addition, the modulation may long outlast the initial stimulus, the occupation of receptor by neurotransmitter.

Several other features of this general scheme deserve mention. First, we have already mentioned that most second messenger pathways involve G protein–activated enzymes. Accordingly, second messenger–mediated receptor–channel coupling might be thought of as just a special case of G protein coupling, but in fact, the temporal and molecular complexities introduced by having a second messenger mediate the response justify a separate category for this coupling mechanism. A second important feature is that amplification can occur at each step of what may be a multistep second messenger cascade (Fig. 12–7), so that the activity of many target ion channels may be modulated by the occupation of a relatively small number of receptors. A corollary of this is that activation of a second messenger system may produce coordinated changes in the activity of more than one kind of ion channel in the cell membrane (Fig. 12–7). Furthermore, while tuning the activity of ion channels, a second messenger can also influence other cellular processes that have nothing to do with ion fluxes. Clearly, these features are not generally characteristic of directly coupled systems.

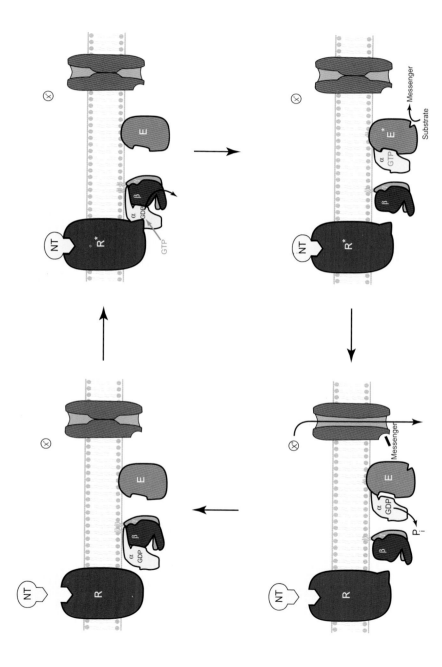

Figure 12–6. Second messenger–mediated receptor–channel coupling. In some cases neither the neurotransmitter (NT) receptor (R) nor the G protein interacts directly with the ion channel. In these cases, an intracellular second messenger influences ion channel activity. Contrast with Figures 11–7 and 12–2. E, enzyme regulated by G protein that produces a second messenger; X$^+$, ion that can flow through the open ion channel; R*, receptor activated by bound NT.

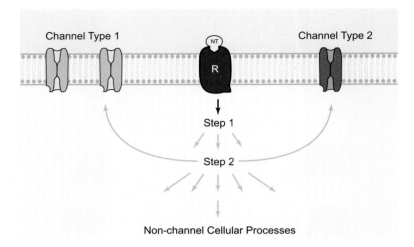

Figure 12–7. Amplification in second messenger cascades. With second messenger coupling the binding of a single neurotransmitter (NT) can activate many channels of a given class, activate several classes of channels, and affect other cellular processes not associated with ion channels.

Different Kinds of Second Messengers

A comprehensive treatment of the vast field comprising second messenger systems is far beyond the scope of this book; indeed, whole books have been written about individual second messengers Therefore, at the risk of omitting some that are favorites of our readers, we will confine ourselves in this summary to a few intracellular signaling pathways that are known to play important roles in the modulation of neuronal excitability. We will first present an overview of the biochemistry of second messenger production and mechanism of action, and then will discuss some specific experimental examples that illustrate the role of these pathways in neuromodulation.

The specific pathways that we will discuss here include

1. the adenylate cyclase/cyclic AMP–dependent protein kinase system; and
2. the phospholipase-mediated turnover of membrane phospholipids, which can lead to the production of the three distinct second messengers: inositol-1,4,5-trisphosphate (IP_3), diacylglycerol (DAG), and arachidonic acid.

Before we discuss these, however, let us consider the general question of how changes in ion channel properties mediated by second messengers might be identified. Changes that last for a long time are, of course, strong candidates, but this criterion is not an unequivocal one. However, the cell-attached mode of the patch clamp technique does provide a convincing test for second messenger–mediated alteration of ion channel properties, independent of what the second messenger might be (Fig. 12–8).

The gigaohm seal between a patch electrode and the plasma membrane prevents the movement of neurotransmitter (and other) molecules

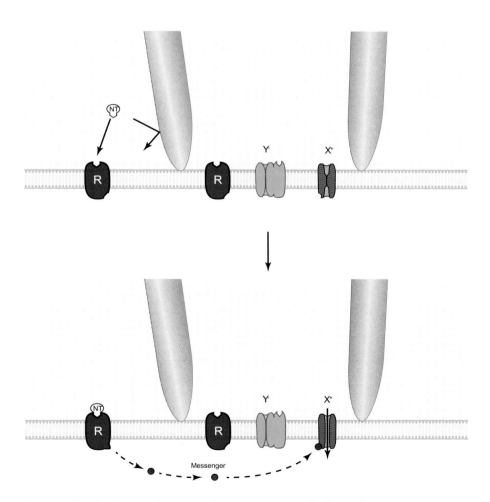

Figure 12–8. A test for second messenger mediation of neurotransmitter effects on ion channels. A neurotransmitter (NT) applied outside a patch electrode does not have access to receptors (R) within the patch. Accordingly, it can affect ion channels within the patch only by interacting with some receptor outside the patch. These receptors can communicate with channels inside the patch only by means of some diffusible intracellular messenger.

between the extracellular bathing medium and the inside of the electrode. Accordingly, transmitters placed in the bathing medium cannot have access to receptors within the patch electrode (Fig. 12–8, top). Therefore, directly coupled ion channels within the patch (green in Fig. 12–8) cannot be activated by a transmitter outside the patch electrode. If the transmitter activates ion channels within the patch, it must do so by binding to receptors outside the electrode. The only way these receptors can communicate with the channels in the patch is via some diffusible intracellular messenger (Fig. 12–8, bottom). An important feature of this general test for second messenger modulation is that it requires no a priori assumptions about the identity of the particular second messenger involved.

The adenylate cyclase/cyclic AMP–dependent protein kinase system. Cyclic AMP was first described by Earl Sutherland and colleagues as the second messenger mediating hormonally stimulated glycogen breakdown in the liver. A surprisingly long time elapsed before it was accepted as the intermediary in some neurotransmitter responses in nerve cells, but many such cyclic AMP–mediated responses have now been described. The major components of this system are illustrated in Figure 12–9a. Cyclic AMP is synthesized from ATP by the enzyme *adenylate cyclase*, which is coupled to a variety of different receptors via a specific G protein. The second messenger then activates the *cyclic AMP–dependent protein kinase*, which is a tetrameric complex of two each of two kinds of subunit. The holoenzyme complex is completely inactive. Binding of cyclic AMP to the *regulatory subunits* causes them to dissociate from the *catalytic subunits*. The free catalytic subunits are active (Fig. 12–9b) and can catalyze the transfer of the terminal phosphate from ATP to the hydroxyl groups of serine or threonine residues in the target protein to produce a *phosphoprotein*. This system is turned off by two kinds of enzymes: *phosphodiesterases*, which break down the cyclic AMP, and *phosphoprotein phosphatases*, which dephosphorylate the substrate proteins (Fig. 12–9a).

As we shall see later in this and subsequent chapters, phosphorylation (and dephosphorylation) of enzymes, ion channels, or other proteins can lead to large changes in their functional properties. The way in which a particular cell responds to an elevation of cyclic AMP will thus depend on its particular spectrum of cell-specific substrate proteins that can be phosphorylated by the cyclic AMP–dependent protein kinase. Some specific examples of different cyclic AMP–mediated modulations in nerve and muscle cells are presented in Chapters 18 and 19. It was thought for a long time that all actions of cyclic AMP in eukaryotic cells are mediated by the cyclic AMP–dependent protein kinase as just

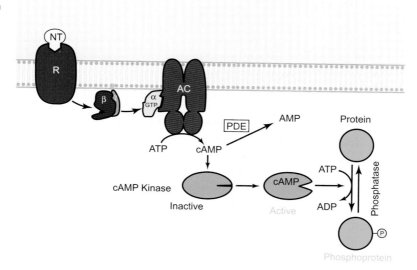

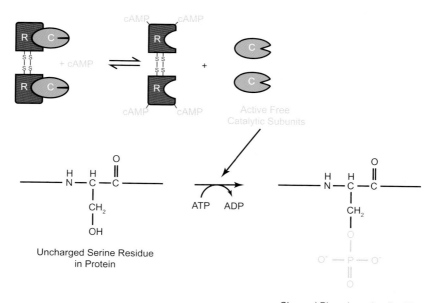

Figure 12–9. The adenylate cyclase/cyclic AMP–dependent protein kinase system. *a*: Cyclic AMP (cAMP) is synthesized from ATP by the enzyme adenylate cyclase (AC) and broken down by a phosphodiesterase (PDE). It activates a cyclic AMP–dependent protein kinase (cAMP kinase). NT, neurotransmitter; R, receptor; P, phosphate group. *b*: The cyclic AMP–dependent protein kinase is an inactive holoenzyme consisting of two regulatory (R) and two catalytic (C) subunits. Cyclic AMP binds to the regulatory subunits, releasing the catalytic subunits that are now active and can catalyze serine (or threonine) phosphorylation of target proteins.

described. However, cyclic AMP also interacts directly with some ion channels and activates them independently of kinase activation. This will be covered in Chapter 13, when we discuss sensory neurons in olfactory epithelia. ●

Turnover of membrane phospholipids. Lowell and Mabel Hokin first demonstrated as long ago as 1952 that acetylcholine and other neuro-transmitters and hormones can stimulate the breakdown and resynthesis (turnover) of the minor membrane phospholipid *phosphatidylinositol* (PI). Almost 25 years passed, however, before it was widely recognized that PI (and other phospholipids) might participate in signal transduction, and since then there has been explosive growth in this field. This may be contrasted with the cyclic nucleotides, which were discovered later but very quickly became established as important second messenger molecules. Why was the PI story so long in developing? It is interesting to speculate that it might have been because lipid biochemistry is so difficult. Only a small, brave band of biochemists was prepared to struggle with these complex, water-insoluble phospholipids, whereas the majority extracted and discarded the lipid from their preparations as quickly as possible so it would not interfere with their study of proteins. In any event, the latent period is well behind us, and it is evident that phospho-lipid metabolism is of fundamental importance in signal transduction in neurons as well as other kinds of cells.

A critical breakthrough came with the demonstration that the phos-pholipid species whose metabolism is stimulated by agonists is not PI itself but rather its doubly phosphorylated derivative, *phosphatidylinositol-4, 5-bisphosphate* (PIP$_2$). As shown in Figure 12–10a, many receptors are coupled (via a G protein) to the membrane enzyme *phospholipase C* (PLC). This acts as a phosphodiesterase to split PIP$_2$ into two products, the water-soluble *inositol-1,4,5-trisphosphate* (IP$_3$), and *diacylglycerol* (DAG), which remains in the membrane. *Arachidonic acid* (AA), which is usually the fatty acid present in the 2 position of the glycerol backbone of PIP$_2$ (see Fig. 12–10b), can be released from DAG through the action of another membrane phospholipase, *phospholipase A$_2$* (PLA$_2$), which may also release AA directly from PIP$_2$. All three of these products—IP$_3$, DAG, and AA—are important second messengers.

The mechanisms by which these three second messengers work are summarized in Figure 12–11. The IP$_3$ diffuses in the cytoplasm and binds to specific receptors on the endoplasmic reticulum (see later discussion). This binding opens a calcium channel in the endoplasmic reticulum mem-brane, and as a result, calcium ions are released into the cytoplasm from storage sites in the lumen of the endoplasmic reticulum (Fig. 12–11a).

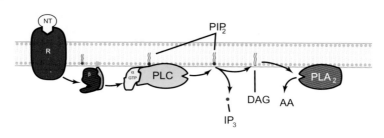

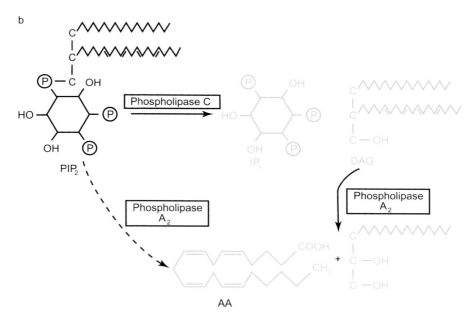

Figure 12–10. Three second messengers from polyphosphoinositides. *a*: Some neurotransmitter (NT) receptors (R) are coupled (via a heterotrimeric G protein) to the enzyme phospholipase C (PLC). *b*: The chemical structures of the second messenger molecules derived from the breakdown of the polyphosphoinositides (see reviews by Gamper and Shapiro, 2007, and Oldham and Hamm, 2008). AA, arachidonic acid; DAG, diacylglycerol; IP_3, inositol trisphosphate; PIP_2, phosphatidylinositol-4,5-bisphosphate; PLA_2, phospholipase A_2; P, phosphate group.

As we shall discuss later, the elevated cytoplasmic calcium can influence ion channel activity and a myriad of other cellular functions.

The DAG, in contrast, remains associated with the membrane. An increase in DAG results in a translocation from cytoplasm to membrane of *protein kinase C* (often called PKC), another member in the family of

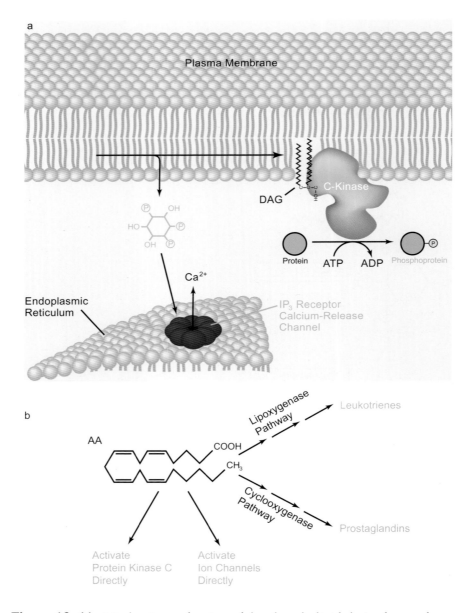

a

Plasma Membrane

DAG

C-Kinase

Protein   ATP   ADP   Phosphoprotein

Ca^{2+}

Endoplasmic
Reticulum

IP_3 Receptor
Calcium-Release
Channel

b

Leukotrienes

Lipoxygenase
Pathway

AA

COOH

CH_3

Cyclooxygenase
Pathway

Prostaglandins

Activate
Protein Kinase C
Directly

Activate
Ion Channels
Directly

Figure 12–11. Mechanisms of action of the phospholipid-derived second messengers. *a*: Diacylglycerol (DAG) activates a protein kinase called *protein kinase C* (PKC or C kinase). Inositol trisphosphate (IP_3) opens calcium channels in the endoplasmic reticulum (drawing modified from Berridge, 1997). *b*: Arachidonic acid (AA) may directly activate PKC or some ion channels. It can also be metabolized to the leukotrienes and the prostaglandins. P, phosphate group.

protein kinases that catalyze the phosphorylation of serine and threonine residues in substrate proteins (some of which are ion channels). Protein kinase C becomes active only when it associates with the membrane and binds DAG (Fig. 12–11a).

In the test tube, protein kinase C can be activated by the addition of DAG together with phosphatidylserine, the latter apparently serving as a substitute for the cell membrane. In addition, a series of compounds called *phorbol esters*, first characterized as tumor-promoting agents, are useful pharmacological tools that can substitute for DAG in activating protein kinase C, both in the test tube and when applied to intact cells. PIP_2 may not be the most important source of DAG; PIP_2 is only a minor component of the plasma membrane, and the amounts of DAG produced in response to an agonist are often too large to be accounted for solely by PIP_2 breakdown. However, some agonists activate a *phospholipase D* (again via a G protein), which catalyzes the release of phosphatidic acid from phosphatidylcholine, the major membrane phospholipid. The phosphatidic acid can then be metabolized further to DAG. This pathway provides a mechanism for producing DAG and activating protein kinase C without concomitant activation of the IP_3 calcium release system.

The actions of AA, the third second messenger product of phospholipid turnover, are less well understood. Arachidonic acid can be metabolized via several different pathways, each of which gives rise to biologically active products, including the prostaglandins and the leukotrienes (Fig. 12–11b). In addition, AA can act as an activator of one form of protein kinase C in the brain.

The complexity of lipid signaling is, in fact, even greater than indicated here. For example, a phosphatidylinositol-1,4,5-trisphosphate (PIP_3) can be produced in cell membranes via the action of an enzyme called *PI-3 kinase*, and this may also be involved in the modulation of membrane ion channels. In addition, receptor-mediated hydrolysis of a complex membrane lipid called *sphingomyelin* produces several products that may act as intracellular messengers. Undoubtedly, there are surprises still to be uncovered in this area.

The Endocannabinoids

The discovery and characterization of a cell surface receptor that binds to Δ^9-tetrahydrocannabinol, the psychoactive ingredient in marijuana, led to a search for endogenous cannabinoids that might interact with the receptor. Two such *endocannabinoids*, anandamide (from a Sanskrit word meaning "internal bliss") and 2-arachidonyl glycerol (2-AG), have been characterized in considerable detail. Both are complex lipid molecules related to arachidonic acid, which we have already discussed as one of the important intracellular second messengers derived from PIP_2 (see Fig. 12–10). 2-AG is synthesized from DAG by the enzyme diacylglycerol lipase-α, and among its metabolites is arachidonic acid. The synthesis of anandamide is more complex and less well understood, but it too can be metabolized to arachidonic acid. In addition to contributing to the production of a second messenger, the endocannibinoids themselves act as neurotransmitters. They play critical roles in such diverse phenomena as mood, memory, reward, and the regulation of metabolism. The CB1 and CB2 receptors that bind to and mediate the actions of the cannabinoids are G protein–coupled receptors that are expressed prominently on presynaptic nerve terminals. The activation of inhibitory G proteins by endocannabinoid-occupied CB1 and CB2 leads to a decrease in calcium entry into presynaptic endings and a consequent reduction in the release of neurotransmitters.

Down-regulation of G protein–coupled receptors. We have mentioned briefly that many neurotransmitter receptors are subject to *down-regulation*, also often called *desensitization*. These terms refer to a progressive decrease in the response to a neurotransmitter during maintained exposure or multiple exposures of the cell to the same transmitter. In some cases, desensitization results from receptor *internalization*, the removal of receptors from the plasma membrane by the pinching off and internalization of membrane vesicles that contain receptor molecules (Fig. 12–12a). This agonist-dependent process is best understood for certain growth factor receptors and may be important for regulating β-adrenergic receptors as well. Internalization may result in degradation of the receptors within the cell, although in at least some cases the internalized receptors can be recycled back to the plasma membrane (Fig. 12–12a).

Several members of the G protein–coupled receptor family can also be down-regulated by phosphorylation. Photoreceptors contain an enzyme called *rhodopsin kinase*, which specifically phosphorylates light-activated rhodopsin and decreases its sensitivity to light (Fig. 12–12b). The agonist-occupied β-adrenergic receptor can be phosphorylated and down-regulated by the cyclic AMP–dependent protein kinase, as well as by a more specific β-*adrenergic receptor kinase* (βARK; see Fig. 12–12b). There are a number of serine and threonine residues that are potential sites for phosphorylation near the carboxyl terminal of the G protein–coupled receptors (on the cytoplasmic side of the membrane). These residues and adjacent amino acid sequences are well conserved in other members of this receptor family, and some of them are also

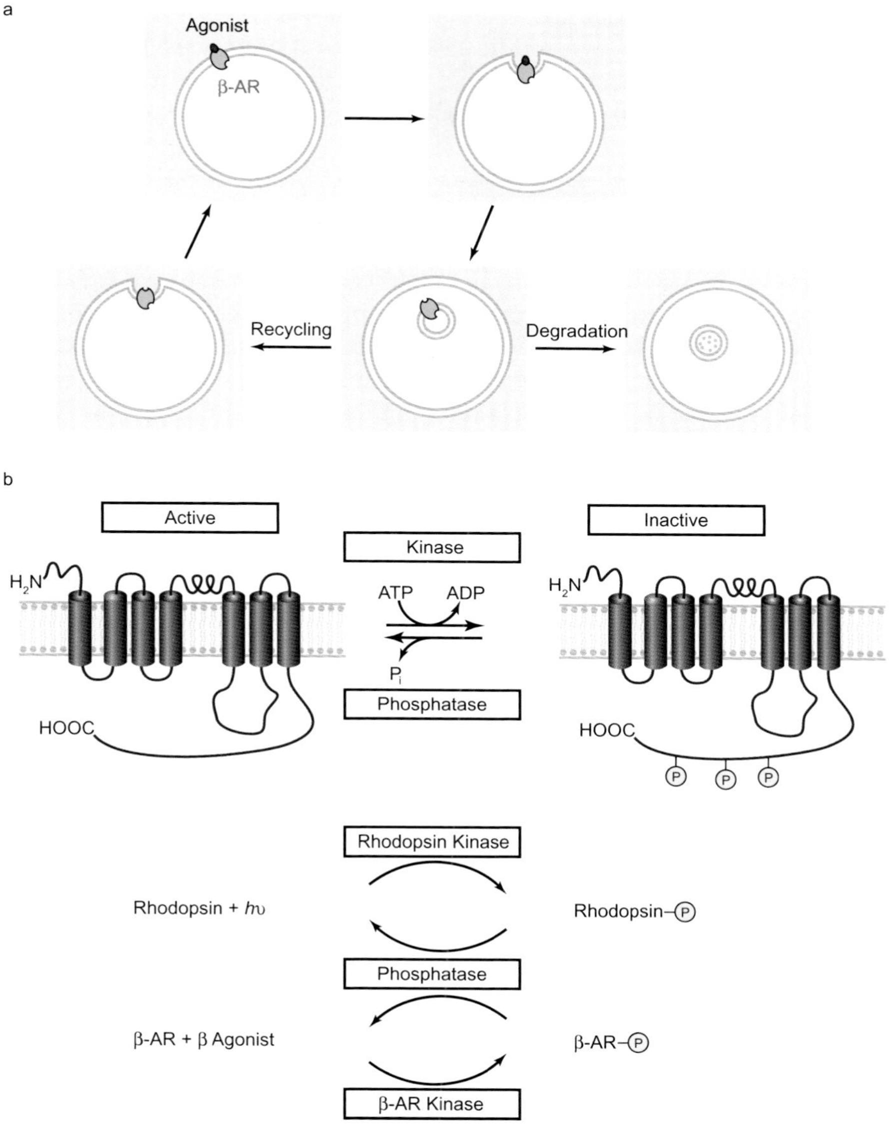

Figure 12–12. Receptor down-regulation. *a*: Agonist-occupied receptors may be internalized by endocytosis, and either recycled to the cell surface or degraded. *b*: Members of the G protein–coupled receptor family, such as rhodopsin or the β-adrenergic receptor (β-AR), may also be down-regulated by phosphorylation. These down-regulations can be reversed by phosphoprotein phosphatases. hυ, photon of light; P, phosphate group.

down-regulated by phosphorylation. Phosphorylation by several different protein kinases is also involved in desensitization of at least one member of the ligand-gated receptor/channel family, the nicotinic acetylcholine receptor, and thus this general mechanism may be widespread.

Features of G protein signaling—a reprise. Modulation of neuronal properties by G proteins, either via direct interaction with ion channels or via second messenger systems, is ubiquitous. Thus it is worth summarizing here some of the key features that are characteristic of G protein signaling. First, it provides a wide temporal range for physiological responses, ranging from tens of milliseconds for direct ion channel modulation to seconds or even minutes for second messenger–mediated responses. A second important feature is that G protein signaling often involves enzymatic cascades that can vastly amplify the response to just a small amount of neurotransmitter or other first messenger. Finally, because a second messenger system may influence several different kinds of ion channels or other effector systems, G protein signaling may allow for the coordinated modulation of multiple effectors to produce a coherent cellular response.

Calcium Signaling

Free cytoplasmic calcium plays a ubiquitous role in neuronal signal transduction processes. Its concentration is some four or five orders of magnitude lower than that in the extracellular space or in intracellular calcium storage sites and is tightly regulated by a variety of pumps and binding proteins. One important source of cytoplasmic calcium is entry across the plasma membrane through voltage-dependent calcium channels; another is release from storage sites in the endoplasmic reticulum, particularly under the influence of IP_3.

Intracellular calcium-release channels. How is IP_3 able to cause calcium release from the endoplasmic reticulum stores (Fig. 12–11)? It does so by binding to an *IP_3 receptor*, a ligand-gated ion channel located in the membrane of the endoplasmic reticulum. The IP_3 receptor/channels are formed as tetramers of large (about 300 kDa) protein subunits, each of which can be divided into three domains (Fig. 12–13a): a ligand-binding domain, a transduction/modulation domain, and a channel domain that includes six membrane-spanning segments and a pore region related to that in the voltage-dependent ion channels (see Chapters 3–7). The IP_3-gated ion channel formed from these subunits can be observed by reconstituting endoplasmic reticulum membranes into artificial phospholipid bilayers

(Fig. 12–13b) or by patch clamping isolated nuclei (the endoplasmic reticulum is continuous with the nuclear membrane). They are cation-selective channels that resemble in many ways plasma membrane ion channels (Fig. 12–13b).

Neurons and other cells also contain another endoplasmic reticulum calcium-release channel, called the *ryanodine receptor*, which is opened by elevated cytoplasmic calcium. Thus calcium that enters the cell via plasma membrane voltage-dependent calcium channels can activate the ryanodine receptor/channel and cause a regenerative release of calcium. It is also possible that calcium released into the cytoplasm through IP$_3$ receptor/channels can activate nearby ryanodine receptor/channels, hence giving rise to complex patterns of intracellular calcium signaling. The subunits of

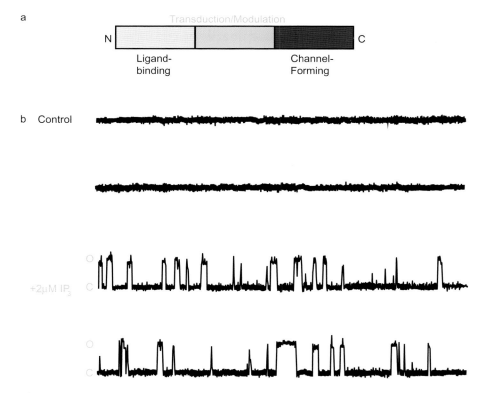

Figure 12–13. Calcium-release channels. *a*: Structural domains of the IP$_3$ and ryanodine receptors. N and C, amino and carboxyl terminals, respectively. *b*: An example of an IP$_3$ receptor/channel reconstituted in an artificial bilayer membrane. The channel is closed (C) and opens (O) only in the presence of IP$_3$ (from an experiment by Barbara Ehrlich and colleagues). See also Mikoshiba (2007).

the ryanodine receptor are similar in overall structure to those of the IP_3 receptor (Fig. 12–13a) but are almost twice as large. There are multiple sites for phosphorylation, by a variety of protein kinases, in both the IP_3 and ryanodine receptors, and they are modulated in complex ways.

Calcium as a second messenger. Calcium plays a central role in the activity of all cells. In nerve cells and other cells that possess calcium channels in their plasma membranes, we have seen that calcium can act as a charge carrier to modulate beating or bursting activity, action potential shape, and other aspects of electrical activity.

Charge transfer, however, may be among the more banal of calcium's many functions (Fig. 12–14). Its entry across the plasma membrane or release from intracellular stores produces such diverse effects as (*1*) triggering of secretion, (*2*) muscle contraction, and (*3*) activation of protein kinases, other enzymes, and calcium-dependent ion channels. Changes in intracellular calcium can be induced directly by certain neurotransmitters, such as those linked to formation of IP_3, or others that activate calcium-permeable ion channels, such as the NMDA receptor channel (see

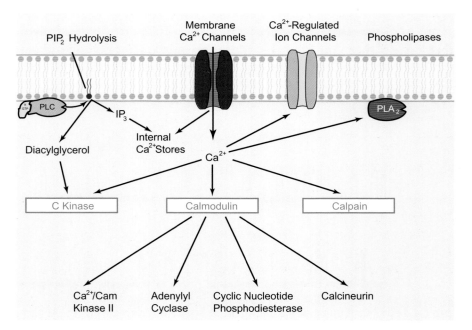

Figure 12–14. Some actions of intracellular calcium ions. Calcium can interact directly with several membrane and cytoplasmic proteins and regulate their properties. Among its cytoplasmic binding proteins is calmodulin, which in its calcium-bound form regulates the activities of many enzymes.

Chapter 11). Let us not forget, however, that changes in this second messenger also occur as a direct result of neuronal activity. Calcium entry through voltage-dependent calcium channels results in transduction of an electrical signal—a change in voltage—into a chemical signal.

The intracellular calcium signal is received and decoded by calcium-binding proteins, including the ubiquitous small protein *calmodulin*, discussed in Chapter 9. Calcium binding induces a large conformational change in calmodulin, exposing a hydrophobic domain that can interact with a large variety of effector proteins (Fig. 12–14). Among these are a variety of ion channels and at least four different types of calcium/calmodulin-dependent protein kinase. One of these kinases, the *multifunctional calcium/calmodulin-dependent protein kinase* (Ca^{2+}/Cam kinase II; see Chapter 9), is present at high concentration in both presynaptic terminals and postsynaptic densities and comprises as much as 1% of total brain protein. Like the cyclic AMP–dependent protein kinase, this enzyme acts on a wide range of substrates, among them proteins that may be involved in synaptic transmission. We described in Chapter 9 the modulation of neurotransmitter release at the squid giant synapse by Ca^{2+}/Cam kinase II in the presynaptic terminal, and its high concentration in postsynaptic densities has led to speculation about its role in receptor and/or ion channel modulation. One feature of this enzyme is that, like many other kinases, it can undergo autophosphorylation. That is, it phosphorylates itself in a calcium/calmodulin-dependent manner. Particularly intriguing is the fact that once several of the subunits are phosphorylated, the enzyme becomes active independent of calcium/calmodulin (Fig. 9–12). Thus a transient calcium signal can produce long-lasting activation of this kinase, a finding that has led to hypotheses that the enzyme acts as a "calcium switch" to produce long-lasting changes in neuronal properties of the sort discussed in Chapter 19.

Calmodulin as an auxiliary subunit of ion channels. Another feature of calcium signaling deserves mention here. It has been known for many years that calcium can influence the activity of some ion channels, such as the calcium-dependent potassium channels that we have discussed in several different contexts and will meet again in later chapters. More recently it has become evident that at least some of these actions of calcium on ion channel properties are mediated by channel-associated calmodulin. For example, the activities of several different classes of calcium-permeant channels (among them voltage-dependent calcium channels, NMDA receptor/channels, and the cyclic nucleotide-gated channels in sensory receptor neurons that we shall discuss in Chapter 13) are inhibited by calcium, acting via calmodulin that is bound constitutively to the channel

(Fig. 12–15, left). In contrast, the intermediate (IK) and small (SK) conductance calcium-dependent potassium channels described in Chapter 7 are activated by calcium, also via constitutively bound calmodulin (Fig. 12–15, right). The emerging theme is that calmodulin, in addition to its many other cellular roles, can act as an ion channel auxiliary subunit that mediates feedback inhibition of calcium entry.

Interactions and commonalities among second messenger systems. The diversity in second messenger systems does not (yet) approach that of the first messengers—the extracellular signaling molecules—or their membrane receptors. However, there is considerable complexity in second messenger systems because the different second messenger systems can interact at various levels. For example, G proteins may participate in the

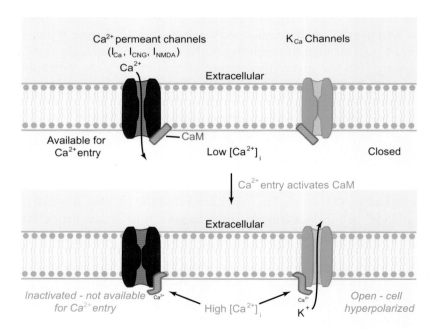

Figure 12–15. Calmodulin confers calcium sensitivity on certain ion channels. Calmodulin (CaM) binds to some channels that allow calcium entry (*top left*). The entering calcium then activates the channel-bound calmodulin, causing channel inactivation and feedback inhibition of calcium entry (*bottom left*). In contrast, calmodulin bound to IK and SK calcium-dependent potassium channels (*top right*) mediates channel opening in the presence of calcium, resulting in membrane hyperpolarization and again feedback inhibition of calcium entry (*bottom right*). (Modified from Levitan, 1999.)

formation of several different second messengers. In addition, one second messenger may influence the activities of enzymes that are involved in the turning on or off of another messenger pathway. To give just several examples, there are calcium/calmodulin-dependent cyclic AMP phosphodiesterases and phosphoprotein phosphatases, and calcium can both stimulate and inhibit adenylate cyclases. In the face of this complexity, it is gratifying to find common features among these pathways. Most second messengers are generated via G protein–mediated enzymatic reactions, and most use protein phosphorylation as their final common mechanism for producing a biological response. Nevertheless, one must always take into account the interactions among second messenger systems in drawing conclusions about their roles in physiological responses. ❧

Other Receptor Families

Although we have focused our attention on the G protein–coupled receptor family in the discussion of second messengers, it is important to realize that there are other receptor families whose activity can also influence the same second messenger pathways. For example, some growth factor receptors, structurally unrelated to the G protein-coupled family, also stimulate the IP_3/DAG pathway directly. The fact that we have not dealt with other receptors in as much depth as the directly coupled or G protein–coupled families should not be taken to imply that other receptor systems are any less important. Rather, it reflects the breadth of the field and the fact that more is known about how these two families regulate neuronal excitability. We have also chosen the directly coupled and G protein–coupled receptors as examples of receptor families in which the relationship between structure and function is best understood.

Neuromodulation Results from Changes in the Properties of Neuronal Ion Channels

Let us revisit the concept of neuromodulation, which we introduced in Chapter 3 (see Figure 3–14), in the context of what we now know about ion channels and molecular transduction mechanisms. Among the earliest examples of modulation of action potential duration (as in Figure 3–14a) was in neurons of the chick dorsal root ganglion (DRG), a way station in the pathway through which sensory information from the periphery reaches the spinal cord. A variety of neurotransmitters can cause a narrowing of action potentials in DRG neurons (Fig. 12–16a). Among the transmitters that produce this effect are the peptides enkephalin and somatostatin, as well as serotonin, GABA, and norepinephrine. In principle, there are two possible ways such a shortening could come about. There might be an increase in the potassium currents responsible for spike repolarization; alternatively, or in addition, the transmitters might directly decrease the calcium or sodium currents that underlie the depolarizing phase of the spike. Voltage clamp experiments strongly support the latter explanation. As shown in Figure 12–16b, the various transmitters cause a decrease in a current with a voltage dependence expected for a calcium current (compare with Fig. 7–1c).

What is the molecular mechanism mediating this shortening of action potentials in DRG neurons? Intracellular application of GDPβS, or pretreatment of the DRG neurons with pertussis toxin, can block the actions of norepinephrine or GABA. In addition, application of a diacylglycerol (DAG) activator of protein kinase C can decrease the calcium current in these neurons, and a protein kinase C inhibitor can block the transmitter-induced

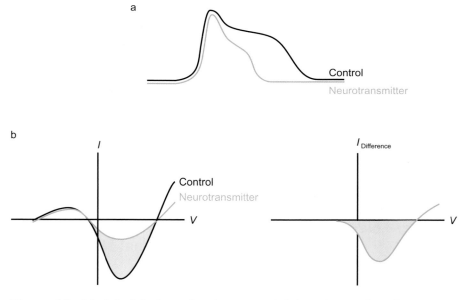

Figure 12–16. Modulation of action potential duration in dorsal root ganglion neurons. *a*: A variety of neurotransmitters can cause a shortening of the action potential in dorsal root ganglion neurons. *b*: Voltage clamp analysis indicates that this results from a decrease in inward current. At the right, the difference current elicited by the neurotransmitter is plotted. *I*, current; *V*, voltage.

spike shortening. Taken together, these results suggest that a G protein–mediated activation of phospholipase C, with the ensuing release of DAG and activation of protein kinase C (see earlier discussion), is responsible for the decrease in calcium current and shortening of the action potential. The physiological consequence of this shortening is likely to be a decrease in transmitter release at the DRG neuron terminals and an attenuation of the amount of sensory information that is allowed to reach the spinal cord. This is of particular interest in the case of enkephalin. An enkephalin-induced decrease in action potential duration, with a consequent decrement in release of the pain pathway sensory transmitter, substance P, from DRG neurons, might account for some of the analgesic actions of enkephalin and other opiate agonists.

Another example of neuromodulation: modulation of synaptic efficacy. It is widely believed that modulation of chemical synaptic efficacy may underlie many important behavioral phenomena, including learning and memory. Indeed, it is now evident that a change in the properties of ion channels can alter synaptic efficacy, without changing action potential amplitude and duration or spontaneous neuronal activity. A rather subtle action of a transmitter on neuronal excitability has been described in at least two kinds of vertebrate neurons: the large B cells of the bullfrog sympathetic ganglion, and rat hippocampal pyramidal cells. In each of these cell types a neurotransmitter alters neuronal excitability by inhibiting a potassium current that is not active at the resting potential. In the sympathetic ganglion neurons, a potassium current called the M-current (I_M) can be inhibited by a variety of neurotransmitters, including the peptides substance P and luteinizing hormone-releasing hormone (LHRH), and muscarinic cholinergic agonists. Recall from Chapter 7 that the channels that carry this current are made up of $Kv7.2$ and $Kv7.3$ (KCNQ2 and KCNQ3) subunits. Because the M-current is not very active at hyperpolarized potentials, application of one of these agonists has only a small effect on the cell's resting potential. However, the response to a depolarizing stimulus, for example, an excitatory synaptic potential, will be very much enhanced because the M-current is not available to oppose the depolarization (Fig. 12–17a). A similar phenomenon is observed in hippocampal pyramidal neurons, in which norepinephrine blocks a calcium-dependent potassium current, thought to be carried by the SK class of calcium-dependent potassium channel, which is responsible for a slow afterhyperpolarization (AHP) that follows a train of action potentials. This current (I_{AHP}) is not active at rest when calcium levels are low, and thus norepinephrine has little or no effect on the cell's resting potential. However, a depolarizing stimulus that produces action potentials will

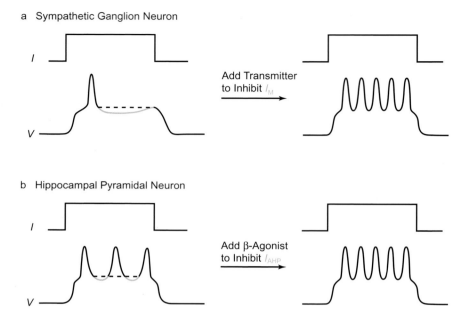

Figure 12–17. Modulation of synaptic efficacy. *a*: In experiments with sympathetic ganglion neurons of the bullfrog, addition of a transmitter that blocks the M potassium current (I_M) alters the cell's response to other depolarizing synaptic stimuli. *b*: Similarly, norepinephrine modulates the response of rat hippocampal pyramidal neurons to other depolarizing synaptic stimuli. I_{AHP} refers to the current responsible for the spike afterhyperpolarization in these neurons.

be more effective in the presence of norepinephrine (Fig. 12–17b), because the calcium-dependent AHP, which normally follows the action potentials and dampens the excitatory response, has been suppressed.

These few examples suffice to emphasize that the action of one transmitter may be modulated profoundly by that of another. Although the effects of a single transmitter on resting membrane properties may appear to be minor, its true influence on the properties of a neuron may be evident only when it is coupled with stimulation of another synaptic or hormonal input.

Phosphorylation and Dephosphorylation of Ion Channels: A Common Mechanism in Neuromodulation

Many neuromodulatory phenomena occur through the actions of protein kinases and phosphoprotein phosphatases. One question that immediately arises concerns the identity of the proteins that are phosphorylated by the kinases. By analogy to alterations in the activities of some enzymes, which are known to result from direct phosphorylation of the enzyme molecules themselves, it seemed possible that ion channels are targets for protein kinases and phosphatases, and that direct phosphorylation of ion channel proteins alters their functional properties. It is now evident that many ion channel proteins are substrates for a variety of different protein kinases and phosphatases. Although undoubtedly other phosphorylated proteins may interact with and influence the activity of ion channels, channel phosphorylation and dephosphorylation clearly are of fundamental importance for neuromodulation.

What Determines the Biological Response

Where does the specificity lie in the response of a target cell to a particular extracellular signal? This is a dilemma that has been with us since it was realized that cyclic AMP can mediate different biological responses in different kinds of cells. We are still without a complete understanding of this issue of response specificity, but important clues have emerged from experiments in which receptors have been expressed in cell types in which they are not normally found. These receptors can couple to the host cell's endogenous transduction mechanisms, and activation by agonist then produces a biological response that is characteristic of the host cell rather than of the exogenous receptor and its ligand. For example, serotonin receptors, which may be coupled to ion channels in the nerve cells from which they are isolated, can cause malignant transformation when they are expressed in fibroblasts and activated by serotonin. Similarly, changes typical of the response to sperm can be evoked by serotonin when one of its receptors is expressed in eggs. These experiments tell us that the specificity resides in the target cell and in the particular biological response system(s) that the cell makes available to interact with the receptor and transduction mechanism.

Summary

Extracellular signals must be recognized by the target cell and transduced into an appropriate biological response. Signal recognition is accomplished by the specific membrane receptors that are coupled to different kinds of transduction mechanisms, which in nerve cells often regulate the activity of ion channels. The simplest receptor–ion channel coupling system, discussed in Chapter 11, consists of the ligand-binding site and channel within a single protein molecule or macromolecular complex.

A coupling mechanism of intermediate complexity involves protein–protein interactions, between a G protein and ion channel, in the plane of the plasma membrane. Finally, many ion channels are coupled to receptors via diffusible intracellular second messengers such as cyclic AMP, diacylglycerol, and IP_3. The purpose of this diversity in the categories of receptor–channel coupling may be to provide a wide temporal range in the responses of neurons to neurotransmitters, hormones, and sensory stimuli.

Diversity also exists *within* the category of second messenger–mediated coupling. There are a variety of second messenger systems that at first glance appear to bear little relationship to one another. However, several of these share a common final mechanism of action on response systems, namely protein phosphorylation via one of several second messenger–dependent protein kinases. Modulation of neuronal excitability by protein phosphorylation often involves direct phosphorylation of the ion channel protein itself or of some closely associated regulatory component.

13

Sensory Receptors

Although most neurons receive input from other neurons, the business of the brain is to act on information from the outside world. Specialized cells have evolved for the receipt of such information. These include cells that are responsible for sight, hearing, touch, taste, and smell, as well as those that signal to the brain the state of internal organs. Most external and internal stimuli are received by three classes of sensory cells: (*1*) those that respond to mechanical stimulation, (*2*) those that are influenced by light, and (*3*) those that sense changes in their chemical environment. In addition, some cells respond to other signals such as changes in temperature. In all cases, the stimulus alters the activity of ion channels, sometimes through the agency of second messengers. Sensory cells therefore provide clear examples of the principles encountered in previous chapters. We will now discuss briefly a few of the varied transduction mechanisms that have been found in each of the major classes of sensory cells.

Receptor Potentials

First let us consider in general terms the kind of output signal that these specialized sensory receptors send to the central nervous system. Some sensory cells are true neurons, with axons that travel from the sense organ to other parts of the nervous system. An external stimulus produces a depolarization or hyperpolarization of the membrane that is known as the *receptor potential*. As shown in Figure 13–1, the time course and amplitude of the receptor potential generally mirror those of the stimulus. If a depolarizing receptor potential is large enough to exceed the action potential threshold, the neuron will fire action potentials at a frequency that

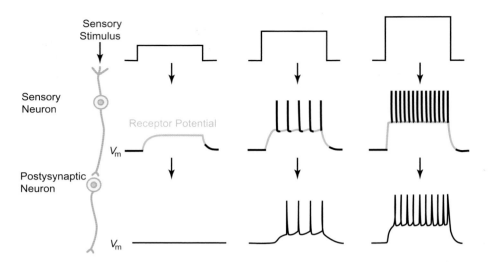

Figure 13–1. Receptor potentials and frequency coding. The size of a sensory stimulus determines the size of the receptor potential. This, in turn, is encoded in the frequency of firing of the sensory neuron and/or its postsynaptic neuron. V_m, membrane potential.

reflects the size of the stimulus. Thus we can see that information about stimulus strength is now encoded in action potential frequency. At the first synapse made by the sensory receptor neuron, the action potential frequency will determine the amount of transmitter released, and this in turn will control the magnitude of the depolarizing postsynaptic response. The latter will be translated back into a particular firing frequency in the postsynaptic cell.

Other sensory cells do not have axons. Although action potentials can sometimes be evoked in such cells, a sensory stimulus normally causes a depolarizing or hyperpolarizing receptor potential that does not cause firing. Instead, the change in membrane potential alters the rate of neurotransmitter release onto a postsynaptic neuron, and action potential frequency coding enters the picture only after this first synapse. As we shall see later, the classic example of this is the hyperpolarization of vertebrate photoreceptors in response to light.

Mechanoreceptors

Many sensory cells that respond to physical movement, termed *mechanoreceptors*, are found in or under the skin. These are true neurons, with their cell bodies in the dorsal root ganglion. Recall that we have already considered the excitability of dorsal root ganglion neurons previously. The business end of these mechanoreceptors is at the peripheral endings of their axons in the skin. Figure 13–2 shows that the morphology of such endings is quite varied. One type of mechanoreceptor ending, located deep below the surface, is the *Pacinian corpuscle*. This consists of the bare ending of an axon, surrounded by "onion skin" layers of connective tissue.

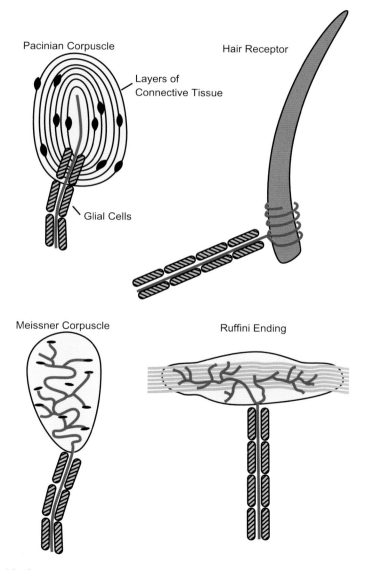

Figure 13–2. Mechanoreceptors in the skin. (Modified from Martin, 1981.)

Meissner corpuscles and *Ruffini endings* are found closer to the surface of the skin and consist of more elaborate branchings of a nerve fiber, also enclosed in connective tissue. Other receptors comprise bare nerve fibers wrapped around the base of a hair follicle. Although in Figure 13–1 we suggested that frequency of firing generally mirrors the amplitude of a sensory stimulus, this is complicated by the fact that many sensory cells *adapt* their response to a *maintained* stimulus. This is particularly true of these different types of mechanoreceptors, whose electrical responses differ in that some fire only transiently when pressure is applied to the skin, whereas others fire persistently throughout a maintained tactile stimulus. These and other receptors account for some of the varied sensations of touch and pressure on the skin. Mechanoreceptors are also found in many other parts of the body. For example, *Golgi tendon organs* and *muscle spindles* are found in joints and muscles, and provide information on the position and length of muscles.

Current work is revealing the molecules that are responsible for the transduction of mechanical movement into changes in the firing rate of mechanoreceptor cells. Two such molecules are *Piezo1* and *Piezo2*, which are membrane proteins that are enriched in cells that respond to pressure from a motor-driven glass rod (Fig. 13–3a). Four Piezo proteins, which have an estimated number of between 24 and 39 transmembrane segments each (Fig. 13–3c), come together to form extraordinarily large protein complexes that appear to be *stretch-activated* ion channels. These open transiently to allow both sodium and potassium ions to flow across the membrane when it is distorted by the pressure pulse, producing the receptor potential. When levels of *Piezo* channels are reduced using genetic manipulations, as has been done in mammalian cells and in sensory neurons of the fruit fly *Drosophila*, the transient current is eliminated (Fig. 13–3b). ❷

Hair cells of the cochlea. *Hair cells* respond to a specialized form of mechanical stimulation and are found in the inner ear of vertebrates. (These cells are entirely unrelated to the cells that innervate hair follicles, as in Fig. 13–2). They are found both in the vestibular organs and in the *cochlea*. In the former they are responsible for transducing information about gravity and movements of the head, whereas in the cochlea, hair cells are the sensory cells of the auditory system. What is particularly interesting about these cells responsible for hearing is that they must not only provide the brain with information about the *intensity* of a sound but must also respond selectively to different *frequencies* of sound waves.

Figure 13–4 shows an image of a row of hair cells. They are elongated in shape with a distinctive arrangement of hairs, termed *cilia*, located at

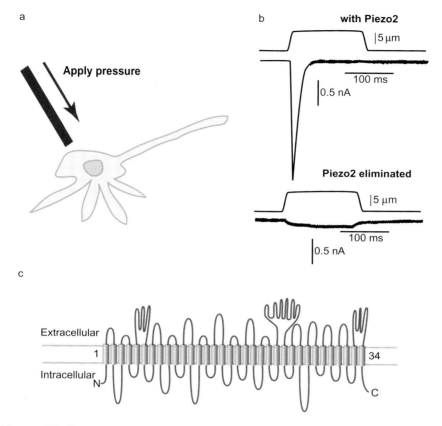

a

Apply pressure

b

with Piezo2

|5 µm

100 ms

0.5 nA

Piezo2 eliminated

|5 µm

100 ms

0.5 nA

c

Extracellular

1   34

Intracellular

N

C

Figure 13–3. Piezo proteins transduce mechanical stimulation into depolarizing current. *a*: Mechanical responses of a neuron can be tested by pushing on its plasma membrane using a motor-driven glass rod. *b*: Work by Ardem Patapoutian and his colleagues showed that, in neurons and cell lines that contain the Piezo2 protein, application of pressure produces a transient inward current that is lost when levels of Piezo2 are decreased. *c*. Proposed transmembrane organization of a single Piezo2 protein. (Modified from Nilius and Honoré, 2012.)

one end (Fig. 13–4, lower panel). The cilia are of two types. Most, but not all, hair cells have one long *kinocilium*, which, in an electron microscope, closely resembles moving cilia such as those in a sperm tail. Its structure is maintained by microtubules that extend the length of the kinocilium, with two central microtubules surrounded by a ring of nine others (Fig. 13–5a). Some hair cells, however, such as those in the mammalian cochlea, seem to do quite well without a kinocilium. The remaining 30–100 cilia, which are present in all hair cells and are of varying lengths, are termed

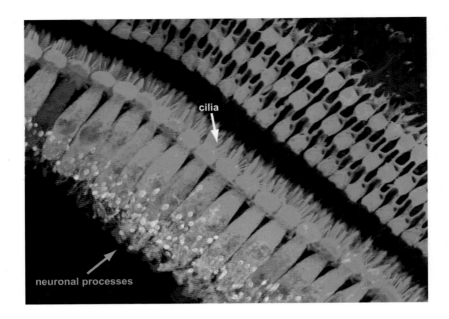

cilia

neuronal processes

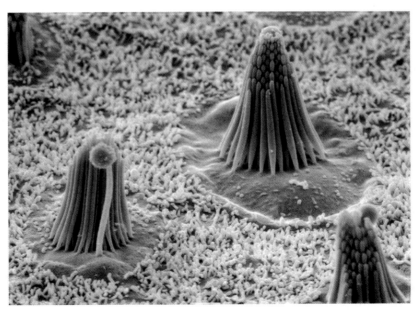

Figure 13–4. Hair cells. The upper image shows a row of hair cells stained in green with their nuclei in blue and, in red, the processes of afferent neurons at their base. The cilia are evident at the tops of the individual cells. This confocal image was made by Sonja Pjott. The lower scanning electron micrograph shows the cilia of hair cells in more detail. This image, which was taken by John Assad and David Corey, is of cells in the saccule of a bullfrog, an organ that senses both sounds and ground-borne vibrations.

stereocilia. These do not contain microtubules but are filled with filaments of actin and proteins that cross-link actin filaments. The entire collection of cilia is termed the *hair bundle*.

There is another important structural feature of the hair bundle in sensory hair cells. Each of the cilia is connected to an adjacent cilium by a very fine filament, termed a *tip link*. As the name implies, these filaments join the tip of one cilium to that of the next (Fig. 13–5b). The point at which tip links are anchored to the membrane of a cilium is believed to be the site at which the ion channels responsible for sensory transduction are located.

Physical movements of the hair bundle cause a rapid change in the membrane potential of a hair cell. When the bundle is moved toward the kinocilium, the membrane depolarizes by 10–20 mV. In contrast, displacement of the bundle away from the kinocilium hyperpolarizes the cell (Fig. 13–5b). The changes in membrane potential are caused by the opening and closing of channels located in the plasma membrane of the stereocilia themselves. It is the mechanical force that pulls or pushes on the tip link filament that is thought to lead to the opening or closing of the channels. These channels are relatively nonselective for cations such as sodium, potassium, and calcium ions. The response of the channels following mechanical displacement of the cilia occurs extremely rapidly, within 20–100 sec. This means that opening of the channels is probably linked *directly* to mechanical deformation of the cilia, rather than through a second messenger system. As we shall now describe, a striking feature of the response of hair cells in the cochlea is that their responses are specifically tuned to different frequencies of sound.

Hair cells are tuned by position in the cochlea and by electrical resonance. Figure 13–6 shows that different cells respond optimally to different sound frequencies. This is illustrated by the *tuning curves* for several different hair cells, which represent the intensity of sound of different frequencies that must be applied to produce a fixed change in membrane potential. A major factor that determines the tuning curve for an individual cell is its position in the cochlea. The cell bodies of hair cells, together with their supporting cells, form a sheet of cells in the cochlea. The tips of the hair bundles are normally in contact with a stiff, carbohydrate-containing sheet, known as the *tectorial membrane*, that lies over the layer of cell bodies. Vibrations caused by sound waves entering the cochlea set up lateral movement of the tectorial membrane relative to the underlying cells. This in turn bends the hair bundles and transforms the mechanical vibration into an electrical oscillation of the membrane potential in the hair cells.

a

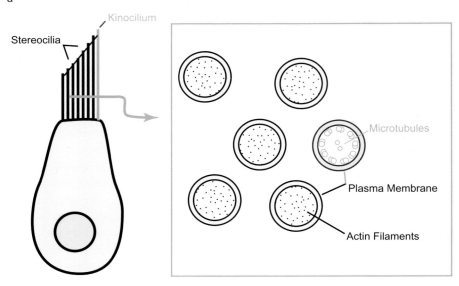

b

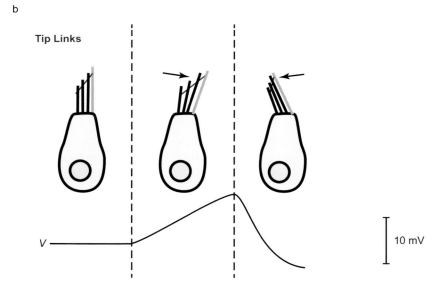

Figure 13–5. Hair cell cilia. *a*: Each hair cell has a single kinocilium and many stereocilia. *b*: Movement of the hair bundle toward the kinocilium depolarizes the cell. Movement in the opposite direction produces a hyperpolarization. *V*, voltage. See Animation 13–1. ◖

The mammalian cochlea is an elongated structure, folded into a spiral that resembles the shell of a snail (Fig. 13–6b). The sheets of hair cells extend from the base of the spiral to its apex. This mechanical design for the cochlea, coupled with the fact that changes in the thickness of the tectorial membrane occur along its length, provides a mechanism for the selective response of hair cells to different frequencies of sound. As was first suggested by the German physicist Helmholtz more than a hundred years ago, and elaborated by Georg von Bekesy in the 1950s, low-frequency sounds cause the greatest vibrations to occur at the apex of the cochlea. In contrast, high-frequency sounds maximally deflect the hair bundles in cells at the base. The frequency to which a hair cell responds optimally is therefore determined by its physical position along the coiled cochlea.

In birds, and in amphibians such as bullfrogs and turtles, another mechanism for tuning has been discovered. The membrane potential of hair cells in these animals undergoes spontaneous oscillations, or can be induced to oscillate after a transient depolarization (Fig. 13–7a). Although the frequency of oscillation can be increased or decreased by applying depolarizing or hyperpolarizing current, each cell has a characteristic frequency at which it oscillates around its resting potential. The characteristic frequency of different hair cells varies from tens to many hundreds of cycles per second. When a cell is stimulated with sound waves at its characteristic frequency, a maximal fluctuation of the membrane potential is evoked (Fig. 13–7b). Higher or lower frequency sounds are much less effective. The oscillations can largely be explained by the activity of only two types of ion channels: calcium channels and calcium-activated potassium channels. The opening of calcium channels causes the depolarizing phase of an oscillation. As calcium enters the cell, calcium-dependent potassium channels begin to activate. When a sufficient number of these have been opened, the membrane hyperpolarizes and calcium entry decreases. As intracellular calcium falls, the potassium channels close, and calcium channels again activate, renewing the cycle.

How is tuning to different frequencies achieved? Calcium current does not differ much from one hair cell to the next. In contrast, the rate at which the calcium-dependent potassium current activates varies enormously between cells (Fig. 13–8). It is these differences in the kinetics of potassium current that account for the fact that hair cells have different characteristic frequencies of oscillation. A particularly intriguing question that has yet to be answered is what determines the kinetics of the potassium channel in different cells. As we saw for the *Shaker* potassium channel in Chapter 5, alternative splicing of RNA from the *Slo* potassium channel gene (which encodes large-conductance calcium-activated potassium channels) produces a variety of channel proteins that differ in their

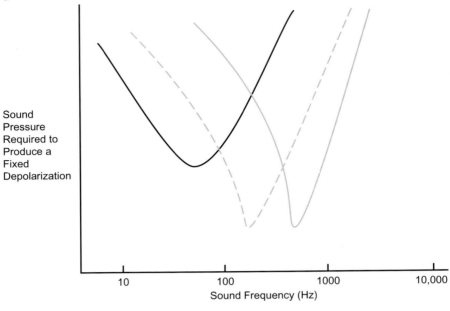

a

Sound
Pressure
Required to
Produce a
Fixed
Depolarization

10   100   1000   10,000

Sound Frequency (Hz)

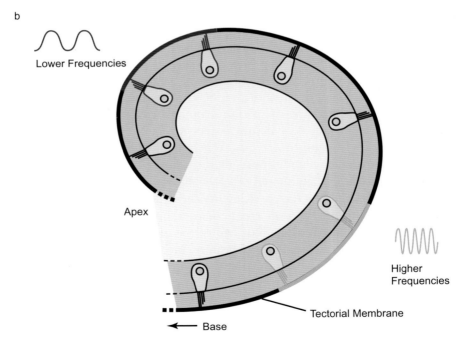

b

Lower Frequencies

Apex

Base

Tectorial Membrane

Higher
Frequencies

Figure 13–6. Tuning of hair cells. *a*: Tuning curves for three different cells. *b*: Spatial localization of cells with different frequency responses along the cochlea. See Animation 13–1.◉

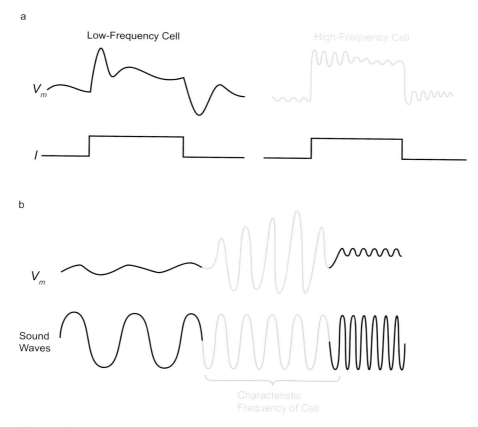

Figure 13–7. Membrane potential (V_m) oscillations in hair cells. *a*: Endogenous oscillations and oscillations evoked by depolarizing currents were revealed in recordings by Fettiplace and his colleagues. *b*: Sound waves at the characteristic frequency of a cell cause the largest fluctuation in membrane potential. Experiments of this kind, using mechanical displacement of the hair bundles as a stimulus, were also carried out by Lewis and Hudspeth (1983).

kinetic properties. Moreover, β-subunits of these channels also modify the rates of opening and closing of these channels. Gradients of different isoforms of the *Slo* channel and of β-subunits have been found to exist in the cochlea of both higher and lower vertebrates, and are likely to account for the differences in the electrical properties of high-frequency and low-frequency hair cells. Such tuning of individual hair cells by electrical resonance accounts for the selective response of cells in the lower range of auditory frequencies, in species such as birds and amphibians. Although it is more difficult to record from hair cells in mammals, there is currently no evidence that such tuning occurs in mammalian species. Within the

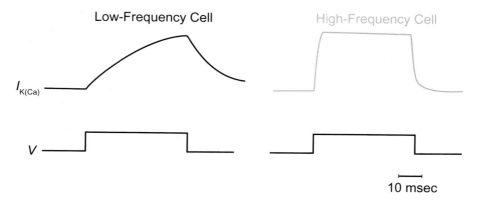

Figure 13–8. Calcium-dependent potassium current ($I_{K(Ca)}$) in hair cells. Voltage clamp recordings by Art and Fettiplace (1987) found more rapidly activating currents in cells with higher-frequency responses.

mammalian brain, however, it is well established that neurons that respond to high sound frequencies have different levels of potassium channels than the neurons that respond to low-frequency sounds.

Outer Hair Cells Provide Mechanical Tuning

There are two types of hair cells in the mammalian cochlea. So-called *inner hair cells* are the actual sensory receptors that transmit information to the brain. There are about 4000 such inner hair cells in a human cochlea. Along the length of the cochlea, lying parallel to this row of inner hair cells, are found three rows of *outer hair cells* (Fig. 13–9a). The properties of these outer hair cells are entirely distinct from those of the inner hair cells. For example, rather than sending information toward the brain, they receive inputs from auditory nuclei in the brainstem. The most striking feature of the outer hair cells is, however, their response to rapid electrical simulation.

In contrast to most other cells, including the inner hair cells, the plasma membrane of outer hair cells is kept relatively rigid by a *cortical cytoskeleton* that lines the inside of the membrane. When viewed with the freeze fracture technique (see Fig. 8–2), the membrane itself is found have a very dense array of intramembranous particles. In response to electrical stimulation, or to application of acetylcholine, this membrane contracts, producing a shortening of the cell (Fig. 13–9b). While at first glance this response appears to resemble that of a muscle cell, there is one key

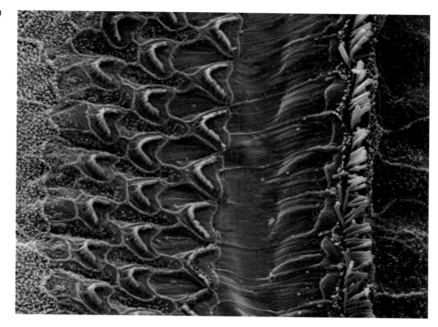

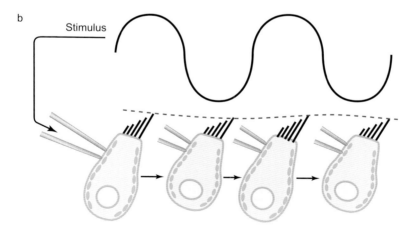

Figure 13–9. Outer hair cells. *a*: An electron micrograph showing a top view of three rows of outer hair cells and one row of inner hair cells. The stereocilia have been colored red on the picture. On the outer hair cells stereocilia are arranged in a V shape on each cell, while those on the inner hair cells are arranged in a straight line (courtesy of Uli Mueller; see Kazmierczak and Müller, 2012). *b*: Alternate contraction and relaxation of an isolated hair cell in response to a sinusoidal stimulus applied through a patch pipette. The contractile protein Prestin in the plasma membrane is shown in blue.

difference. Changes in the length of an outer hair cell occur extremely rapidly in response to changes in voltage; they can be followed by making movies of changes in the shape of cells in response to changes in voltage driven by music or other auditory stimuli (Video 13–1). ⊙ Thus the contractions of such a cell can follow stimulation at frequencies of 3000–8000 Hz or higher. The speed of this response entirely rules out a contractile mechanism such as that found in muscle cells.

A protein called *prestin* is found at a very high density in the lateral wall of the outer hair cells, but is absent from the nonmotile inner hair cells (Fig. 13–10a). This is a key component of the contractile apparatus of outer hair cells and is, in fact, the fastest motor protein known. Prestin is a member of a family of proteins that transport chloride ions across membrane, and it binds cytoplasmic chloride ions, although it is unclear whether prestin itself can transport chloride (Fig. 13–10b). Nevertheless, when prestin is bound to chloride, changes in voltage produce contractions of the membrane which in turn produce movements of the overlying

a   b

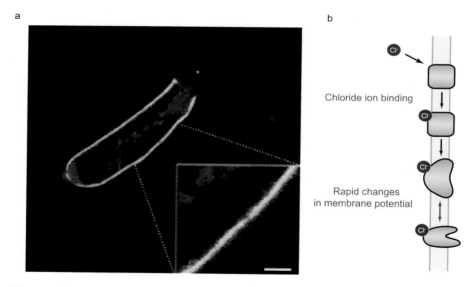

Chloride ion binding

Rapid changes
in membrane potential

Figure 13–10. Location of prestin in outer hair cells. *a*: Outer hair cells were stained with an antibody against the prestin protein (seen as yellow/green) and also with a dye that stains all cell membranes (red). Prestin is detected only in the plasma membrane of the cell body but not in the stereocilia or in the cuticular plate at the top of the cell that separates the lateral walls of the cell body from the stereocilia (Yu et al., 2006). *b*: A model for prestin action. The binding of chloride ions converts the protein into a form that can rapidly sense changes in membrane voltage, which cause contraction/expansion of the protein in the plane of the membrane (Song and Santos-Sacchi, 2013).

tectorial membrane in which the hair cells are embedded. This provides further tuning of the tectorial membrane to different frequencies of sound. Information about this tuning, however, is relayed to the central nervous system by the less numerous inner hair cells.

Photoreceptors

The way that visual stimuli are translated into electrical activity of a photoreceptor cell in the retina is better understood than any other sensory event. The process termed *phototransduction* actually occurs in a variety of cells. For example, birds possess *extraretinal receptors*, which are found in the pineal gland within the brain itself. These allow a bird to sense changes in the light of its environment, even in the absence of a functioning retina. In the retina of vertebrates there exist two cell types, the *rods* and the *cones*, which have different sensitivities and respond to different frequencies of light. The cone cells can further be subdivided into cells that preferentially sense different colors. However it is rods, from a variety of species, that have been the favored cell type for studying visual transduction.

Rods, rhodopsin, and cyclic GMP. Figure 13–11 shows the structure of a rod photoreceptor. The cell has two parts. The rod *outer segment* is elongated and contains a stack of flattened discs made from internal membranes. This is connected by a thin bridge to the remainder of the cell, the *inner segment*, that contains the nucleus, the mitochondria, and the presynaptic terminal that synapses onto other neurons in the retina. It is the outer segment that is the business end for visual transduction. Within the internal membranous discs is found the light-sensitive protein *rhodopsin*. This is made up of an *opsin* protein, bound to a light-sensitive molecule or *chromophore* termed *retinal*. As was mentioned in Chapter 12, rhodopsin belongs to the family of G protein–coupled receptors, and it is the first protein of this family to have its three-dimensional structure determined by electron microscopy and by X-ray crystallography (Fig. 13–12). The chromophore molecule retinal may exist in a number of different forms, of which 11-*cis*-retinal and all-*trans*-retinal are the two major isomers (Fig. 13–13). On its own, neither opsin nor retinal absorbs visible light. In combination, however, absorption of a photon of light causes an isomerization of retinal from the 11-*cis* form to the all-*trans* form.

This light-dependent isomerization of retinal then causes a structural rearrangement of the protein. Rhodopsin that has been activated in this

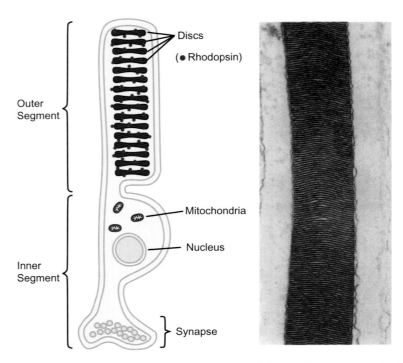

Figure 13–11. A photoreceptor. The drawing (*left*) is of an entire rod photoreceptor. The micrograph (*right*) shows only the outer segment of a salamander cone (courtesy of Dr. John E. Dowling).

way is termed *meta*-rhodopsin. For all of the subsequent steps in visual transduction, it is useful to think of this molecule as analogous to a receptor that has just bound its neurotransmitter. In fact, the structure of the opsin protein is very similar to that of neurotransmitter receptors such as the β-adrenergic receptors. Recall that receptors such as the β-adrenergic receptor act through GTP-binding proteins. Not surprisingly, therefore, the steps that follow the production of *meta*-rhodopsin involve a second messenger through the agency of a GTP-binding protein called *transducin* (G_T) (Fig. 13–14).

When *meta*-rhodopsin binds to transducin, GDP is replaced by GTP, and the α_T subunit of transducin is liberated from its complex with the βγ subunits. The target of the newly liberated α_T is an enzyme in the membranous discs, a *phosphodiesterase*, that cleaves the second messenger cyclic GMP to 5'-GMP. Even in the dark, the levels of cyclic GMP in the outer segments are maintained by a balance between its rate of synthesis through guanylate cyclase and degradation by the phosphodiesterase. The action of α_T, formed after exposure to light, is to stimulate the phosphodiesterase,

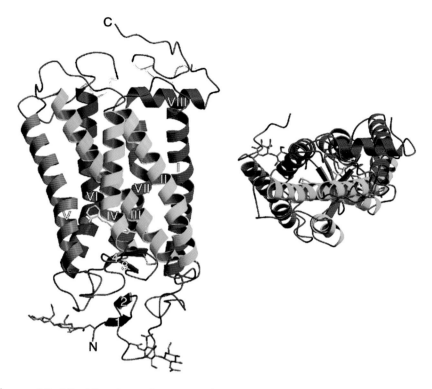

Figure 13–12. The three-dimensional structure of rhodopsin, determined by X-ray crystallography. *Left:* Organization of the transmembrane regions viewed from the plane of the membrane. *Right:* A view of the protein from the cytoplasmic side of the membrane (Palczewski et al., 2000).

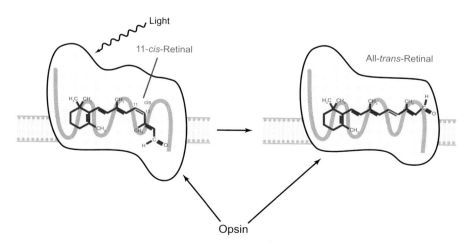

Figure 13–13. Rhodopsin is made up of an opsin protein, bound to the light-sensitive molecule retinal. The light-induced isomerization of retinal was discovered by George Wald.

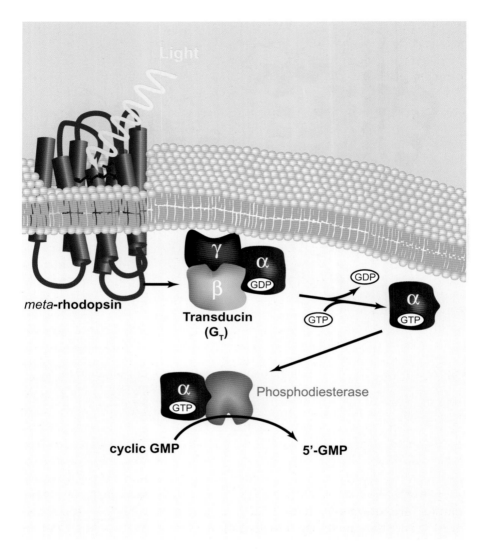

Figure 13–14. Exposure to light produces a decrease in cyclic GMP levels in photoreceptors. *Meta*-rhodopsin activates the GTP-binding protein transducin (G_T), causing dissociation of GTP-bound α_T-subunit. This, in turn, activates cyclic GMP phosphodiesterase.

producing a *drop* in the levels of cyclic GMP. This drop occurs within about 100 msec of the onset of a light flash, sufficiently fast to account for a visual response. In many respects, photoreceptors are built backward. When excited by light, they respond by *dropping*, rather than raising, their concentration of the second messenger cyclic GMP.

This cascade of reactions that follows the formation of *meta*-rhodopsin produces a very significant amplification of the signal generated by light. It has been estimated that a single molecule of *meta*-rhodopsin, which is formed by the action of a single photon of light, diffuses in the membrane and activates several hundred transducin molecules before it is rendered inactive (see discussion later in this chapter). The subsequent stimulation of the phosphodiesterase by α_T provides further amplification such that a single photon of light can lead to the destruction of more than 10^5 molecules of cyclic GMP.

Cyclic GMP-gated channels. The change in cyclic GMP levels can be translated into an electrical response in the photoreceptor because of the existence of a class of channels that are directly activated by cyclic nucleotides such as cyclic GMP. These channels, termed *CNG channels*, for Cyclic Nucleotide–Gated channels, are close relatives of the *Shaker* family of potassium channels we encountered in Chapter 5. In contrast to *Shaker* channels, however, the pore of the GNG channels is relatively nonselective for cations. Binding of cyclic GMP to the channel occurs near the carboxyl terminus (Fig. 13–15a), at a sequence that closely resembles the cyclic nucleotide binding domains in the cyclic AMP- and cyclic GMP-dependent protein kinases. Binding of cyclic GMP at these sites, however, leads to direct activation of the channel rather than triggering an enzymatic reaction. This can be demonstrated by making an inside-out patch recording on membrane from the rod outer segments. When cyclic GMP is added to the cytoplasmic face of the patch, a large increase in conductance, attributable to the opening of the channels, can be measured (Fig. 13–15b).

To understand how GNG channels work in photoreceptors, we must first consider the electrical properties of a rod at rest in the dark. The CNG channels are found selectively in the plasma membrane of the outer segments, where they are tethered to discs of membranes that contain rhodopsin through a tethering protein complex termed *peripherin-2/ROM-1* (Fig. 13–16). Because of the abundance of sodium ions in the extracellular fluid, the major ion that enters the outer segments through the nonselective CNG channels is sodium. The opening and closing of these channels is not very dependent on membrane potential. From our discussion of ion channels, we know that a cell with a preponderance of such sodium channels would be expected to have a very positive resting potential. The effect of the rod sodium channels is, however, counterbalanced by potassium channels. The interesting thing about these potassium channels is that they are found in a very different part of the cell, the membrane of the *inner* segment that includes the nucleus and synaptic terminal. Because there is good

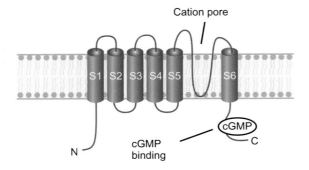

a

Cation pore

S1 S2 S3 S4 S5   S6

cGMP
binding

N

cGMP
C

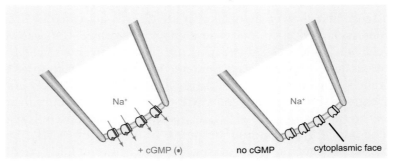

b   Inside-Out Patch Recordings

Na⁺

+ cGMP (●)

Na⁺

no cGMP   cytoplasmic face

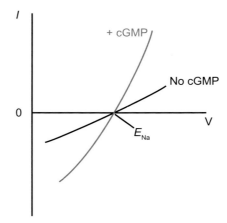

I

+ cGMP

No cGMP

0

E_{Na}

V

Figure 13–15. Cyclic GMP opens sodium-permeable channels. *a*: Presumed membrane organization of the cyclic GMP-gated cation channel. Compare with that of the *Shaker* potassium channels (see Fig. 5–8). *b*: When cyclic GMP is added to the cytoplasmic face of an inside-out patch of membrane from a rod photoreceptor, the *I–V* curve changes to reflect an increase in sodium conductance. This finding was obtained by Evgeniy Fesenko and colleagues (1985). E_{Na}, sodium equilibrium potential.

electrical continuity between the inner and outer segments, the mean membrane potential is kept fairly negative as a result of the open potassium channels. This spatial distribution of channels, however, creates a circulating current, termed the *dark current*, which flows in through the outer segment sodium channels, through the bridge into the inner segment, and out through the potassium channels (Fig. 13–16).

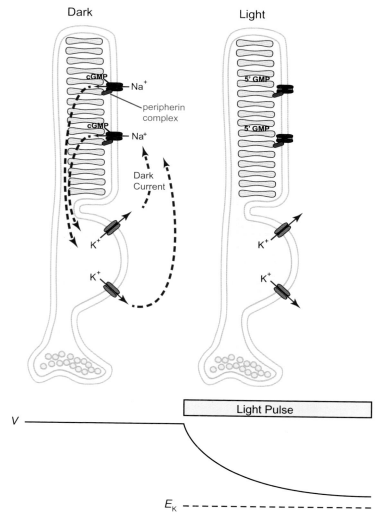

Figure 13–16. The dark current. In the dark, current flows in through CNG channels in the outer segment of a rod photoreceptor and out through potassium channels in the inner segment. A pulse of light closes the CNG channels, resulting in hyperpolarization of the rod. E_K, potassium equilibrium potential.

The effect of shining light on a rod is to shut down many of the CNG channels in the outer segment. This produces a marked decrease in the dark current (Fig. 13–16). As a result, the potassium channels, which remain open in the inner segment, produce a *hyperpolarization* of the cell toward E_K, and a *reduction* in the rate of spontaneous transmitter release at the terminal. The closure of the CNG channels can be directly attributed to the drop in cyclic GMP levels in the outer segment. The decrease in transmitter released from the photoreceptors onto other neurons in the retina is interpreted as an increase in light signal. ❽

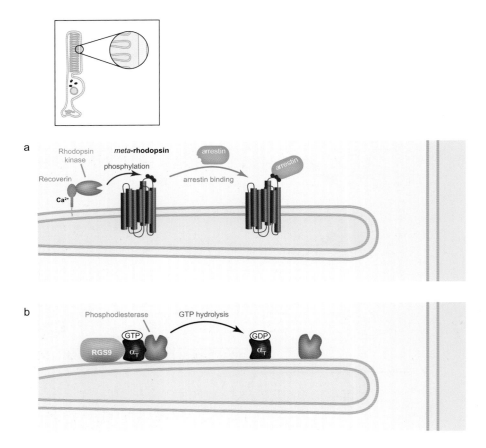

Figure 13–17. Termination of the light response. *a*: Inactivation of *meta*-rhodopsin by phosphorylation followed by arrestin binding. *b*: Inactivation of the α subunit of transducin by GTP hydrolysis. (Modified from Burns and Pugh, 2010.)

Terminating the light response. The analogies between visual transduction and neurotransmitter action can be taken further when one considers how the response to a flash of light is terminated. Although *meta*-rhodopsin is unstable, and left to its own devices would spontaneously dissociate into opsin and all-*trans*-retinal, this occurs too slowly to be useful in visual transduction. At least two biochemical reactions are required for normal vision. First, in Chapter 12 we saw that rhodopsin can be phosphorylated by a rhodopsin kinase that resembles β-ARK, the kinase that phosphorylates the β-adrenergic receptor, leading to its down-regulation. Rhodopsin kinase preferentially phosphorylates *meta*-rhodopsin. This phosphorylation promotes the binding of *meta*-rhodopsin to *arrestin*, a 43 kDa protein whose function is to prevent the *meta*-rhodopsin from activating further transducin molecules, thereby terminating the light response (Fig. 13–17a). After phosphorylation, the all-*trans*-retinal dissociates from rhodopsin, leaving the opsin protein, which must bind another 11-*cis*-retinal before it can again be activated by light. The overall level of rhodopsin kinase activity is controlled by a calcium-binding protein called *recoverin*, which normally inhibits the kinase and may play a role in adjusting the sensitivity of the eye to different ambient levels of light (Fig. 13–17a). A second biochemical process is required to stop phosphodiesterase activity. A protein termed *RGS9* (**R**egulator of **G**-protein **S**ignaling **9**) acts on the GTP-bound α_T transducin subunit, accelerating the hydrolysis of GTP to GDP and thus preventing further activation of the phosphodiesterase by α_T (Fig. 13–17b). This second process appears to be the key rate-limiting step for ending the response to a visual stimulus. ❷

Chemoreceptors

The olfactory system. The task of a *chemoreceptor* cell is to signal to the nervous system a change in its chemical environment. Major use of chemoreceptors is made in those parts of the body specialized for taste (the *gustatory sense*) and smell (*olfaction*). The latter sense is particularly remarkable in the specificity with which it can distinguish different odorant molecules. Figure 13–18a shows the structures of three molecules that, at first glance, appear generally similar in shape and size. The sensations that they elicit, however, leave no doubt that they stimulate different patterns of activity in neurons of the olfactory system.

Figure 13–18b shows the anatomy of olfactory chemoreceptors in a vertebrate. These are neurons that are aligned in sheets within the nasal epithelium. Their axons travel to the brain, where they synapse within the olfactory bulb. Each receptor neuron has a dendrite that extends toward a layer of mucus lining the nasal cavity. There it forms a dendritic "knob" from which fine cilia extend into the mucus. It is this part of the cell that responds to odorant molecules that are borne in with the air circulating over the mucus.

It is perhaps not surprising that the response of the olfactory system to odorants results from a process directly analogous to neurotransmission.

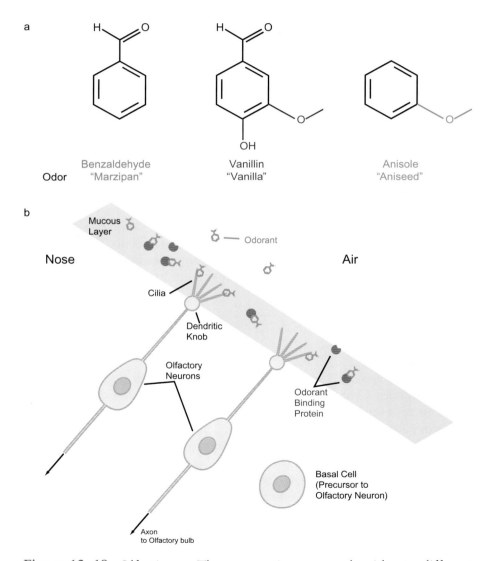

Figure 13–18. Olfaction. *a*: Three aromatic compounds with very different odors. *b*: Transport of odorants to olfactory neurons.

After all, odorants are simply chemicals that must interact with receptors on the sensory cells. Nevertheless, the olfactory system has to be able to respond appropriately to a vast number of possible odorants. It is estimated that we can distinguish between 5000 different odorants in the environment and can detect considerably more. To accomplish this, evolution has provided a very large number of different receptor molecules, the *olfactory receptor* proteins.

Like rhodopsin in the visual system, olfactory receptor proteins, termed *ORs*, have seven hydrophobic membrane-spanning domains, and are members of the G protein–coupled receptor family discussed in Chapter 12. Most tissues and parts of the brain need only a relatively small number of such receptors to allow them to respond appropriately to hormones and neurotransmitters. There are, however, as many as 400 different olfactory receptor proteins, each encoded by its own gene. The information used to encode these olfactory proteins takes up approximately 1% of the genome of vertebrates. There is sufficient similarity in the amino acid sequences of different olfactory receptors to define them as a family of closely related proteins. Because some family members are more closely related than others, it is also possible to assign each member of this enormous family to one of a number of specific subfamilies. Despite these similarities, however, there are several regions within the protein that are highly diverse in their sequence in different family members. The greatest diversity is found in the third, fourth, and fifth membrane-spanning domains (Fig. 13–19). It is these regions that are believed to bind specific odorant molecules.

The olfactory receptors are distributed among the olfactory sensory neurons, such that each individual sensory cell appears to express only one olfactory receptor. It has been possible to express the genes for some of the receptors in non-neuronal cells, and also in olfactory neurons, to test their response to odorants. It appears that the response of each receptor is rather specific in that it recognizes only a very small subset of possible odorants. For example, one well-studied receptor, OR-I7, recognizes aldehydes with chains of 5–11 carbon atoms, such as octanal, which has a pleasant, orange-like odor. ❽ It is, however, much less sensitive to the slightly smaller molecule, heptanal, which has a more "grassy unripe fruit" odor, and it is completely insensitive to most other odorants. Thus it is possible that, in the olfactory epithelium, a very low concentration of a specific odorant could activate one, and only one, specific receptor. Higher concentrations of odorants, however, probably activate several different receptors. The nature of the odor is thus encoded by the particular combination of sensory neurons that are activated by the odorant.

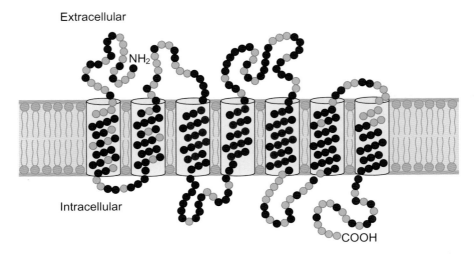

Figure 13–19. An odorant receptor protein. Individual amino acids are shown as circles. Residues that show the greatest variability between different odorant receptor proteins are shown in blue, and those in transmembrane domains that interact with odorants are shown in red. (Modified from Ressler et al., 1994.)

There are clear parallels between the activation of chemoreceptor cells and photoreceptors. Just as transducin is specific to photoreceptors, there exists a GTP-binding protein, termed G_{olf}, which is specific to the olfactory neurons. For olfaction, however, cyclic GMP appears not be involved. Instead, exposure to many odorants results in the activation of adenylate cyclase and the production of cyclic AMP or in the formation of inositol trisphosphate, a product of the activation of phospholipase C (see Fig. 12–10). The elevation of cyclic AMP levels activates a channel that is a member of the nonselective GNG family (Fig. 13–15a), producing a slow depolarization, with inward currents developing over several hundred milliseconds following application of odorants. Thus opening of the channel is achieved by the binding of cyclic AMP directly to the channel protein. The olfactory channel differs from the channel in rods, however, in that it can be opened by both cyclic AMP and by cyclic GMP; that in rods is highly specific for cyclic GMP.

Another point of similarity between olfactory and visual transduction is found in the way that the light-sensitive substance retinal and the odorant molecules are brought to the receptor cells. In the visual system, retinal is synthesized from the related molecule retinol, a highly lipophilic molecule. This is transported to the retina in a form that is bound to a carrier protein termed *retinol-binding protein*. This protein is closely related in its structure to *odorant-binding protein*, a dimer of two identical 19 kDa subunits found in the mucus around the olfactory cilia. The major role of odorant-binding protein is to bind and to concentrate airborne odorants and then to ferry them to the tips of the cilia (Fig. 13–18b).

One particularly interesting aspect of olfactory neurons is that, unlike most other neurons, they are not permanent. In the retina, we saw that photoreceptors are continually renewing their outer segments. Renewal in the adult olfactory system is accomplished by destroying *entire* olfactory neurons and replacing them by new neurons. The new neurons arise from precursor basal cells (Fig. 13–18b) in a manner similar to the formation of neurons during embryonic development. The axons of the newly formed olfactory neurons must then navigate their way to their postsynaptic targets in the adult olfactory bulb.

ORs have a say in telling the axons of olfactory neurons where to go. The axons of olfactory receptor neurons project from the nasal epithelium into the *olfactory bulb*, within the central nervous system (Fig. 13–20). Here they make connections with two types of neurons, mitral cells and tufted cells, both of which send their output further into the brain. These synaptic connections are made within large, spherical structures that have a diameter of 100–200 µm; these structures are termed *glomeruli*. Knowing the identity of the ORs has enabled the construction of "wiring diagrams" showing the pathways by which a single receptor connects to the brain. For example, genetic strains of mice have been developed in which all of the axons of neurons that make one specific olfactory receptor can be stained and their connections traced. Thus it has been found that the olfactory receptor neurons that bear the same receptor are not clustered together, but rather are found in a mosaic-like pattern over the surface of the epithelium. The axons from these neurons, however, converge in glomeruli in the olfactory bulb (Fig. 13–20). There are about 2000 glomeruli in the olfactory bulb of a rodent, and there are about 1000 different ORs. Thus axons from neurons containing a specific OR generally terminate in only two glomeruli that are dedicated to that receptor, one on each side of the olfactory bulb.

How does the axon of a specific olfactory neuron find its appropriate glomerulus? The general molecules involved in axon guidance and synapse formation will be covered in Chapters 16 and 17. The olfactory system may, however, have an additional twist. Genetic studies have shown that changing the particular OR that a receptor neuron makes can alter the target that its axon finally chooses. Thus, by some means that is not yet understood, the activity of ORs may function not only in sensory transduction, but may play a role in the selection of synaptic connections. &

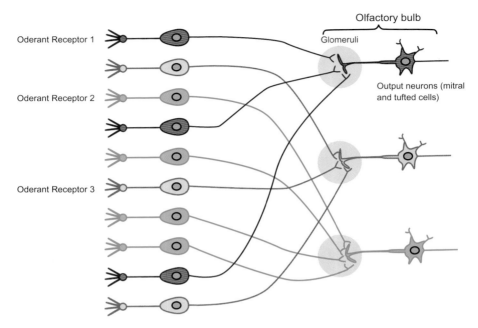

Figure 13–20. Connections of olfactory receptor neurons to the olfactory bulb. The outputs of neurons that are in different parts of the olfactory epithelium but that have the same OR converge onto the same glomerulus.

Taste Receptors

Another important class of chemoreceptive cells is found in the lingual epithelium of the tongue. These do not need to distinguish among as varied a selection of chemical stimuli as the olfactory neurons. Figure 13–21a shows a picture of a taste bud, with its receptor cells. In contrast to the olfactory cells, the taste receptors do not have axons but are innervated by afferent fibers that relay gustatory information to the brain. The sensing of stimuli occurs in finger-like projections, or *microvilli*, at the apical surface of the taste buds. Taste receptors are responsible for the basic types of taste sensations in food: sweet, salty, bitter, and umami (a "meaty" taste produced by substances such as glutamate). Each of these types of sensations has its own receptor cells, with their own type of molecular machinery (Fig 13–21b).

The sensation of sourness appears to be produced by the direct action of protons on ion channels in the microvilli of taste receptors cells dedicated to sensing acid. A key sensor for acid is a TRP channel constructed of channel subunits *PKD2L1* (also called TrpP3) and *PKD1L3*. We

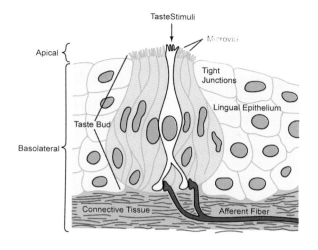

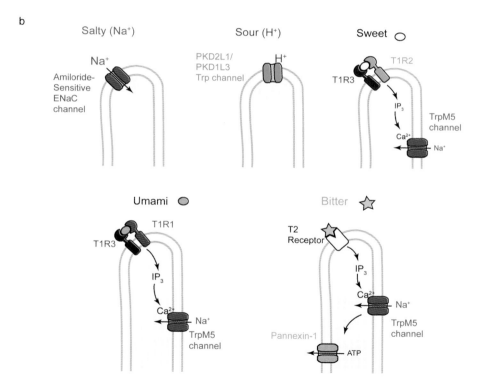

Figure 13–21. Taste transduction. *a*: Taste receptor cells within a taste bud (adapted from Kinnamon, 1988). *b*: Mechanisms that transduce different taste stimuli. The release of ATP through pannexin-1 is thought to occur for all the G protein–coupled protein responses, but for simplicity is shown only in the diagram for bitter taste.

encountered TRP channels in Chapter 7; the letters *PKD* stand for Polycystic Kidney Disease, as this family of TRP channel subunits was first discovered to be defective in this disease. The sensation of saltiness is, in contrast, produced by the entry of sodium ions into receptor cells dedicated to detecting salt through *epithelial sodium channels (ENaC)* that are sensitive to the drug *amiloride*. We also encountered ENaC channels earlier in our Web text on mechanosensitive channels in the nematode. ❽

In contrast to salt and acid, which are first detected by ion channels, sweet and umami stimuli are first detected by G protein–coupled receptors termed *T1* receptors on the microvilli (Fig. 13–21b). These receptors resemble the metabotropic glutamate receptors that we covered in Chapter 12 in that they have long extracellular N-terminal domains that are required for ligand binding. There are three members in the T1 family—T1R1, T1R2 and T1R3—and these combine such that a T1R2+T1R3 dimeric receptor is the receptor for sweet taste, while the umami receptor is a complex of T1R1 with T1R3. Transduction of bitter-tasting food is carried out by a second family of G protein–coupled receptors, termed *T2* receptors, which has 25 members in humans. In contrast to the T1 family, T2 receptors more closely resemble rhodopsin and ORs, which have only a short N-terminal domain, and in which ligands can bind directly to transmembrane domains.

As with vision and smell, the taste system has its own G protein α subunit, called *Gα-gustducin*. Both T1 and T2 receptors activate Gα-gustducin. The βγ subunits released from Gα-gustducin activate phospholipase C to produce IP_3, which then liberates calcium from internal stores (see Fig. 12–12). Thereafter, the transduction process is a little unorthodox and differs from that of photoreceptors or olfactory receptors. The released calcium activates another TRP channel termed *TRPM5*, causing depolarization of the cell (Fig. 13–21b). The neurotransmitter used by the taste receptor cells is simply ATP, which, following the depolarization, appears to be released through gap junction hemichannels formed by pannexin-1. We encountered both hemichannels and pannexins in Chapter 8.

Other Sensory Receptors

An additional sensation that has not yet been covered is that of being hot or cold. A key discovery that led to our current understanding of this sensation was made by the discovery of a receptor for *capsaicin*, the key ingredient in chili peppers that is responsible for both the pain and the pleasure of eating very spicy foods. This receptor turned out to be yet another TRP channel, TRPV1, which is present in dorsal root ganglion neurons. Like the mechanoreceptors we discussed at the start of the chapter, these neurons send their processes out to the skin. Increases in temperature shift the voltage-dependence of the channel to negative potentials, causing them to open at the resting potential and allowing sodium influx sufficient to trigger firing. Several other TRP channels, TRPV2, V3, and V4, are now known to also respond to increases in temperature, covering the range of sensations from pleasant warmth to very painful heat. A different TRP channel, TRPM8, which opens on cooling rather than heating, is found on neurons that respond to cold temperatures.

Another sensory pathway is activated by insect bites, or agents that cause painful local inflammation of tissue. While this is often ascribed to the sense of touch, it is in fact mediated by chemoreceptors. The immediate stimulus causes the release of active substances from non-neuronal cells in the skin. These substances include peptides such as *bradykinin*, and lipid molecules such as the arachidonic acid metabolic products called *prostaglandins*. These act directly on the processes of sensory neurons that have their cell bodies in the dorsal root ganglion. **&**

Summary

Sensory cells have evolved pathways that allow ion channels to be regulated by external stimuli such as movement, light, or chemicals. In some cases, such as in photoreceptors and olfactory and taste receptors, the means by which the external stimulus is transduced is reasonably well understood. Such cells appear to handle information in ways similar to those used by neurons that deal with information coming from a presynaptic pathway, by altering the levels of second messengers such as cyclic nucleotides, which then open or close ion channels in the plasma membrane. In contrast, in mechanoreceptors, movement is directly linked to the gating of ion channels. The study of sensory receptors has also uncovered some unexpected twists in signaling pathways. Ion channels that are opened directly by cyclic GMP and cyclic AMP have been found in photoreceptors and olfactory receptors. Some auditory hair cell receptors are exquisitely tuned to specific frequencies of stimulation, rather than solely to the amplitude of a sound wave. It is likely that such mechanisms were discovered first in sensory cells because of the intense interest neurobiologists have in these cells that tell us about the outside world. On further searching, some of these mechanisms have also been found in nonsensory neurons.

IV

Behavior and Plasticity

The previous two parts of this book dealt with the mechanisms that neurons use for intracellular and intercellular information transfer. In this final part we will discuss *behavior*, the output of the nervous system. We will also cover *plasticity*, changes in properties of individual neurons and the patterns of connections among them that lead to changes in behavior. Our discussion of plasticity will focus first on *development*, a time of rapid and dramatic changes in neuronal properties. Although our emphasis throughout this book has been on adult neurons, we begin with development because many of the cellular and molecular mechanisms that contribute to developmental plasticity are also relevant to plasticity in the adult. Chapter 14 deals with the early events that determine whether an immature cell develops into a neuron or into some other cell type. It also covers the factors that govern *neuronal survival*, both during the formation of the nervous system and in response to damage to the brain later in life. This is followed in Chapter 15 by an account of the many molecules that regulate the growth of neuronal precursors and their *differentiation* into adult nerve cells. Once an immature neuron begins to extend an axon, this axon must find its way, often over very long distances, to its synaptic target. The role of various protein molecules in this process of *pathfinding* is summarized in Chapter 16. Such molecules play an important role in guiding neuronal processes through the extracellular matrix and over the surfaces of other cells. Specific cell surface molecules also tell neurons when they have reached an appropriate synaptic target and can stop migrating. In Chapter 17 we discuss *synaptogenesis*, the formation of the complex chemical synapse that occurs when an axon reaches a suitable target. During development and also in the adult, synapses can be continually formed and eliminated. The mechanisms that contribute to the *selective maintenance* of some synapses therefore are also presented in this chapter.

The ways in which the *intrinsic excitability* of neurons within interconnected *networks* can generate behavior are the subject of Chapter 18. Certain small groups of neurons that can mediate surprisingly complex behavioral outputs have been analyzed in detail. These analyses have provided insights into the way individual neurons are uniquely tailored to their role in a network and the way changes in their cellular properties alter the output of the network. Computational models based on biological data can be used to investigate this organization and to make quantitative predictions that can be tested by further biological experiments. Finally, in Chapter 19 we end the book with a discussion of *learning and memory*. These fundamental features of animal behavior, the ability to modify behavior as a result of experience (learning) and to maintain the new behavior, often for as long as the animal lives (memory), have fascinated scientists for many years. Recent advances in our understanding of neuronal properties and their modulation (see Chapters 11 and 12), and of the selective stabilization of synaptic pathways (see Chapter 17), are providing new insights into the cellular and molecular mechanisms of learning and memory.

The Birth and Death of a Neuron

R adical changes in the structure and connections of neurons occur during the development of the nervous system. Immature neurons are subject to chemical and mechanical influences that cause them to migrate to various locations in the nervous system, to extend axonal and dendritic processes toward other cells, and then to make and break synaptic connections with these cells before a final pattern of branching and connections is established. While a full account of the different stages of neuronal development is beyond the scope of this book, we will address the theory that neuronal plasticity in the adult animal may utilize mechanisms that are active during development, a period of profound plastic changes in neuronal structure and function. Many of these mechanisms may also contribute to the maintenance of neuronal form and connections in the adult. In this chapter, we shall therefore first give a brief and general account of the normal course of neuronal development, and discuss some of the molecules involved in the earliest steps of the formation of the nervous system. We shall cover how new neurons come into being, both during development and in the adult. Finally, we discuss the mechanisms that lead to the death of a neuron. The following three chapters (Chapters 15, 16, and 17) will cover molecules that determine the existence and properties of select groups of neurons, the growth of axons, and the formation of synaptic contacts.

Cell Determination

Early in the development of a vertebrate embryo there exist three layers of cells. The cells destined to become neurons are found in the most external layer, termed the *ectoderm*. Immediately under the ectoderm lies a layer of cells termed the *mesoderm*. The first step in neuronal development is the determination of cells in the ectoderm to become neuronal precursors; this is often called *neural induction* (Fig. 14–1). In the nervous system of vertebrates, this appears to result from the action of *neural inducers*—diffusible molecules—released from nearby cells in the mesoderm. After this stage, the cells in the ectoderm can develop only into neurons, glial cells, or a limited number of other cell types.

Some of the most critical experiments for our understanding of neural induction have come from studies of the African frog *Xenopus laevis*. In normal development, only a small part of the external ectoderm layer, that situated at the most dorsal (top) side of the embryo, develops into neural tissue. The remainder of the ectoderm develops into *epidermis*, the

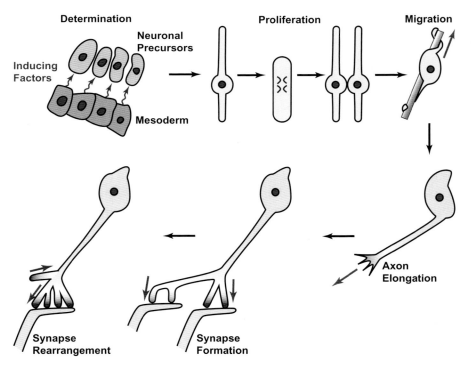

Figure 14–1. Some of the stages in the development of neurons discussed in this chapter and in Chapters 15–17.

external skin of the animal. This occurs both in an intact embryo and in a culture dish. The *animal cap* is a layer of ectoderm that stretches from the dorsal to the ventral side of the embryo. When any part of the animal cap is dissected out from the embryo and maintained in a culture dish, its cells develop into epidermis (Fig. 14–2).

In the 1920s the biologist Hans Spemann found that when a small, specific part of the developing embryo, which has now come to be termed the *Spemann organizer*, is grafted onto the ventral part of a recipient embryo, it causes the ectodermal cells at that position to develop a secondary nervous system. Again, this experiment can be recreated in a culture dish. If the animal cap of ectodermal cells is cultured together with the Spemann organizer, its cells take on the appearance and molecular properties of neural tissue (Fig. 14–2). The Spemann organizer is an important source of neural inducers. A variety of these molecules have been found, and they have been given names such as *Noggin, Chordin, Follistatin,* and *Cerberus.* Application of any one of these to the ectodermal cells induces the formation of neural tissue.

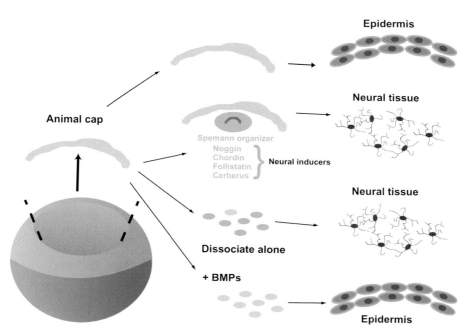

Figure 14–2. Different developmental paths of ectodermal cells when they are cultured alone, with the Spemann organizer, or with bone morphogenetic proteins (BMPs).

How do these neural inducers work? An important clue came from the finding that if ectodermal cells are completely dissociated from each other, before culturing, they eventually take on many of the characteristics of neurons or glia (Fig. 14–2). In such solitary confinement, they send out long processes from their cell bodies and synthesize proteins that are normally restricted to the nervous system. This discovery led to the suggestion that, when left to their own devices, ectodermal cells prefer to develop into neural tissue. In a tightly packed aggregate of cells such as the animal cap, however, the cells are exposed to factors that inhibit the formation of the nervous system and, instead, urge them to form epidermis. Such neural inhibitors have indeed been identified. They belong to a class of signaling molecules termed *bone morphogenetic proteins* (BMPs).

BMPs Are Neural Inhibitors

Bone morphogenetic proteins belong to a large family of factors termed the *transforming growth factor-β (TGF-β)* family. Like the neuropeptides that we considered in Chapter 8, the BMPs are first synthesized as large precursor proteins, from which the active BMP molecule is cleaved. They assemble into dimers and are then secreted from cells by a constitutive pathway (see Fig. 8–5). More than 20 BMPs are known. As we shall see in Chapter 15, they play important roles in the development of the nervous system at later stages, in addition to their ability to divert ectodermal cells from a neural fate, which we will discuss now.

On contacting an ectodermal cell, a BMP binds to a receptor comprised of two different subunits, the type I and type II BMP receptor subunits (Fig. 14–3a). Both subunits are protein kinases that phosphorylate proteins at serine residues. When a BMP is bound, the type II receptor phosphorylates its partner the type I receptor, which in turn phosphorylates a cytoplasmic protein termed *Smad1*. This protein is one of a number of *R-Smads* (receptor Smads), proteins involved in ferrying messages from receptors of the TGF-β family at the plasma membrane to the nucleus of the cell. (The term *Smad* is a hybrid word derived from the names for these signaling molecules in flies and worms, where they were first discovered). The phosphorylated Smad1 protein then binds another cytoplasmic protein, termed *Smad4* (sometimes termed a *Co-Smad*), allowing the Smad1–Smad4 complex to enter the nucleus of the cell and influence transcription. The end result of activation of the Smad pathway in ectodermal cells is to make them into epidermis.

The pathway from a cell that secretes BMPs to its receptors on nearby cells is not a straightforward one. Once BMPs have been released, they

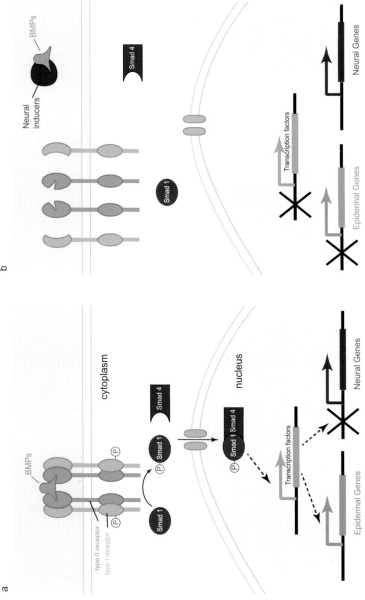

Figure 14–3. Signaling pathways for the bone morphogenetic proteins (BMPs). *a*: BMPs act though Smad proteins to activate epidermal genes and to suppress neural genes. *b*: In the presence of neural inducers, the BMP signaling pathway is inhibited, leading to the activation of neural genes (Chang and Hemmati-Brivanlou, 1998; Bond et al., 2012).

diffuse through the extracellular space, where they to bind to proteins that are normally resident in this space. It is here, in the external environment of the cells, that the neural inducers do their work. Inducers such as Noggin bind BMPs directly in the extracellular space, preventing the access of the BMPs to their normal target cells (Fig. 14–3b). In the absence of the activated Smad signaling pathway, the nucleus of the ectodermal cells carries out a program of activity designed to make a neural cell.

What Does It Mean to Be a Neuron?

How does inhibition of the Smad pathway lead to the production of neurons or other cells of the nervous system? At the heart of the difference between a cell that is going to become a neuron and one that will eventually be an epidermal cell is their pattern of *gene expression*. Inhibition of the BMP/Smad pathway must lead to the activation of the genes that encode proteins required for the construction of a nervous system, and to the inhibition of those specific to the epidermis.

The regulation of neural-specific genes, such as those for specific ion channels or neurotransmitter receptors, is a multistep process. Once the Smad1/Smad4 complex reaches the nucleus, as occurs in ectodermal cells exposed to BMPs, it binds to other proteins and to specific sequences of DNA, altering the transcription of genes (Fig. 14–3a). Most of the genes initially influenced by the Smad pathway are, however, not those that encode proteins with day-to-day functions for epidermal or neural cells. Rather, it is the synthesis of *transcription factors* that is first modified. Transcription factors are proteins that, once synthesized in the cytoplasm, quickly return to the nucleus, where they bind other regions of DNA that regulate the formation of epidermal or neural proteins. The BMP/Smad pathway leads to the synthesis of a particular set of transcription factors that stimulate epidermal genes and inhibit the transcription of neural genes. When the BMP/Smad pathway is inhibited by neural inducers such as Noggin, the epidermal proteins are no longer made, and other transcription factors, resident in the nucleus, allow the synthesis of neural proteins to take over. By controlling the synthesis of transcription factors, BMPs and neural inducers are able to exert orderly control over the decision to become an epidermal cell or a neural cell.

Table 14–1 provides a list of some of the transcription factors that control neural development. One important transcription factor that plays a key role in the decision of whether to become a neuron or other cell type is called *REST*, for **R**epressor **E**lement 1 (RE-1)-**S**ilencing **T**ranscription factor. This protein binds to a 21-nucleotide stretch of

Table 14–1 Examples of Transcription Factors That Determine the Fate of Stem Cells and their Progeny

Type of Transcription Factor	Biological Role	Examples
Proneural	Basic helix-loop-helix (bHLH) proteins that, when expressed, commit the cells to exit the cell cycle and commit to becoming neurons. They also activate the Notch signaling pathway in adjacent progenitor cells (see Chapter 15).	Mash1, Neurogenins 1–3, Math1
Patterning	The expression of these transcription factors subdivides the developing nervous system into different regions.	(1) Homeodomain (HD) proteins such Pax6, Otx2, Nkx6.1, Irx1–6, which specify cells along the dorsoventral axis of the neural tube (2) Homeodomain proteins such Otx, Gbx, En, and Hox family proteins that specify cells along the anterior–posterior axis of the neural tube. Also Phox2b and Phox2a proteins that make adrenergic neurons (see Chapter 15). (3) The bHLH protein Olig2 which specifies cells along the dorsoventral axis
Stem cell renewal	When expressed, these prevent stem cells from differentiating into neurons or glia. They oppose the actions of the proneural transcription factors.	TLX (*tailless*), a nuclear hormone receptor Sox2 Hes1, Hes5 (*hairy*, enhancer of *split*), which are bHLH transcription factors
Factors that prevent the expression of neural-specific genes	These act as silencers to prevent the synthesis of messenger RNA for many neural proteins.	REST (RE1-silencing transcription factor)

DNA termed the *RE-1 silencer element*, which is present in the regulatory regions of the genes for many of the proteins found selectively in mature neurons. By interacting with a host of other nuclear proteins, REST selectively suppresses the synthesis of messenger RNA for the neural proteins. Thus REST is one of those transcription factors that prevent cells from becoming neurons. In contrast to REST, the transcription factors that turn on genes required for formation of the nervous system are encoded by *proneural* genes. One class of transcription factors particularly important for the formation of the nervous system is the *basic helix-loop-helix* factors, or *bHLH* transcription factors, so called because of a conserved pair of helices in the structure of these DNA-binding proteins.

Once the simple choice of whether to make a nervous system has been made, there remains the trickier question of whether to make neurons or glia, and whether to make a cerebral cortex, a spinal cord, or an auditory brainstem. Such decisions are governed by secreted signaling factors, termed *morphogens*, which locally organize tissues. Morphogens confer positional information to each cell in a growing organism and contain specific instructions for cell proliferation and fate. The BMPs are one example of a class or morphogens; and others are discussed later in the chapter. Helping to make these decisions is another class of important transcription factors that contain conserved regions termed *homeodomains*. These divide the growing nervous system into discrete regions that will eventually become different brain regions. The particular mix of bHLH and homeodomain transcription factors that become activated in a cell may be the key determinant of what type of neuron that cell will eventually become. The next chapter will describe in more detail how some of these patterns get established. ❧

Cell Proliferation

During and following the actions of factors that determine the developmental fate of a particular group of neuronal precursors, the cells begin to divide (proliferation in Fig. 14–1). During development of the vertebrate central nervous system, the embryo invaginates such that the location in which the neuronal precursor cells proliferate is adjacent to internal fluid-filled ventricles. This location is the *germinal* zone. As the cells divide, they undergo a series of changes in shape that are characteristic of dividing cells in most epithelia (Fig. 14–4a). These are, in sequence:

1. The cells send out processes that span the thickness of the germinal zone (G_1 phase).

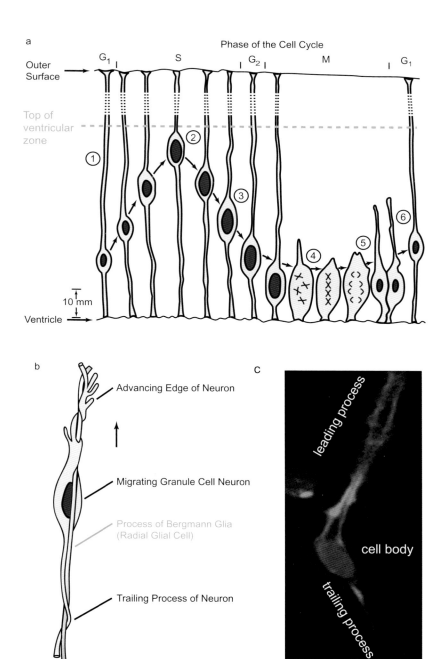

Figure 14–4. Proliferation and migration. *a*: Movements of cell nuclei during proliferation (Jacobson, 1978). *b*: The anatomical work of Pasco Rakic (1971) demonstrated the migration of neurons in the cerebellum along radial glial fibers. *c*: Fluorescent video image of a cerebellar granule cell migrating along a radial glial cell (Solecki et al., 2009). The cytoskeletal protein actin in the processes of the neuron is stained green and its nucleus is stained red. The radial glial cell itself cannot be seen in this image. See Video 14-1. ⭕

2. As the cells synthesize DNA, the nuclei of the cells move along the processes away from the ventricles (S phase).
3. The nuclei return along the processes toward the ventricular surface.
4. The processes of the cells retract.
5. The cells divide.
6. The two daughter cells re-extend processes to span the germinal zone, and the cycle begins again.

After several such cycles, the cells either lose the ability to divide further and become *postmitotic*, or they migrate to another region of the developing nervous system where they undergo several more divisions before ceasing to proliferate. In the vertebrate nervous system, an immature cell that has undergone its final division and is destined to become a neuron is termed a *neuroblast*. There is some semantic confusion here in that, in invertebrates, the same term *neuroblast* is used to describe cells that can proliferate but whose progeny are destined to become neurons.

Stem Cells

In the developing brain, immature cells continually assess their environment for the presence of factors such as BMPs, neural inducers, and other cells. As a result of each of these interactions, the developing cell makes a new set of proteins that determine the ultimate function of that cell in the adult brain. Like Peter Pan, however, some cells refuse to grow up. They persist from the embryo to the adult as cells that have not followed the pathways that make epidermis, neurons, glia, or other mature cell types. Such cells are termed *stem cells*.

Stem cells have the capacity for unlimited *self-renewal*. Each time a stem cell divides, it produces two cells, one of which is identical to the original stem cell (Fig. 14–5). Depending on the environment, the other daughter cell may die, become another stem cell, or become a progenitor cell. In some cases, which of these events actually occurs depends in part on a protein called *Numb*, which is distributed asymmetrically as the cells divides. Numb inhibits *Notch* signaling, a process described in the next chapter. The daughter cell that ends up with the Numb protein typically becomes a *progenitor* cell, which then follows environmental cues to produce highly differentiated descendants such as neurons or other mature cells. In fact, the progeny can usually produce many different types of differentiated cells. There is, however, no looking back. Once the progenitor cells have embarked on a course of development, they can never produce another stem cell. In contrast, the daughter cell without the Numb protein remains an uncommitted stem cell for the rest of its life, and may continue to divide to produce more stem cells.

The central function of stem cells may be to provide a never-ending source of progenitor cells for the replenishment of adult tissues. Stem cells may have different properties in different tissues. Very early in the development of the embryo, there exist stem cells that have the capability of producing progeny that end up in every tissue of the mature organism. Such *embryonic stem (ES) cells* are termed *pluripotent*. In contrast, stem cells in the brain are thought to be capable of generating only a limited number of cell types that include both neurons and glia. As a result, such stem cells are often referred to as *multipotent*.

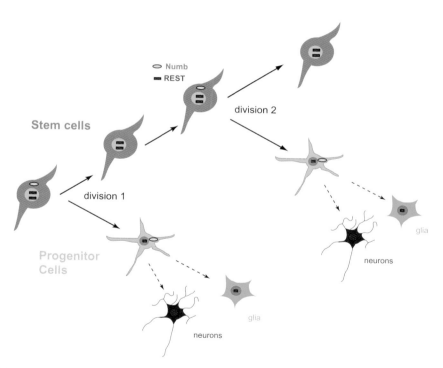

Figure 14–5. Stem cells. Division of stem cells produces more stem cells, as well as cells that eventually differentiate into mature cells of the nervous system. In some cases, after division, the cells that retain the Numb protein become neuronal progenitor cells. Cells that eventually become neurons lose expression of the REST silencer transcription factor.

Cell Migration

Once a progenitor cell has been generated, it follows environmental cues to eventually produce highly differentiated descendants such as neurons or glia. These environmental cues include direct physical interactions with other cells and factors released from nearby cells. If a cell develops into a neuron, factors released by other cells also influence the type of neurotransmitter it synthesizes and the specific mixture of receptors, ion channels, and other proteins that determines the characteristics of the fully differentiated neuron. Some of these factors will be covered in Chapter 15.

After their final division, a few cells may already be situated close to their eventual location in the nervous system. The majority, however, migrate (Fig. 14–1) considerable distances to arrive at their final destination. One particularly important stem cell type that is generated at this early stage is the *radial glial cell* (Fig.14–4b,c). This cell has two very important functions. Although termed a "glial" cell for historical reasons, it is, in fact, a dividing stem cell with the capacity to give birth to new neurons. At the same time, it extends cytoplasmic processes across the full thickness of the cortical wall, providing a track along which neurons can migrate. In contrast to the cell depicted in Figure 14–4a, radial glial cells maintain this process during the entire cell cycle. While many types of radial glial cells die or differentiate into astrocytes when their job in neural development is over, some persist into adulthood. Examples of these are *Bergmann glia*, which are produced in the germinal zone of the developing cerebellar cortex (Fig. 14–4b). Their radial fibers provide a pathway for the migration of neurons from the external surface (the external granule layer) down into the developing cortex. The cells that travel along these glial fibers eventually form a layer of granule cell interneurons in the mature cerebellar cortex.

Neurons can still be born in the adult brain. In the adult brain, *neurogenesis*, the process of making new neurons, occurs only in a few select areas. As is the case in the developing embryo, stem cells for this process are located adjacent to the fluid-filled ventricles, in the *subventricular zone*. Some of the new neurons born in this area then migrate along a pathway known as the *rostral migratory stream* and eventually settle in the olfactory bulb (Fig. 14–6a,b). These new neurons establish full functional connections with other cells and become fully integrated into the neural circuits that are responsible for the sense of smell. Another site of stem cells that provide an ongoing source of neurons is the *subgranular* zone, a layer of the hippocampus (Fig. 14–6c). These cells provide for the replenishment of neurons in the *dentate gyrus* of the hippocampus. As we

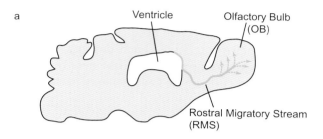

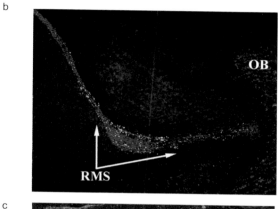

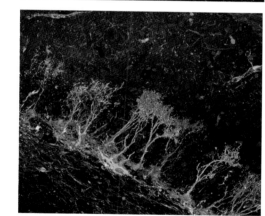

Figure 14–6. The rostral migratory stream. *a*: Path of migration of newly born neurons in the adult brain from ventricles to the olfactory bulb (Gage, 2000). *b*: A section of adult mouse brain labeled with bromodeoxyuridine (green-white) to show newly formed neurons migrating toward the olfactory bulb (OB). Dyes have also been used to show mature neurons (red) and the rostral migratory stream (RMS, blue) (van der Kooy and Weiss 2000). *c*: An image of neural stem cells in the hippocampus made by Grigori Enikolopov and Ann-Shyn Chiang. The stem cells are shown as green and the nuclei of neurons are stained red.

shall see in Chapter 19, this region is thought to be important for the acquisition of new memories. It has been estimated that, in the mammalian brain, one new neuron is added each day for every two thousand existing neurons. The ongoing production of new neurons in these key areas of the brain suggests that new neurons may perhaps be required for some forms of learning.

The fact that the mammalian brain can make new neurons in certain regions, and that these neurons can be integrated into the normal circuitry of the brain, also offers hope for the treatment of diseases such as Parkinson's disease, which are caused by the death of entire populations of neurons. It is possible to take a fully differentiated human cell, such as a skin fibroblast, and to "reprogram" it so that it regresses to an embryonic stem cell–like state. This is accomplished by choosing an appropriate mix of transcription factors, such as some of those in Table 14–1, and using genetic approaches to force the cell to synthesize these factors. Such cells are termed *induced pluripotent stem (iPS) cells*. Moreover, again, with the right mix of transcription factors, it is possible to take a differentiated human cell and convert it directly into a multipotent *induced neural stem cell* (termed an *iNSC*) or even directly into a neuron, bypassing the progenitor stage. The latter is then termed an *induced neuron (iN)* cell. The implantation of iNSC or iN cells into patients may eventually provide a mechanism for restoring the normal complement of neurons in areas that have suffered neuronal loss. ❽

Cell Death During Development

As important as the birth of new neurons is for the proper development of the nervous system, many neurons also have to die to make things right. Entire populations of neurons may arise during development only to die before a stable mature nervous system is formed. Very elegant examples of how cell death is used to build a simple nervous system have been described in invertebrates such as leeches, nematodes, and grasshoppers. For example, the nervous system of the grasshopper comprises a series of segmental ganglia along the length of the body. These ganglia are not identical and contain different numbers of neurons. Initially, the pattern of cell division is similar in all of the segments. Later, the "unwanted" cells in a given segment simply die, leaving the remaining cells to establish the appropriate pattern of connections. This phenomenon has been termed *programmed cell death*.

In vertebrates as many as three-quarters of the neurons destined for a specific neuronal pathway may die during early embryonic development. In some cases, survival of a neuron may require exposure to some diffusible *trophic factor* released by its postsynaptic or presynaptic target. Thus removal of a target, for example, a limb bud in a chick embryo, causes the atrophy of sensory neurons that normally would innervate the limb. Conversely, implanting an extra limb partially prevents the loss of cells in the sensory ganglia. In other cases, the survival of a group of neurons may depend on exposure to a specific factor at a critical time in development. This process of dying, termed *apoptosis*, is associated with a specific set of biochemical reactions that differ from those that occur during death from injury. In particular, the nucleus and the cytoplasm of the affected cells shrink, and the DNA of the cell is cleaved at specific sites (Fig. 14–7). Fragments of the dying cells are rapidly removed by *phagocytosis*, a process in which these fragments are taken up into neighboring cells or into *macrophages*, blood cells whose function it is to remove debris.

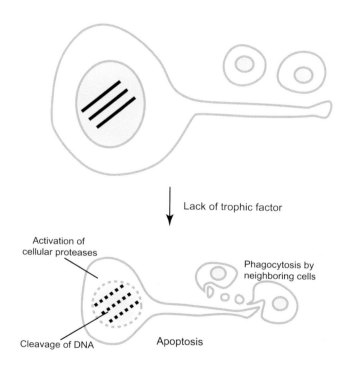

Figure 14–7. Programmed cell death. Lack of trophic factors leads to proteolysis of cellular proteins, cleavage of DNA, and removal of cell fragments by adjacent cells. See Animation 14–1. ◖

Insights into apoptosis have come from the identification of numerous proteins encoded by "death genes" in the nematode *Caenorhabditis elegans*, whose development proceeds through a very stereotyped pattern of selective cell death. The first of these genes identified, termed *ced-3*, encodes a protease that becomes activated in the cytoplasm of a cell destined to die, resulting in the destruction of cellular proteins and the breakup of the nuclei. Mammals have a number of proteases that are counterparts of the ced-3 enzyme; these are termed *caspases*.

Life and Death Decisions Are Made by Mitochondria

Clearly, caspases are dangerous enzymes. Yet they exist even within mature neurons and must normally be maintained in an inactive state. The activation of these proteases occurs only when an irrevocable decision has been made that it is appropriate for a cell to die. This decision is generally made by a second class of "death genes"—those that encode the Bcl-2 family of proteins. The Bcl-2 family of proteins can function as ion channels, and they carry out their work in the outer membrane of mitochondria (Fig. 14–8a). In neurons, the greatest density of mitochondria is found at synaptic endings.

To understand how the Bcl-2 proteins regulate apoptosis, we must review some basic properties of mitochondria. These organelles have two membranes, a highly invaginated inner membrane and a smooth outer membrane. Between the two membranes is the *intermembrane space*. The best-known function of mitochondria is to make the ATP that powers all cellular functions. An important component in the pathway that uses oxygen to drive ATP synthesis is the enzyme *cytochrome c*, which normally resides in this intermembrane space. In response to external signals that induce apoptosis, cytochrome c is released from the intermembrane space into the cytoplasm, where it binds and activates caspases, triggering apoptosis. The first enzyme directly activated by binding to cytochrome c is *caspase 9*. This enzyme in turn activates another caspase, *caspase 3*, leading to a chain of biochemical reactions that destroy cellular DNA and other cellular constituents (Fig. 14–8a). In neurons, the external signal that triggers this sequence of events is typically the withdrawal or loss of trophic factors. We shall cover the biological actions of several of these neuronal factors in Chapter 15.

The Bcl-2 family of proteins appears to be a key regulator of the permeability of the outer mitochondrial membrane. Bcl-2 proteins come in two flavors: those that promote apoptosis and those whose function is to *prevent* programmed cell death, termed *antiapoptotic* proteins. The Bcl-2

a

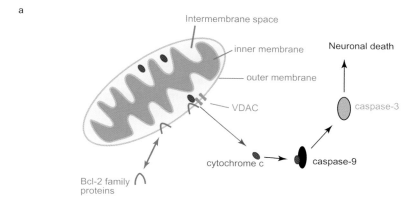

b

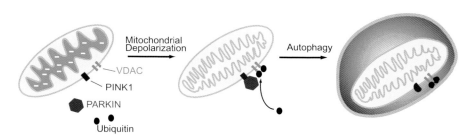

Figure 14–8. Role of mitochondria in cell death. *a*: Regulation of the permeability of the outer mitochondrial membrane by Bcl-2 family proteins. Release of cytochrome c from the mitochondria into the cytoplasm activates caspases, leading to apoptosis. See Animation 14–1.⬤ *b*: Actions of Parkin and PINK1 in autophagy of mitochondria. VDAC, voltage dependent anion channel.

protein itself is antiapoptotic, as is its relative Bcl-x$_L$. Proapoptotic members have names such as *Bax, Bid*, and *Bad*. Some of these proteins have been shown to be ion channels. In contrast to the plasma membrane channels covered in the first few chapters of this book, however, the Bcl-2 channels are not integral membrane proteins, and they are not inserted into membranes by a secretory pathway (see Fig. 8–5). Instead, they bind to cellular membranes directly and insert their ion channel pore region across the membrane. In fact, while some of the Bcl-2 family proteins are bound to the outer mitochondrial membrane at all times, others are usually bound to proteins in the cytoplasm but can shuttle between the cytoplasm and the mitochondrial membrane. This occurs in response to changes in the external factors that induce apoptosis. ⊗

When activated by external signals, the proapoptotic Bcl-2 family members, such as Bax, form a channel with an exceptionally large pore that allows the passage of cytochrome c across the outer mitochondrial membrane, leading to the activation of caspases (Fig. 14–8a). In mice in which the genes for either caspase 9 or caspase 3 have been deleted, there is an excessive overall production of neurons because of the lack of neuronal apoptosis. Paradoxically, this excessive production leads to the death of the animals before birth. Of course, it is essential that cell death occur only in "unwanted" neurons. In some way that is not yet fully understood, signaling pathways that activate antiapoptotic Bcl-2 proteins render the mitochondria resistant to the action of the proapoptotic proteins and prevent death of the cells. Thus the development of the nervous system represents a battlefield in which different external factors vie for control of the permeability of the mitochondrial membranes of the neurons.

When a Neuron Dies Before Its Time

A number of well-known neurological diseases occur due to the untimely death of neurons in the adult brain. These include *Parkinson's disease* and *Alzheimer's disease*. In addition, adult neurons may die following a blow to the head or after a stroke, caused by the bursting of blood vessels in the brain or a cutoff of blood flow to part of the brain. In each of these cases the mechanism that leads to neuronal death is different, and each provides some insights into what is required for normal brain function.

Parkin, α-synuclein, and Parkinson's disease. This disease results from the premature death of neurons that make the neurotransmitter dopamine in the basal ganglia, an area that controls movements. It affects approximately 1% of the population over age 65. Before the neurons die, they make large, spherical aggregates called *Lewy bodies* within their cytoplasm. The major protein within these aggregates is *α-synuclein*, whose normal function is not yet understood. While the cause of most cases of Parkinson's disease is not yet known, some cases can be attributed to mutations in the gene for specific proteins such as α-synuclein itself, or for the proteins *Parkin* and *PINK1*.

As part of the general maintenance of neurons, new mitochondria are continually being produced while old or damaged ones are destroyed. Because, as we have seen, mitochondria are laden with dangerous molecules, they have to be disposed of very carefully. Parkinson's disease may result in part from an inability to destroy mitochondria without damaging the entire neuron. The membrane potential across the inner mitochondrial membrane is normally very negative ($\sim$180 mV) but becomes much more positive as the function of a mitochondrion fails. This depolarization of the inner membrane is sensed by the protein kinase PINK1, which is located in the outer membrane, triggering the recruitment of the cytoplasmic protein Parkin to the outer membrane (Fig. 14–8b). Parkin then carries out its function of covalently transferring *ubiquitin* molecules to other outer mitochondrial membrane proteins such as a mitochondrial ion channel termed the *VDAC* channel. This is followed by total envelopment of the entire mitochondrion by another membrane, a process termed *autophagy*. The autophagosome moves the mitochondrion to the lysosome, where it is safely disassembled and destroyed. When this process fails because of errors in the Parkin/PINK1 pathway, the neuron is likely to die. Why cell death in Parkinson's disease occurs preferentially in dopaminergic neurons that control movements is, however, not yet fully understood.

Amyloid precursor protein and Alzheimer's disease. The devastating dementia of Alzhemier's disease, which typically begins only late in life, is accompanied by loss of neurons and loss of functional synaptic contacts. In addition, the appearance of brain tissue under the microscope is changed in two striking ways (Fig. 14–9a). First, *amyloid plaques*, also known as *senile plaques*, are found in the extracellular spaces between the neurons. These are large, insoluble aggregates of a protein termed *β-amyloid*. Second, in the cytoplasm of neurons, aggregates of a microtubule-associated protein named *tau* give rise to *neurofibrillary tangles*.

Alzheimer's disease provides another example of why it is so important to dispose of cell components very carefully. The key component in this case is *amyloid precursor protein*, or *APP*. The normal function of this protein, which spans the plasma membrane, is not fully understood; but evidence indicates that during development it participates in the processes of neuronal survival and axonal pathfinding, covered in this and the next two chapters (Chapters 15 and 16). APP can be sliced into pieces by three different enzymes (Fig. 14–9b). *α-Secretase* and *β-secretase* cut APP at two sites in its extracellular domain. In contrast, *γ-secretase* cuts APP within the membrane-spanning region. The latter enzyme is a complex of membrane proteins, one of which is called *Presenilin*. There are two known forms of human Presenilin (Presenilin-1 and -2), and mutations in either one of these can result in an early-onset hereditary form of Alzheimer's disease.

Cleavage of APP by either α- or β-secretase releases soluble APPα and APPβ fragments into the extracellular space. Sequential cleavage by β-secretase and then γ-secretase, however, releases β-amyloid, which then aggregates into the insoluble amyloid plaques. Even before they form the

a

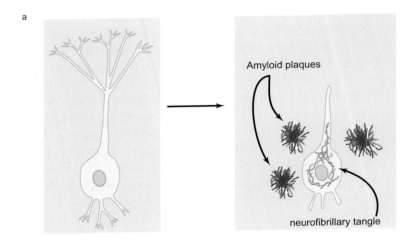

b

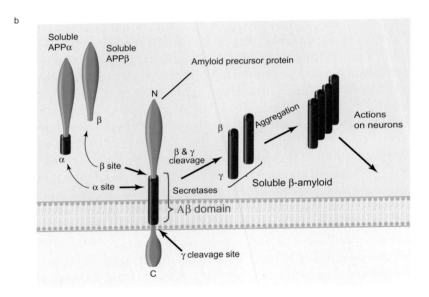

Figure 14–9. Biochemical events during Alzhemier's disease. *a*: Formation of amyloid plaques and neurofibrillary tangles. *b*: Processing of amyloid precursor protein (APP) by secretase enzymes.

plaques, oligomers of β-amyloid bind to the surface of neurons and, directly or indirectly, interfere with a very wide range of neuronal functions, including neurotransmitter action and axoplasmic transport. One consequence is a great increase in the phosphorylation of *tau*, a protein that binds and stabilizes microtubules in the axon and regulates transport along the microtubules. This hyperphosphorylation causes the normally soluble tau protein to form the insoluble neurofibrillary tangles and to accumulate in parts of neurons such as the cell body and dendrites, which normally have very little tau.

Despite a great deal of work on the proteins that malfunction in Alzheimer's disease, we are still a long way from understanding the roles of these proteins in normal neuronal development or how to fix the problems that occur when these biochemical pathways go wrong. The trigger for most cases of Alzheimer's disease is not known, although important clues have come from studies of families in which the disease is inherited. Genetic studies of several other *neurodegenerative* diseases have also identified molecules that make life-or-death decisions on the fate of adult neurons. These include *Huntington's disease*, which impairs motor movements and cognitive functions and in which a protein called *huntingtin* is mutated. *Amyotrophic lateral sclerosis* (ALS, also called motor neuron disease or Lou Gehrig's disease) can be caused by mutations in *superoxide dismutase*, which, like some of the proteins in Alzheimer's disease, cause this normally soluble enzyme to misfold into insoluble aggregates inside neurons. Finally, inherited *spinocerebellar ataxias*, which cause degeneration of the cerebellum, brainstem, and spinal cord, can be caused by mutations in *ataxins*, whose normal role in development is not yet known, as well as by mutations in the gene for the Kv3.3 voltage-dependent potassium channel. A major challenge for future research is to unravel how these proteins regulate neuronal development and survival, and to devise strategies for overcoming the effects of the mutations.

Excitotoxic Cell Death

Programmed cell death, a normal aspect of development, also occurs to a limited extent in the adult brain, so as to compensate for the continual formation of new neurons and glia from stem cells. Neurons can, however, also die by other, more pathological mechanisms. Neurons become damaged and may die when deprived of oxygen, a condition termed *hypoxia*. This occurs during *ischemia*, when blood supply to part of the brain becomes impaired, causing a lack of both oxygen and glucose. Loss of neurons as result of hypoxia and ischemia causes the disabilities that follow a stroke, and are a major cause of death in many countries. In contrast to the orderly, slow biochemical changes that precede apoptosis, neuronal death in response to lack of oxygen and glucose can be very rapid. Such cell death has been termed *necrosis*, and it is usually accom-

panied by swelling of the cells, rather than the physical shrinkage characteristic of programmed death.

   A multitude of undesirable cellular events occur within neurons following deprivation of oxygen (Fig. 14–10). A fall in ATP levels results in
an inability of cells to maintain their normal resting potentials and ionic
gradients. The depolarization that follows causes the release of the excitatory neurotransmitter glutamate. The loss of ATP also compromises the
activity of the uptake mechanisms that normally pump extracellular glutamate back into the cytoplasm. This leads to *excitotoxicity*, a major
cause of neuronal death. Excess glutamate during excitotoxicity produces
abnormally high levels of intracellular sodium and calcium ions, which
enter cells through voltage-dependent channels and through the NMDA
subclass of glutamate receptors. Several biochemical processes have been
implicated in this rapid form of neuronal death, including the activation
of calcium-dependent proteases and the destruction of the cytoskeleton.

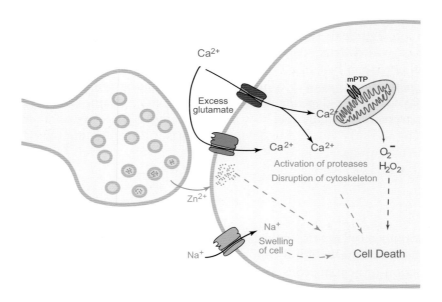

Figure 14–10. Excitotoxic cell death. Elevated levels of extracellular glutamate lead to the influx of calcium and sodium. Cell swelling, activation of the
mitochondrial permeability transition pore (mPTP), and the destruction of cellular components can lead to rapid necrosis. Apoptotic mechanisms may also
be triggered, leading to delayed cell death if neurons survive the early excessive
stimulation.

In addition, the excess calcium that enters mitochondria activates *the mitochondrial permeability transition pore, or mPTP.* (This term is completely unrelated to the PTP, or post-tetanic potentiation, we encountered in Chapter 9). In contrast to the BCl-2 family channels that open in the outer mitochondrial membrane, the mPTP is a large conductance channel in the inner membrane that, when fully opened by calcium ions, equilibrates the matrix of the mitochondrion with the cytoplasm, effectively disabling the mitochondrion.

Even if a neuron manages to survive these early events it may still be doomed. Some of the events occurring during excitotoxicity can produce more delayed reactions, and can even trigger the biochemical reactions of apoptosis, resulting in delayed cell death. Calcium entry into the mitochondria may also trigger the formation of *reactive oxygen species* such as superoxide ions and hydrogen peroxide, which contribute to cell death (Fig. 14–10). Another agent that has been suggested to be a participant in the execution of neurons is the zinc ion. At many excitatory synapses, a proportion of presynaptic vesicles are filled with high levels of zinc (Fig. 14–10). It has been suggested that, when they are released during normal neurotransmission, these zinc ions play a role in shaping postsynaptic responses. During the excessive amounts of secretion and depolarization that occur during ischemic brain damage, however, large amounts of released zinc are taken up by the postsynaptic neurons, where they accumulate in the cell bodies. Such accumulation of zinc leads to opening of the mPTP and to the death of the postsynaptic neurons.

Summary

The development of the nervous system requires the participation of a variety of factors that influence neuronal determination, proliferation, migration, and differentiation. The earliest steps in the formation of a neuron involve the actions of factors such as the bone morphogenetic proteins and neural inducers. Acting on cells that still have the potential to develop into many different types of cells, these factors control the synthesis of transcription factors and determine whether the complement of genes that becomes activated corresponds to those required for building a neuron. The birth of new neurons occurs at a high rate early in development, but in some brain regions persists in adults. The normal formation of the nervous system also requires the programmed death of many neurons. Decisions as to whether a specific neuron survives or perishes during development are made by factors that control the permeability of the outer mitochondrial membrane. In the adult nervous system, death of neurons can be brought about by excessive excitatory stimulation, or by malfunctions in biochemical pathways that can sometimes be linked to mutations in specific genes, such as those that produce neurodegenerative diseases such as Parkinson's disease or Alzheimer's disease. The interplay between neuronal birth and death is fundamental for the formation and function of the nervous system.

Neuronal Growth and Trophic Factors

The previous chapter briefly covered some of the major events that occur during development of a neuron. We saw that the neural inducers, together with the bone morphogenetic proteins, play a key role in the early steps of making a nervous system. We also covered some of the molecules that determine whether a neuron survives or dies during the process of development or in response to injury. Clearly, there are many other factors that lead to the formation of the many varied types of neurons, with different morphologies, diverse patterns of synaptic connections, and different electrical properties. These cells coexist with other cell types, such as glia or sensory cells. The resultant properties of any specific mature neuron result from its exposure during development to the action of many different external factors. These include molecules secreted by other cells far away from the neurons, as well as proteins on the surface of neighboring cells with which the growing neuron makes contact. This chapter will cover more of the molecules that act on developing neurons and their precursors.

General Actions of Growth Factors and Trophic Factors

Table 15–1 lists some of the factors that influence the fate and specific properties of cells in the nervous system. Molecules in the first group may generally be considered pure *morphogens*, or molecules whose concentration surrounding a cell determines the eventual identity of the cell. We have already encountered the bone morphogenetic proteins (BMPs) in this category. The colorful names of some of the others, such as *Sonic Hedgehog* or *frizzled*, exist because they were first used to describe mutant animals in which some aspect of normal development had been perturbed. A favored animal for such genetic experiments is the fruit fly *Drosophila*. We have already seen, in Chapter 5, how a study of mutant flies termed *Shaker* led to the characterization of a family of potassium channels. The names of the mutants are now used to refer to the normal genes and to the protein factors or receptors they encode. Other names are hybrids. For example, the term *Wnt* is a cross between the *Drosophila* mutant *Wingless* and a homologous mouse gene *int-1*.

Table 15–1 Neuronal Growth Factors and Trophic Factors

Ligands	*Receptors*	*Transduction*
Morphogens		
Bone morphogenetic proteins (BMPs, Nodal)	Type I and II BMP receptors	Serine kinase receptors, Smads
Wnts (soluble secreted glycoproteins)	Frizzled, Dally	G protein–coupled transduction
Delta, Jagged (Notch pathway ligands)	Notch	Hairy, Enhancer of Split
Sonic hedgehog (Shh)	Patched, Smoothened	Cubitus interruptus (Gli) transcription factors
Growth factors		
Fibroblast growth factors (FGFs)	FGF receptor	Receptor tyrosine kinase
Epidermal growth factor family ligands (EGFs)	EGF receptor	Receptor tyrosine kinase
Neurotrophins (NGF, BDNF)	Trks	Receptor tyrosine kinase
Cytokines		
Ciliary neurotrophic factor (CNTF), leukemia inhibitory factor (LIF)	Cytokine receptors	JAK-STAT pathway
Steroid hormones		
Estrogen, glucocorticoids	Estrogen receptor, glucocorticoid receptor	Binding of receptor–steroid complex to DNA
Other factors		
Retinoic acid (RA)	RA nuclear receptors	Binding of receptor–RA complex to DNA

Molecules called *growth factors* are also required for the proliferation of some of the cells that will eventually become components of the brain. When stem cells are exposed either to *epidermal growth factor (EGF)* or to *β-fibroblast growth factor (β-FGF)*, they stimulate the cell divisions that are required for growth of the nervous system. The receptors for EGF and β-FGF belong to the receptor tyrosine kinases, which we covered in Chapter 12. As their names suggest, these factors were first discovered as proteins that stimulate cell division in non-neuronal cells but have turned out to act on neurons as well. This chapter will cover the actions of these and the other factors listed in Table 15–1.

Before we discuss the properties of some of these molecules and their receptors, it is necessary to make some general comments about the way these factors can influence the properties of a cell:

(1) *A growth factor can act either instructively or selectively.* Exposure of a precursor cell to a growth factor may specifically activate genes that are required for the development of a mature neuron or glial cell. Some of the TGF-β family proteins have been shown clearly to function in this manner. Such an action is termed *instructive*. In contrast, many factors act by promoting the survival of cells that already express specific properties. Such factors regulate the apoptotic and antiapoptotic pathways covered in the previous chapter and are said to act by a *selective* mechanism. As we shall see later in this chapter, many of the actions of the *neurotrophins*, such as NGF and BDNF, can be explained by this mode of action. Instructive and selective actions are, however, not exclusive, and the same growth factor can use both mechanisms at different times during development.

(2) *Timing is everything.* The effects of growth factors and their signaling pathways depend on the developmental stage of the target cell. In Chapter 14, we saw that the competition between neural inducers and BMPs determines which cells of the ectoderm will develop into the nervous system. Once this important decision has been made, however, the BMPs (and the inducers) can instruct the developing cells to become particular types of neurons or other cell types. We shall encounter these actions again very shortly when we discuss how the forebrain comes to differ from the spinal cord, and why motor neurons form in the ventral but not dorsal regions of the spinal cord.

(3) *The relative amount of different factors is critical.* Development is shaped by the coordinated action of many factors that originate in different locations. Some factors listed in Table 15–1 that induce

a cell to make a developmental decision are located on the plasma membrane of adjacent cells, while others are secreted by non-neuronal cells located outside the developing nervous system. Spatial gradients are known to exist for many of these factors. As a result of these gradients, cells in different locations are exposed to very different ratios of developmental factors.

Anterior–Posterior Patterning

Gradients play a key role in *anterior–posterior* patterning. This process ensures that the forebrain develops at the anterior end of the nervous system and that the spinal cord is formed at the caudal or posterior end. As we have seen in Chapter 14, BMPs and neural inducers are essential for the formation of neural tissue. Slightly later in development, inducers such as Noggin and Chordin, which antagonize the actions of BMPs, are essential for the normal formation of the anterior parts of the nervous system; mutant animals in which these inducers are lacking fail to develop a normal forebrain. In the more posterior regions of the developing nervous system, additional factors are at play. For example, there exists a gradient of several members of the FGF family (termed *FGF-3* and *FGF-8*) in which their concentration is greatest at the posterior end of the immature nervous system (Fig. 15–1). Inhibition of the action of FGF during development results in embryos that have defects in trunk and tail regions. Application of FGF to forebrain cells that have been dissected away from the rest of the nervous system induces them to activate genes that are normally characteristic of hindbrain.

The actions of FGF alone cannot explain all anterior–posterior patterning. Two other factors, *Wnt3A* and *retinoic acid* (Fig. 15–1), may also be important for producing the more posterior regions of the nervous system. *Wnts* are large, secreted glycoproteins of over 300 amino acids. Their membrane receptors are called *Frizzled* receptors, and, like many of the neurotransmitter receptors described in Chapter 12, they have seven transmembrane-spanning segments. A second receptor termed *LRP5* (for Low-density lipoprotein Receptor-related Protein 5) is also required (Fig. 15–2). When activated by Wnt binding, they signal through several different cytoplasmic pathways. The central signaling molecule, however, is *β-catenin*, a protein that has two roles. First, by binding to *cadherins*, which we will encounter in Chapter 16, it regulates how strongly a cell adheres to its neighboring cells. Second, it can move between the cytoplasm and the nucleus to regulate the transcription of genes (Fig. 15–2). When Frizzled is not bound to a Wnt, β-catenin in the cytoplasm is

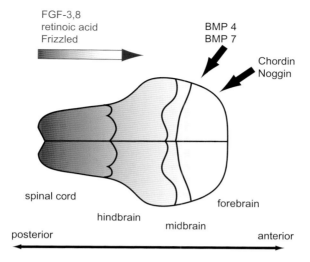

FGF-3,8
retinoic acid
Frizzled

BMP 4
BMP 7

Chordin
Noggin

spinal cord

forebrain

hindbrain

midbrain

posterior

anterior

Figure 15–1. Anterior–posterior gradients of secreted factors shape the formation of forebrain, midbrain, hindbrain, and spinal cord.

continually being degraded in a "destruction complex" of several proteins that include the proteins *axin, APC* (for **A**denomatous **P**olyposis **C**oli), and *glycogen synthase kinase 3β (GSK 3β)* (Fig. 15–2a). It is phosphorylation of β-catenin by the latter kinase that leads to its destruction by the ubiquitin system we encountered in Chapter 14.

When a Wnt binds to Frizzled and LRP5, a series of molecular rearrangements occurs. A cytoplasmic protein called *dishevilled* is recruited to the receptor complex. Axin is then attracted away from the destruction complex to the receptors, causing the destruction complex to disassemble. As a result, β-catenin is no longer degraded but can move from the cytoplasm to the nucleus, where it interacts with other transcription factors to stimulate the transcription of specific genes (Fig. 15–2b).

The Frizzled receptor is expressed in a graded fashion, with highest levels at the posterior end of the embryo (Fig. 15–1). In frog embryos, application of the Wnt family member Wnt3A can transform anterior neural tissue into posterior tissue, and inhibition of Wnt signaling prevents development of the posterior nervous system. Conversely, in mice, genetic manipulations that lead to overproduction of a Wnt (Wnt8c) result in complete loss of the anterior forebrain.

Retinoic acid is a small molecule closely related in structure to retinal, the light-sensitive molecule used in phototransduction (see Chapter 13). While *retinol* (also termed *vitamin A*), the precursor of retinoic acid, is widespread, retinoic acid is synthesized from retinol only by

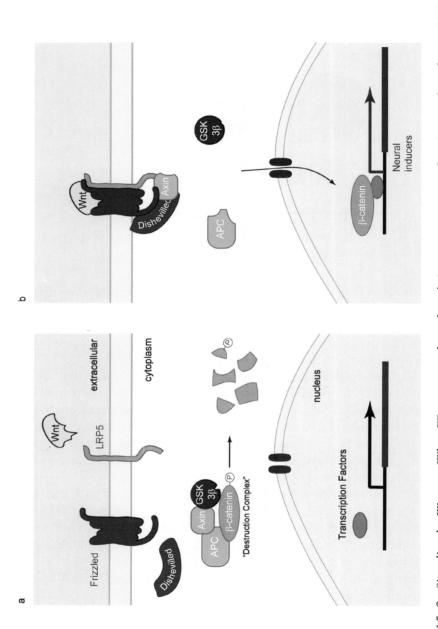

Figure 15–2. Signaling by Wnts. *a*: When Wnts are not bound to their receptors, a cytoplasmic complex of axin, APC, and GSK 3β binds and phosphorylates β-catenin, leading to its destruction. *b*: On binding of a Wnt to the Frizzled and LPR5 receptors, the destruction complex falls apart and β-catenin enters the nucleus.

certain tissues and diffuses from these tissues to act on nearby targets. Its molecular action is very similar to that of steroid hormones, which are discussed later in this chapter. Retinoic acid readily crosses cell membranes and binds a receptor that directly binds to specific sites on DNA (termed *RAREs*, for **R**etinoic **A**cid **B**inding **E**lements) in the nucleus of the target cells. When applied to intact embryos, retinoic acid inhibits the formation of the proteins that mark the anterior parts of the nervous system and promotes the activation of genes characteristic of posterior regions. Thus, this small acid molecule, in combination with Wnts and FGFs may contribute to anterior–posterior patterning of the nervous system.

Dorsal–Ventral Patterning

Gradients also contribute to *dorsal–ventral* patterning. Within the spinal cord, motor neurons are located ventrally, as is the *floor plate*, a structure composed of epithelial cells that extends the entire length of the spinal cord and into the brain (Fig. 15–3a). As we shall see in Chapter 16, cells in the floor plate play an important role in guiding axons to their targets. Other types of neurons are located in the dorsal parts of the spinal cord. Located even more dorsally, between the developing spinal cord (also termed the *neural tube*) and the ectoderm, is the *neural crest*, which, as we shall see later, is a collection of cells that eventually make neurons whose cell bodies lie outside the central nervous system (the sympathetic and parasympathetic nervous systems), as well as other cell types.

A gradient for the factor *Sonic Hedgehog*, often abbreviated to *Shh*, is instrumental in determining the different properties of cells along the dorsal–ventral axis (Fig. 15–3a). This factor is first synthesized in the *notochord*, a non-neural tissue that lies beneath the spinal cord. If the notochord is transplanted to the dorsal side of the developing spinal cord, motor neurons, as well as the floor plate, develop on the dorsal rather than the ventral side. Conversely, if the notochord is eliminated, the ventral part of the cord fails to develop and cells from the dorsal regions take over the ventral area. This occurs because the developmental fate of the ventral cells is normally ensured by the diffusion of Sonic Hedgehog from the notochord. At later stages of development, Sonic Hedgehog is also synthesized in cells of the floor plate, further reinforcing the gradient of this factor from the ventral to the dorsal nervous system.

Sonic Hedgehog is a member of the *Hedgehog* family of secreted proteins. Our understanding of how it functions has come in large part from studies in the fruit fly *Drosophila*. Hedgehog proteins are unusual in that

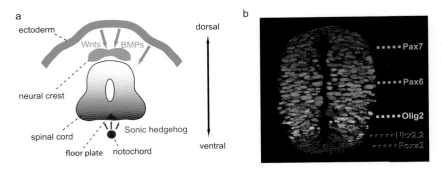

Figure 15–3. Dorsal–ventral gradients of secreted factors determine the formation of the neural crest and different regions of the spinal cord. *a*: A gradient of Sonic hedgehog arises from the notochord and decreases toward dorsal regions. Gradients of Wnts and BMPs are generated in the opposite direction. *b*: A chick embryo spinal cord that has been immunostained for different transcription factors along the dorsal–ventral axis of the developing chick spinal cord (from the work of Gwenvael le Dreau and Elisa Marti). We encountered some of these transcription factors in Chapter 14 (see Table 14–1).

they are covalently bound to the lipid cholesterol as well as to the fatty acid *palmitic acid*, both of which help the protein to associate with lipid membranes. The activated cholesterol-bound Shh protein binds a 12-transmembrane segment membrane receptor termed *Patched* (Fig. 15–4a). A key player in the subsequent response to Shh is *Smoothened*, a member of the seven-transmembrane G-coupled receptor family that, in the absence of Shh, is typically found on intracellular vesicles. In the absence of Shh in the external medium, Patched suppresses the activity of Smoothened and prevents it from being inserted into the plasma membrane. A second key set of proteins required for Shh to carry out its work is the *Gli* family of transcription factors. In the absence of Shh, these Gli proteins are bound to a multiprotein complex containing several protein kinases, including the cyclic AMP–dependent protein kinase, casein kinase I, and glycogen synthase kinase 3β. As described earlier for β-catenin, phosphorylation by these enzymes results in proteolytic destruction of the Gli transcription factors. This destruction can be complete, or it can result in fragments of the full-length Gli proteins that enter the nucleus and repress the transcription of target genes (Fig. 15–4a).

Once Sonic Hedgehog binds to its Patched receptor, there is a dramatic role reversal at each level (Fig. 15–4b). Patched, along with Shh, is now endocytosed into internal vesicles. As a result, Smoothened is no longer inhibited and enters the plasma membrane. Once there, Smoothened sequesters the complex containing the protein kinases away from the Gli

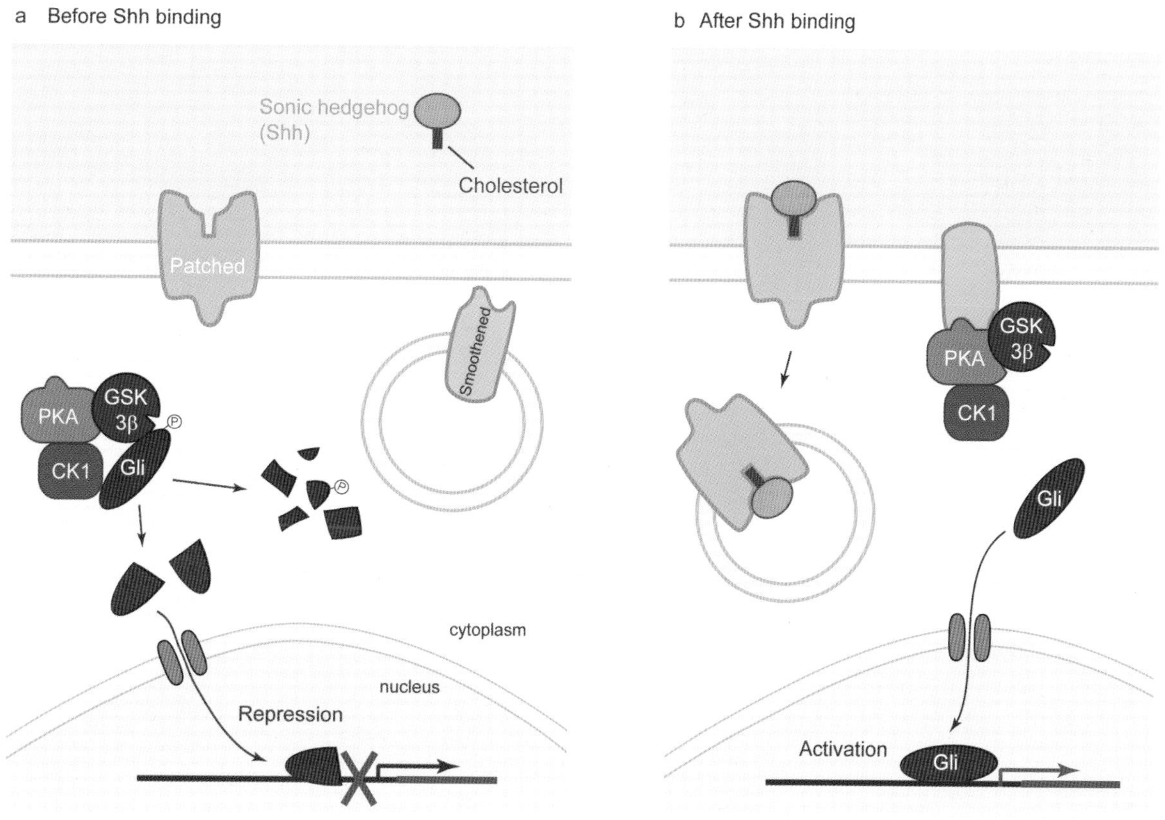

a Before Shh binding

b After Shh binding

Figure 15-4. The Sonic Hedgehog signaling pathway. *a*: When Shh is not bound to its Patched receptors, Smoothened proteins are kept on internal membrane and their activity is suppressed. Gli transcription factors undergo phosphorylation that results in their degradation to inactive fragments or to factors that enter the nucleus to suppress transcription. *b*: On binding Shh, Patched is internalized, and Shh is inserted into the plasma membrane and removes the protein kinase complex from Gli proteins that then activate transcription.

protein, which can no longer be degraded. The full-length Gli proteins then enter the nucleus and activate transcription of the genes they regulate.

While a gradient of Sonic Hedgehog from the notochord shapes the destiny of the ventral spinal cord, a similar scenario is played out on the dorsal side of the developing spinal cord. A variety of factors, including BMPs, are released form the overlying epidermis and from the *roof plate*, a dorsal group of cells analogous to the floor plate. Some of these, termed *BMP2, BMP4, and BMP7*, are required for normal development of neurons in the dorsal spinal cord and of neural crest cells. For example, BMP2 directly instructs cells of the neural crest to develop into neurons. Some Wnts, including Wnt3, which plays a role in anterior–posterior patterning, also promote the formation of neural crest cells and cells of the dorsal spinal cord.

During the development of the spinal cord, each cell along the dorsal–ventral axis is exposed to opposing signals from Sonic Hedgehog and the BMPs/Wnts. Depending on their position in the gradient, the cells encounter different ratios of these factors and activate a different set of transcription factors in their nuclei (Fig. 15–3b). These transcription factors, in turn, direct the synthesis of a slightly different set of proteins. As a result, the cells "know their place" and become neurons or other cells appropriate for their location.

Local Cell–Cell Interactions Shape the Development of Neurons

Within each small part of the nervous system many different types of cells may coexist. For example, the adult cerebral cortex consists of a complex meshwork of astrocytes, excitatory neurons, and inhibitory interneurons. While gradients of growth factors are clearly important in the development of large regions of the nervous system, the way a cell decides to take on an identity different from that of its immediate neighbor requires direct physical contact between the two cells. We shall now give a brief account of a *Drosophila* mutation, termed *Notch*, that has provided major insights into neuronal differentiation in all species. Unlike the soluble factors just discussed, Notch signaling requires direct cell-to-cell contact.

The **Notch** *locus.* During normal development of *Drosophila*, a strip of cells along the ventral midline of the embryo comes to form the *neurogenic region* (Fig. 15–5a). The cells in this region develop either into *neuroblasts*, which give rise to neurons along the ventral cord of the fly, or into *dermoblasts*, which eventually form the epidermis over the ventral cord. (Remember that *Drosophila* is an invertebrate, and that neuroblasts therefore divide to give rise to neurons.) Under normal conditions, about one-quarter of the cells in the neurogenic region form neuroblasts. However, in a class of mutant flies first noted early in the last century, the cells that normally become dermoblasts turn into neuroblasts, resulting in hypertrophy of the nervous system and loss of the ventral epidermis.

The defect in these animals was localized to a stretch of DNA termed the *Notch* locus. A complete deletion of the DNA in this region produces the change in the fate of the precursor cells for the epidermis. This in turn causes the death of the embryo. Other mutations of the DNA in this region are not always lethal but result in a variety of abnormalities of development. The *Notch* locus encodes a large protein of 2703 amino acids (Fig. 15–5b). The predicted structure of this protein has some features of both a growth factor and a receptor within the same molecule. It has one stretch of hydrophobic amino acids that spans the plasma membrane. Near the amino terminal of the protein there is a sequence of 38 amino acids that contains six cysteine residues. Variants of this 38–amino acid sequence are repeated in tandem throughout most of the 1700 amino acids that lie on the extracellular side of the membrane. Each of these repeated sequences is very similar to a region that exists in EGF and related growth factors. The intracellular domain of Notch does not resemble any of the receptors we have considered thus far, but contains a new set of repeated domains that are essential for its normal function. These six intracellular repeats resemble the protein *ankyrin*, a protein of red blood cells that anchors the cytoskeleton to a transporter in the plasma membrane, and are therefore termed *ankyrin repeats*.

Notch does indeed function as a receptor. The prototype ligand for this receptor is a protein termed *Delta*, and mammals have multiple Notch ligands, called *Delta-like* and *Jagged*. In flies, deletion of the gene encoding Delta produces the same alterations in development as those resulting from the loss of Notch. Unlike most of the ligands we have considered thus far, however, Delta is also a membrane-bound protein (Fig. 15–5b). Like Notch, the Delta protein has several EGF-like repeats in its extracellular domain. It also contains a region rich in cysteines termed the *DSL* motif (named for **Delta-Serrate-Lag-2**; the names of three different proteins that contain this motif), which is essential for binding of Delta to Notch. Unlike Notch, however, Delta is simply a membrane-bound ligand and does not appear to be a receptor. Thus its short intracellular domain is not required for normal signaling.

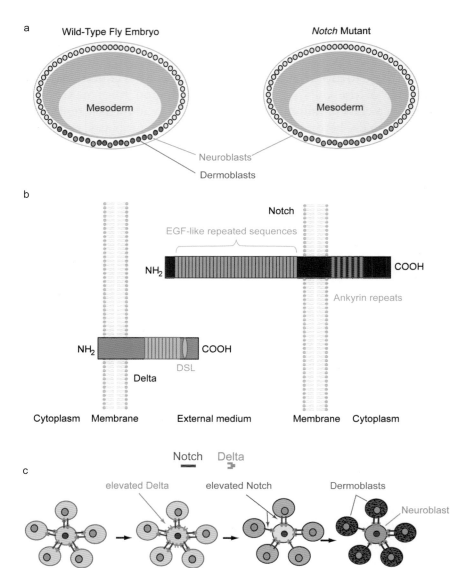

Figure 15–5. The *Notch* mutation. *a*: Cross section of a normal fly embryo and of a Notch mutant, with its overabundance of neuroblasts. *b*: The sequence of the Notch protein was deduced by Spyros Artavanis-Tsakonas and his colleagues (Artavanis-Tsakonas et al., 1995). Notch binds the Delta ligand on adjacent cells. *c*: Lateral specification by Notch–Delta interactions. Adjacent cells that are initially equivalent differentiate into dermoblasts or neuroblasts through changes in the relative levels of Notch and Delta in their plasma membranes.

Initially, all precursor cells in the epidermis are equivalent and have both Notch and Delta in their plasma membranes. The Notch receptors in each cell bind Delta ligands on neighboring cells (Fig. 15–5c). When Notch receptors and Delta ligands are balanced in this way, there is a stalemate in which each cell prevents its neighbor from differentiating. With time, however, it appears that biochemical feedback mechanisms amplify small natural fluctuations in the activity of these signaling proteins. Thus cells with high levels of Delta *lower* the activity of their own Notch receptors but become surrounded by cells with *raised* Notch activity. The cells with low Notch activity then develop into neuroblasts, while their neighbors with high Notch activity become dermoblasts. This process of interaction between adjoining cells has been termed *lateral inhibition* or *lateral specification*.

The signaling pathway that has been adopted by the Notch receptor is one that relies on multiple proteolytic cleavage steps. Even before it starts to function as a receptor it is split into two parts by cleavage of the extracellular EGF-like sequences from the rest of the molecule. The two parts, however, remain very closely glued together by noncovalent bonds (Fig. 15–6). When it binds a ligand such as Delta on its neighboring cell, further cleavage takes place in the external space, causing the EGF-like sequences that bound to Delta to be pulled back into the neighboring cell through endocytosis. Yet another cleavage is then carried out within the membrane-spanning region of Notch by *γ-secretase*, an enzyme we encountered in Chapter 14 in connection with amyloid precursor protein (see Fig. 14–9b). This last cleavage frees the intracellular domain of Notch into the cytoplasm, after which it enters the nucleus to bind a protein colorfully termed *Suppressor of Hairless (Su(H))*. Together with other proteins, such as *mastermind* protein, the Notch intracellular domain then acts as a transcription factor to activate the genes required to form dermoblasts. The mammalian homologs of Su(H) and mastermind have been termed *RPB-J* and *Mastermind-like proteins (MAML)*, respectively (Fig. 15–6).

Homologs of Notch and its ligands are found in all species. They are found in the membranes of many dividing cells, and their role is not specific to the formation of the nervous system. Rather, they seem to be required for many situations in which interactions between adjacent cells result in a switch in the direction of their development. In the very early stages of making a nervous system in vertebrates, as is the case in fruit flies, Notch-dependent lateral inhibition between neighboring cells in the germinal zone helps to decide which cells become neurons and which stay as undifferentiated stem cells (see Fig. 15–5). Later in development, newly generated neurons that migrate along radial glial stem cells activate Notch

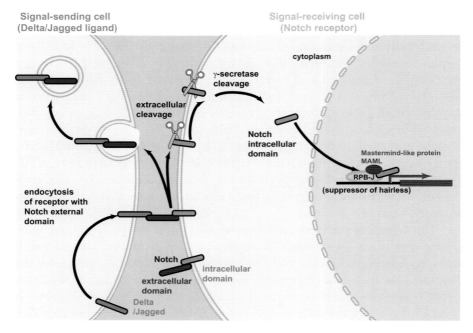

Signal-sending cell
(Delta/Jagged ligand)

Signal-receiving cell
(Notch receptor)

cytoplasm

γ-secretase
cleavage

extracellular
cleavage

Notch
intracellular
domain

Mastermind-like protein
MAML

RPB-J
(suppressor of hairless)

endocytosis
of receptor with
Notch external
domain

Notch

intracellular
domain

extracellular
domain

Delta
/Jagged

Figure 15–6. Steps in the Notch signaling pathway. (Modified from Ables et al., 2011.)

in the glial cells (see Fig. 14–4b). This ensures that the radial glia continue to provide a physical pathway for migration and remain capable of dividing to provide further neurons. Even later, Notch–ligand interactions between cells that touch each other can determine which cells become excitatory or inhibitory neurons, and in the developing retina, which cell becomes a rod, a cone, or a neuron. ❷

The Neurotrophins

By engendering the activation or inhibition of specific transcription factors in the nucleus of their target cells, the cell–cell interactions that we discussed earlier instruct the nucleus of a cell to follow a particular developmental path. We have seen that gradients of soluble factors also promote the synthesis of proteins that determine the eventual phenotype of a neuron. A major role for certain soluble factors is, however, to prevent the apoptotic death of subtypes of cells. In particular, a family of proteins termed the *neurotrophins* is important for neuronal survival during development and for the maintenance of neurons in adult life. The first example and the most thoroughly characterized of these factors is *nerve growth factor* (NGF).

Nerve growth factor. The first clues to the existence of NGF came from experiments by E. Bueker in 1948 that demonstrated that if a muscle tumor is implanted into the body wall of an embryo, the tumor becomes innervated by neurons from the sensory and sympathetic ganglia along the spinal cord. This innervation is accompanied by a great increase in the size of the ganglia that project to the tumor. These experiments were followed up by Rita Levi-Montalcini and Victor Hamburger, who provided evidence for the existence of a soluble substance that promotes the growth of neurons in these ganglia. To characterize this substance, Levi-Montalcini developed a simple bioassay. She placed small pieces of tissue containing sensory neurons or neurons from the sympathetic nervous system into a culture dish. The addition of cells from the muscle tumor was found to produce a very dramatic stimulation of the growth of *neurites* out of the explant (Fig. 15–7a). (The term *neurite*, or *neuritic branch*, is used to describe either an axon or a dendrite, and is a particularly useful term when it has not been established whether the growing process is in fact an axon or a dendrite.)

Subsequent experiments by Levi-Montalcini and Stanley Cohen showed that certain tissues other than muscle tumors could also provide the neurite-inducing factor. In an attempt to characterize the molecular properties of NGF, Cohen treated a crude preparation of the factor with snake venom, which contains a phosphodiesterase that should degrade nucleic acids but leave proteins intact. Surprisingly, the snake venom alone was found to be active in inducing neurite outgrowth and was shown subsequently to contain high amounts of NGF. In mice, the submaxillary salivary gland, an anatomical homolog of the snake gland that secretes venom, was also found to be very rich in NGF. The salivary gland was the source from which the protein was eventually purified.

The biological activity of NGF on sympathetic and sensory neurons resides in a complex of two identical peptide chains that have a molecular weight of 13,259 each (Fig. 15–7b). Direct measurements of the levels of NGF and of NGF messenger RNA in various tissues have shown that NGF is synthesized at a low rate by a variety of tissues innervated by sympathetic neurons. The reason for the atypically high levels in salivary glands and in the saliva of mice is not known. NGF levels may also become elevated upon injury to a nerve, when it appears to be synthesized by Schwann cells. (It has also been claimed that NGF levels increase in the blood of people falling in love.) ❧

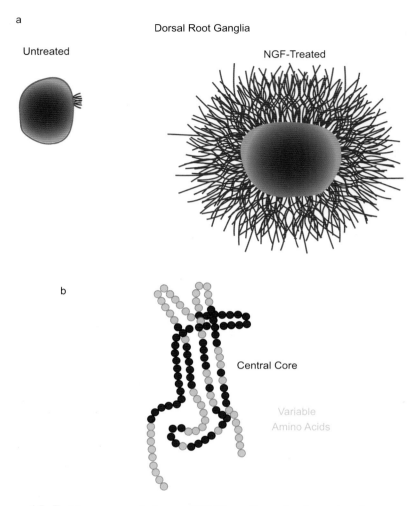

Figure 15–7. Nerve growth factor (NGF). *a*: Growth of neurites from sensory ganglia is induced by NGF. *b*: Structural studies have established that the β subunit of NGF has a central core with external loops. The organization of the peptide chain of β-NGF is shown; the amino acids that are most different in the other neurotrophins are shown in blue.

Actions of NGF during development. Despite its name, NGF is not a true growth factor, because it does not cause the mitosis of neurons or their precursor cells. Nevertheless, NGF plays a key role in the normal development of sympathetic neurons and in their maintenance in adult animals. For example, injection of antibodies to NGF into embryonic or newborn mice results in the complete abolition of the sympathetic nervous system. Injection of antisera to other growth factors does not have as specific an effect. Destruction of the sympathetic nervous system also occurs in older animals that are deprived of biologically active NGF by repeated injections of anti-NGF antisera, although a more prolonged period of application is required than in the embryo or newborn. Conversely, injection of NGF itself into immature animals causes an enlargement of the sympathetic ganglia. Generally, similar results are obtained with sensory neurons, although it appears that antisera to NGF are able to induce a depletion of sensory neurons only when applied early in embryonic development.

On the basis of these sorts of experiments, the hypothesis has emerged that target cells destined to be innervated by sympathetic or sensory neurons provide NGF to ensure the survival of their synaptic inputs, and that NGF acts to inhibit the mechanisms that would normally lead to cell death in the input cells. Studies that measured the amounts of messenger RNA for NGF during development have shown that major synthesis of NGF begins only *after* sympathetic neurons have reached their targets. Dependence on NGF for survival also begins at this time.

BDNF and the other neurotrophins. Although sympathetic and sensory neurons are the major neuronal types sensitive to NGF, only a small number of other types of neurons also respond to this factor, which suggests that the survival of most other neurons may be dependent on other related molecules. A second neurotrophin was discovered as a factor that produces more prolonged survival of sensory neurons and was named *BDNF*, for **B**rain-**D**erived **N**eurotrophic Factor. Molecular cloning then revealed two other closely related members of this neurotrophin family that have been termed *NT-3* and *NT-4*. All of these have a structure very similar to that of NGF. Each has a central core region that is very similar to that of β-NGF, but differs from the others in the structure of the loop regions (Fig. 15–7b) and at the amino and carboxyl terminals.

In contrast to NGF, the other neurotrophins are widely distributed in the brain, both in neurons and in glial cells, as well as in peripheral tissues such as muscle. Correspondingly, they are able to promote the survival of a wide range of neurons, including many motor neurons and other central neurons such as dopamine-containing neurons from the substantia nigra (see Fig. 10–6). In addition, they exert a profound influence on axonal outgrowth and the morphology and electrical activity of neurons (Table 15–2). Moreover, as we shall in Chapter 19, even in the adult brain, the synthesis and secretion of neurotrophins is enhanced by increases in neuronal activity, making them key molecules that regulate the function of mature synapses.

Table 15–2 Actions of the Neurotrophins

Neuronal survival	Prevention of the death of neurons
Nerve growth	Stimulation of the elongation of both axons and dendrites
Nerve sprouting	Stimulation of the sprouting of both axons and dendrites of adult neurons (see Chapter 16)
Anabolic actions	Increasing the size of neuronal somata
Differentiation	Inducing the synthesis of proteins required for a neuronal phenotype
Modulation of transmission	Increasing the synthesis of neurotransmitters, neuropeptides, and their synthesizing enzymes
Electrical properties	Altering the activity and levels of ion channels

The Trk Receptors for the Neurotrophins

NGF and the other neurotrophins produce their effects on neurons by binding to two types of receptors that span the plasma membrane, the *Trk* receptors and a glycoprotein termed *p75NTR* (Fig. 15–8). There are three known Trk proteins—*TrkA, TrkB,* and *TrkC.* The actions of NGF are mediated specifically by TrkA. As expected, therefore, the TrkA receptor is found only on a few types of neurons, primarily sympathetic and sensory neurons and a small number of other neurons in the brain. In contrast, both TrkB and TrkC are found on the majority of neurons. TrkB binds both BDNF and NT-4, while NT-3 acts primarily through TrkC. ❽

A second membrane protein to which all of the neurotrophins bind is p75NTR, sometimes termed *LANR* (**L**ow-**A**ffinity **N**eurotrophin **R**eceptor) (Fig. 15–8). This protein is unrelated in structure to the Trk proteins, but is a member of another family of receptors, the *TNF* (**T**umor **N**ecrosis **F**actor) receptors. By itself, p75NTR is unable to mediate any of the actions of NGF or the other neurotrophins. Indeed, somewhat surprisingly, stimulation of p75NTR by NGF in the absence of the Trk receptors actually triggers, rather than inhibits, apoptosis. Nevertheless, the presence of this low-affinity receptor in neurons appears to potentiate the TrkA signaling pathway and is important for some of the actions of NGF. For example, in animals in which the p75NTR gene has been deleted, much higher than normal concentrations of NGF are required to ensure survival of sympathetic neurons. Moreover, such animals have fewer sensory neurons than normal, and the axons of many of their sympathetic neurons fail to reach their normal targets.

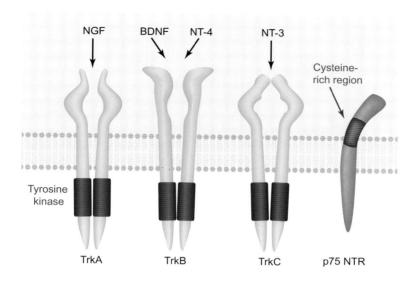

Figure 15–8. Neurotrophins bind Trk receptors, as well as the low-affinity neurotrophin receptor p75NTR.

Signaling pathways for the neurotrophins. Because the effects of the neurotrophins are so diverse, it comes as no surprise that they engage a variety of different signaling pathways. Some of their actions are rapid and very local. For example, application of NGF to a restricted part of a sensory neuron produces effects on the morphology of its neurites that occur within 30 seconds and are confined to the local area of application (see Chapter 16). As shown in Table 15–2, however, many neurotrophin actions result in changes in the rate at which specific proteins are synthesized by the neuron. In many cases, this requires that a signal be sent from the axon terminals, where the neuron encounters the neurotrophin, back to the cell body, where gene transcription and protein synthesis occur. This is accomplished through internalization of the neurotrophin by endocytosis into a *signaling endosome*, which is then carried back to the soma by retrograde transport (Fig. 15–9).

Trk receptors are homodimers with extracellular domains that form a tight binding site for a neurotrophin and with intracellular domains that contain *tyrosine kinase* enzyme activity. The first step that occurs on binding of a neurotrophin is that the tyrosine kinase activity is stimulated, resulting in the autophosphorylation of the receptor molecule. Each activated kinase domain causes phosphate groups to be transferred from ATP

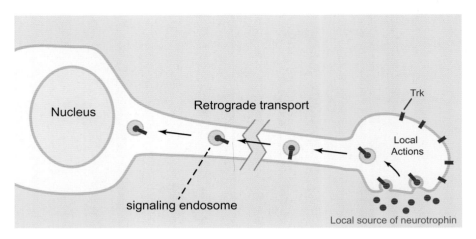

Figure 15–9. Sites of neurotrophin action. After binding to Trk receptors in the plasma membrane, neurotrophins can act locally, or they can be ferried back to the nucleus by retrograde transport in signaling endosomes.

to tyrosine residues on the other chain of the Trk receptor. This sets in motion at least four signaling events, listed next.

(1) The Akt pathway. The very important action of neurotrophins in protecting neurons from apoptosis is mediated by the enzyme PI_3 *kinase* (phosphatidylinositol-3-kinase), followed by the activation of a cytoplasmic serine/threonine kinase termed *Akt*. The initial steps in this pathway are triggered by the tyrosine residues that become phosphorylated when a Trk binds its ligand. These phosphorylated tyrosines serve as points of attachment for proteins that contain structures termed *SH2 domains*. Such domains are found in many signaling proteins, and their function is simply to bind phosphotyrosine. One SH2 domain–containing protein that binds to phosphotyrosines on the TrkA receptor is the *Shc* protein, which, after binding, itself becomes phosphorylated by the receptor (Fig. 15–10a). The Shc protein acts as the first part of a multisegment bridge that links the receptor to several other proteins. The second part of the bridge is an "adaptor" protein termed *Grb2*. This adaptor itself contains an SH2 domain, which allows it to bind to the tyrosine phosphorylated Shc protein. Grb2 also contains two *SH3 domains*. Like SH2 domains, SH3 domains enable signaling molecules to link to each other. Instead of binding to phosphotyrosines, however, SH3 domains recognize specific amino acid sequences that contain a proline and several hydrophobic amino acid residues.

A third adaptor component, Gab1, then binds to one of the SH3 domains and links the chain to PI3 kinase. This enzyme causes the

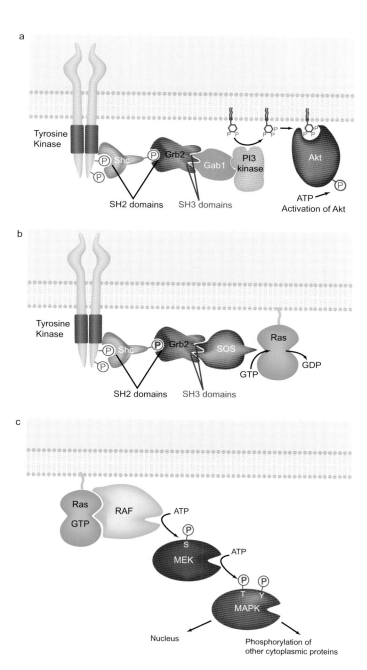

Figure 15–10. The Akt and the ras/raf/MEK/MAPK pathways. *a*: Pathway linking a Trk neurotrophin receptor to activation of PI3 kinase and Akt. *b*: Pathway linking activation of a Trk receptor to the exchange of GTP for GDP on the ras protein. *c*: In its GTP-bound form, ras recruits the raf kinase to the plasma membrane and triggers a cascade of phosphorylation events, including the phosphorylation of MEK and MAPK.

phosphorylation of phosphatidylinositol and PIP_2, which, as we saw in Chapter 12, are parts of the diacylglycerol (DAG)/IP_3 signaling pathway. The actions of PI3 kinase, however, constitute a signaling pathway that is completely distinct from the DAG/IP3 system (Fig. 15–10a). The lipid second messengers produced by PI3 kinase bind to the Akt kinase, docking it close to the plasma membrane. This then allows the transfer of phosphate from ATP to Akt by yet other kinases. Akt activated in this way then influences the activity of Bcl-2 family proteins in neuronal mitochondria. As we saw in Chapter 14, these proteins have the ability to make life-or-death decisions within a cell, and stimulation of this Trk signaling pathway protects cells from apoptosis.

   (2) *From* ras *to* raf *to MEK to MAPK.* While activation of Akt protects NFG-treated neurons from dying, many of the subsequent effects of neurotrophins on neurite outgrowth and gene transcription involve a protein known as *ras*. The ras protein is a 21 kDa phosphoprotein that binds GTP and is homologous to other GTP-binding proteins such as the rab protein (Chapter 8) and the G protein α subunits that couple neurotransmitter receptors to the formation of second messengers (Chapter 12). As with other GTP-binding proteins, it acts as an on/off switch. When bound to GDP it is inactive. When this GDP is exchanged for GTP, ras sets in motion a cascade of events involving the sequential activation of a chain of protein kinases. The ras pathway is triggered by the binding of another protein Sos ❽ to the second SH3 domain on Grb2 (Fig. 15–10b). This allows Sos to contact ras and to promote the exchange of GDP on ras for GTP, thereby putting *ras* into an active state.

   Like some other GTP-binding proteins, ras is linked to the membrane by a lipid tail. In its GTP-bound state, ras recruits a protein kinase termed the *raf kinase* from the cytoplasm to the plasma membrane. On moving to the plasma membrane, the raf kinase becomes activated and phosphorylates another protein kinase, termed *MEK*, at serine residues, thereby activating this second kinase (Fig 15–10c). MEK is unusual among protein kinases in that it has the ability to add phosphates to both serine/threonine and to tyrosine residues. In turn, the activated MEK enzyme phosphorylates yet another protein kinase, *MAPK* (for **Mitogen-Activated Protein Kinase**). This phosphorylation occurs on a threonine *and* a tyrosine residue. It is after this point in the kinase cascade that the multiplicity of effects of the neurotrophins can be understood. In addition to phosphorylating numerous cellular proteins, the activated MAPK enzymes also enter into the nucleus, where they act on transcription factors that alter the rate at which the messenger RNA for specific proteins is synthesized.

(3) Trk receptors trigger PLC signaling. We saw in Chapter 12 that some neurotransmitters stimulate phospholipase C, leading to the production of IP$_3$ and DAG. This same pathway is also activated by neurotrophins. One isoform of phospholipase C, *PLC-γ1*, is activated by binding directly to phosphotyrosines on the Trk receptors.

(4) Direct Trk interactions with ion channels. Some actions of neurotrophins can be very rapid indeed. For example, the application of BDNF to hippocampal neurons activates sodium channels within milliseconds. This has led to the proposal that Trk receptors may be intimately bound to certain ion channels, allowing very rapid modulation of excitability.

PC12 cells and the mechanism of NGF action. Given the many different biochemical events that are triggered by neurotrophins, it can be very tricky to unravel which pathway is responsible for which specific change in the properties of a neuron. The *PC12* cell line was particularly useful in studies of these signaling pathways. These cells were derived from a tumor of adrenal chromaffin cells, termed a *pheochromocytoma*. Like sympathetic and sensory neurons, chromaffin cells are derived from the neural crest, and the TrkA and p75 receptors are present in the plasma membranes of PC12 cells. Treatment of the cells with NGF stimulates neurite outgrowth. Evidence that neurite outgrowth requires the ras/raf/MEK/MAPK pathway came from experiments using a mutant form of ras that is diminished in its ability to hydrolyze GTP. Such a mutant ras protein would be expected to be chronically activated. In fact, injection of this mutant ras protein directly into the cytoplasm of PC12 cells induces morphological differentiation similar to that induced by NGF (Fig. 15–11). Injection of the unmutated, and therefore unactivated, normal ras protein does not produce this effect. The injection of antibodies to the normal ras protein was found to block the effect of NGF on PC12 cells.

The kinases that act downstream of ras can also induce neurite outgrowth in PC12 cells. For example, an activated form of the raf kinase can be introduced into these cells by inserting the DNA encoding the enzyme into the genome of a retrovirus. On infection with the retrovirus, the cells begin to synthesize the activated enzyme and to extend neurites, mimicking the effect of NGF. Similar results are obtained on injection of DNA encoding a mutated MEK enzyme that is active even when not phosphorylated (Fig. 15–11). Moreover, it is possible to make another form of mutated MEK, which is inactive but blocks signaling by competing for the normal cellular raf kinase. Injection of DNA encoding this inhibitor MEK prevents the action of NGF on neurite outgrowth.

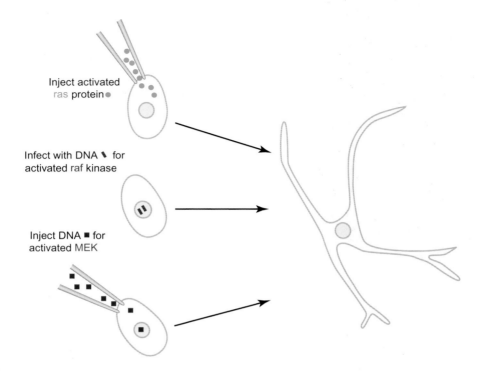

Figure 15–11. A role for the ras/raf/MEK pathway in neurite outgrowth. Injection of activated ras protein, or introduction of DNA for activated raf or MEK enzymes, which leads to the synthesis of the activated proteins within PC12 cells, causes the extension of neurites.

Cytokines and the JAK–STAT Pathway

Another class of molecules, structurally unrelated to the neurotrophin family, has also been found to be required for the survival of many types of neurons and to influence their cellular properties. These molecules, first found to influence the differentiation of cells in the blood and the immune system, are termed *cytokines*. In contrast to the neurotrophins, the cytokines act on membrane receptors that are not themselves tyrosine kinases. Instead, the cytokine receptors stimulate the activity of soluble cytoplasmic tyrosine kinases termed *JAKs* (Janus kinases). There are four known JAK kinases, each of which has seven conserved regions termed *JH domains* (Fig. 15–12a) and can themselves undergo phosphorylation in the JH1 domain that contains the active site of the enzyme.

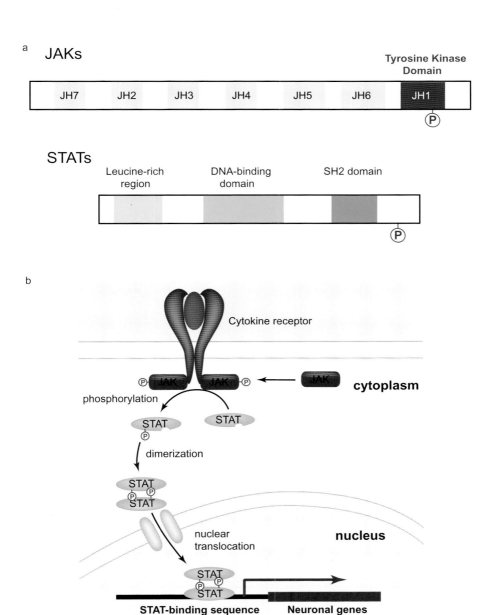

Figure 15–12. The JAK–STAT signaling pathway. *a*: Structures of the JAK kinases and their STAT substrates. *b*: Following binding of cytokine to its receptor, STATs are phosphorylated, leading to activation of specific genes.

One molecule that acts through the cytokine pathway is *CNTF* (Ciliary Neurotrophic Factor), which was first described as a factor required for the survival of neurons of the ciliary ganglion of chicks and is a potent survival factor for many types of neurons. CNTF can also induce the differentiation of astroglia from precursor cells. On the surface of cells, CNTF binds to a protein termed *CNTFα* (CNTF-receptor α) which forms a complex with two other proteins, *LIFRβ* (Leukemia Inhibitory Factor Receptor β) and *gp130* (Fig. 15–13). On activation by CNTF, the receptor complex is able to bind a JAK kinase, which becomes activated and phosphorylates substrate proteins knows as STATs (Signal Transducers and Activators of Transcription) (Fig. 15–12a). Seven such STAT proteins are currently known. These proteins contain SH2 domains that allow them to bind other proteins that are phosphorylated on tyrosine residues, as well as a region rich in leucines. Once a STAT is phosphorylated, these regions allow it to form a dimer with another STAT and then enter the nucleus. There they bind to DNA at regions termed *STAT-binding sequences*, which are situated upstream of genes that become activated on exposure of the cell to CNTF (Fig. 15–12b).

Factors that make a cholinergic neuron. One of the best-studied examples of how cytokine signaling influences the properties of neurons has come from work with cells that form the sympathetic nervous system. These cells migrate from the neural crest (Fig. 15–14a), a column of undifferentiated cells at the dorsal margin of the neural tube that we first encountered when discussing the development of the spinal cord (Fig. 15–3). Most neurons of the sympathetic nervous system synthesize the enzymes required for the production of norepinephrine, which they use as a neurotransmitter. A subset of sympathetic neurons, however, synthesizes and secretes acetylcholine as their major transmitter.

The development of the biochemical machinery to produce either the adrenergic or the cholinergic transmitter system is not determined by the position in the neural crest from which the cells migrate but rather by a factor that the cells encounter during their migration away from the neural crest. The first close encounter is when the precursor cells reach a blood vessel, the dorsal aorta, where they are exposed to BMPs (Fig. 15–14b) that activate the synthesis of new transcription factors *Mash1*, *Phox2b*, *Phox2a*, *Gata2/3*, and *Hand2* (some of these are also listed in Table 15–1). These in turn drive the synthesis of proteins that are required to make a functioning norepinephrine-secreting neuron. The second encounter is when the cells reach their eventual target. Targets that eventually receive a cholinergic input act on the incoming noradrenergic cells

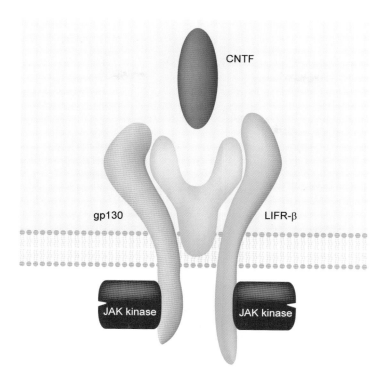

Figure 15–13. The CNTF receptor. This receptor is a complex of three proteins that, when bound to CNTF, activates JAKs, cytoplasmic tyrosine kinases.

to stop the synthesis of tyrosine hydroxylase, the rate-limiting enzyme in the pathway for synthesis of norepinephrine (see Chapter 10). Instead, the neurons now synthesize choline acetyltransferase and make acetylcholine their transmitter.

The switch from noradrenergic to cholinergic neuron can easily be observed in a culture dish. When immature adrenergic neurons are isolated in culture, they maintain the ability to make norepinephrine. If, however, these neurons are cultured with the cell types that normally receive a cholinergic input, the cells switch to making acetylcholine (Fig. 15–14b). There are several different cytokines, secreted into the medium from different tissues, each of which can produce this result. One of these, which can be purified from the medium secreted by heart cells, is *LIF* (Leukemia Inhibitory Factor). A second factor is CNTF. Neither of these, however, appears to be a critical factor in normal development, and normal cholinergic neurons develop in animals that are lacking either of

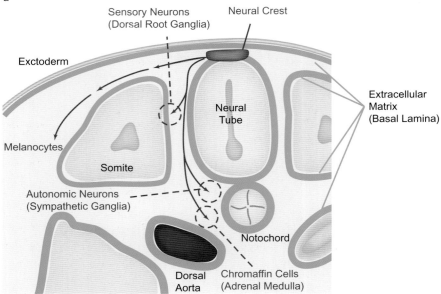

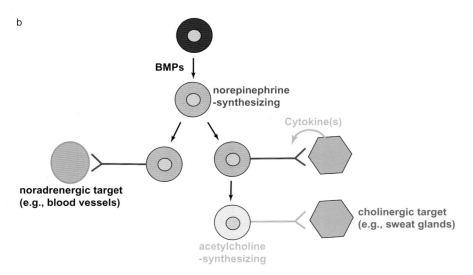

Figure 15–14. The neural crest. *a*: A cross section through the trunk region of a chick embryo. Cells that migrate from the neural crest region develop into sensory, sympathetic, and parasympathetic neurons, as well as other cell types. Also shown is the extracellular matrix of the basal lamina (see Chapter 16). (Modified from Sanes, 1983.) *b*: After BMPs allow neural crest precursors to develop into noradrenergic neurons, cytokine signals from a target tissue can convert them to cholinergic neurons.

these factors. Nevertheless, the real factor is likely to be some cytokine or combination of factors that signals to the nucleus through the JAK–STAT pathway. Genetic knockout of gp130 or blockage of LIPRβ completely prevents the development of the cholinergic neurons. ❷

Other Trophic Molecules

Thus far in this chapter, we have considered some of the factors that determine the survival of sensory and sympathetic neurons. All neurons are exposed to a variety of factors that keep them alive and maintain their individual identity. Because motor neurons in the spinal cord are relatively easy to isolate and to count, additional molecules have been identified in studies with these cells. Approximately half of newly born motor neurons eventually die before the mature nervous system is formed. Evidence that survival of the remainder depends on a target-derived factor came from the finding that surgically implanting an extra limb during development prevents their death by giving them a target and a job to do.

One factor that promotes the survival of motor neurons and of other neurons, such as dopaminergic neurons in the central nervous system, is *GDNF* (Glial cell line–Derived Neurotrophic Factor). GDNF comprises a ~40 kDa dimer of two glycoproteins. It is found in target muscles as well as in Schwann cells, and it acts by binding to a multiprotein receptor that contains *GFRα1* (GDNF Family Receptor α1) and a tyrosine kinase termed *Ret*. Other factors are *cardiotrophin-1*, also found in embryonic skeletal muscle, and *CLC/CLF* (Cardiotrophin-Like Cytokine/Cytokine-Like Factor 1), as well as *IGF-1* (Insulin-like Growth Factor-1), which is able to stimulate the sprouting of axons. Genetic elimination of any of these factors results in the death of subsets of neurons, although no one factor eliminates the development of all motor neurons.

Two trophic factors are better than one. The picture that emerges from studies with these trophic factors is that the many other cells that a young motor neuron and its axon encounter as they migrate to their destination contribute factors to keep the motor neuron alive and shape its identity. These other cells include astrocytes and the myelinating Schwann cells that envelop the axons of motor neurons. Moreover, some factors have a *paracrine* mode of action (that is, they act on an adjacent cell), while other are present within motor neurons themselves and are therefore *autocrine* factors (they act on the same cell type that releases the factor). Some factors act only for a short period during development, while others are required to keep the adult motor neurons alive and functioning.

Proper development of a neuron is likely to require the synergistic action of multiple factors from each of the several sources, and occasionally one factor may be able to substitute for another. For example, BDNF is found in skeletal muscle and its Trk B receptor is present on motor neurons. CNTF is secreted by Schwann cells and is taken up by motor neurons. External application of either of these factors helps to protect motor neurons from cell death. Total loss of BDNF or CTNF does not, however, alter the number of cells that die during the normal period of motor neuron apoptosis (although some motor neurons come to depend on CNTF for their existence much later in development).

Steroid Hormones

For many cells in the nervous system, appropriate development depends on the action of hormones secreted by other organs in the body. Some of these are listed in Table 15–3. In contrast to peptide hormones, these hormones are lipid soluble and readily enter the brain from the blood. They also cross cell membranes without the need for specific carriers or receptors in the plasma membrane. There is evidence that each of these hormones influences the nervous system, based primarily on the finding of receptor proteins for these hormones in the cytoplasm of neurons or glial cells. In most cases, the different receptor proteins are localized to specific groups of neurons, rather than being uniformly distributed throughout the nervous system.

Neurons and glia are themselves also capable of synthesizing small amounts of steroids. These are made either from precursor steroids that enter the brain from the gonads or adrenal gland or directly from cholesterol within the brain. Such locally produced steroids are termed *neurosteroids*.

How do steroid hormones work? The classical mode of action of these hormones is through a class of receptor proteins that, in the presence of the hormone, directly bind DNA in the nucleus, allowing the transcription of specific genes. The general structure of these receptor proteins is shown in Figure 15–15a. There is a high degree of homology in the regions of DNA binding and of hormone binding in the different receptors for this class of hormones. Figure 15–15b shows part of the sequence of the DNA-binding region of the receptors, which is rich in the amino acid cysteine. These are believed to form a coordination complex with zinc

Table 15–3 Steroid Hormones Important for Neuronal Development

Class	Example	Major Source
Androgens	Testosterone	Gonads
Estrogens	β-estradiol (estrogen)	Gonads
Progestins	Progesterone	Gonads
Glucocorticoids	Corticosterone	Adrenal gland
Mineralocorticoids	Aldosterone	Adrenal gland
Neurosteroids	Allopregnanolone	Brain

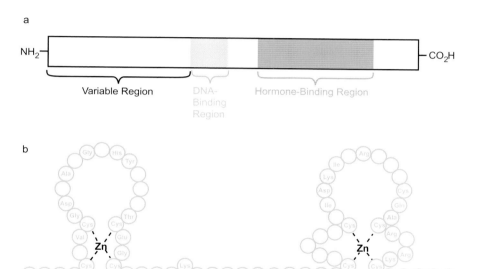

Figure 15–15. Steroid receptors. *a*: An outline of the three distinct regions found in these receptors. *b*: Amino acid sequence of a DNA-binding region. The named amino acids are found in nearly all receptors of this family. *c*: DNA sequence within the prolactin gene required for binding the estrogen receptor (Evans, 1988).

ions, producing two zinc fingers. A *zinc finger* is a structural feature that allows a protein to bind DNA.

When a cell is exposed to hormone, the receptors bind the hormone and undergo a structural change that allows the DNA-binding regions to interact with specific short sequences in the DNA of the cell. These DNA sequences have been termed *hormone response elements* (HREs). They are found in the proximity of genes coding for proteins whose expression is regulated by the steroid/thyroid family of hormones. Although the exact position of the HRE relative to the structural gene encoding the protein itself may vary in different genes, the sequence of the HRE is very similar for each of the genes regulated by a specific hormone. The sequence of the HRE in the gene for prolactin, a peptide whose synthesis is regulated by the steroid hormone estrogen, is shown in Figure 15–15c.

In addition to this now-classic pathway for activation of gene transcription, steroid hormones also produce additional "nongenomic" actions. For example, estrogen receptors are found at the plasma membrane, where, after interacting with estrogen, they form protein complexes with GTP-binding proteins, metabolic glutamate receptors, and tyrosine kinases. These events at the membrane activate a second set of cellular signals, which, acting in concert with the genomic actions, alter the shape and excitability of neurons in ways that are required for the behaviors such as *lordosis*, a stereotyped reproductive behavior described in the section "Activational effects of steroid hormones."

Finally, some neurosteroids can act very rapidly. We have already stated that neurons are capable of making their own steroids. It appears that the major use of such homegrown steroids is to act extremely rapidly on neighboring neurons in ways that are akin to the actions of neurotransmitters. Moreover neurosteroids such as allopregnanolone (Table 15–3) bind to neurotransmitter receptors such as $GABA_A$ receptors, greatly potentiating inhibitory transmission and reducing neuronal excitability.

Organizational effects of steroid hormones. Examples of the neuronal action of external hormones have come from studies of the role of the gonadal steroids in the control of reproductive behaviors. Differences exist between the brains of males and females in the size and synaptic connections of several well-defined groups of neurons. One such group in rat brain is the *sexually dimorphic nucleus of the preoptic area* (SDN-POA), which is five times larger in males than in females. The increased number of neurons in males results from the action of testosterone during development. Thus, if a female rat is treated with testosterone during late embryonic development and over the first 10 days after

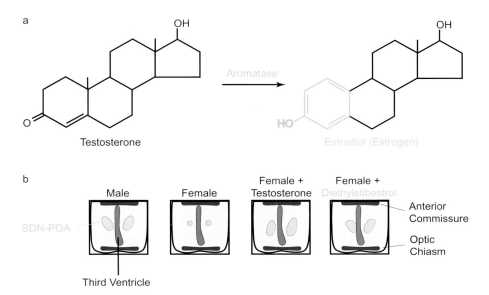

Figure 15–16. Steroid hormones. *a*: Sections through the sexually dimorphic nucleus of the preoptic area (SDN-POA) were prepared by Roger Gorski (1989) and his colleagues, using normal male and female rats and steroid-treated female rats. *b*: Aromatization of testosterone to estrogen.

birth, the SDN-POA develops as in a male (Fig. 15–16a). Interestingly, this action of testosterone is not mediated by a receptor specific for testosterone. Enzymes exist within the neurons that convert testosterone into estrogen. This process is termed *aromatization* (Fig. 15–16b). Thus estrogen is actually the active hormone within the cells, and treatment of immature females with compounds such as diethylstilbestrol, a highly active analog of estrogen, also causes the male pattern of development of the SDN-POA.

The action of a steroid hormone to alter this developmental pathway is an example of what have been termed *organizational* effects of steroid hormones. In particular, the action of the hormone is permanent and restricted to a critical period during development. For example, treatment of female rats with testosterone later than 6 days after birth cannot alter the size of the SDN-POA.

Activational effects of steroid hormones. The role of steroid hormones is not confined to the organization of specific neuronal pathways during development. The hormones also regulate the properties of many mature neurons and thereby influence the onset of specific animal behaviors.

Again, such effects are particularly obvious in reproductive behaviors and have been termed *activational* effects. For example, the morphology and extent of dendritic branching in neurons that innervate muscles involved in copulation remain sensitive to changes in testosterone in adult male rats. Another well-studied effect is the action of estrogen and progesterone in the onset of lordosis behavior in female rats. During this behavior, the female rat assumes a characteristic body posture that indicates sexual receptivity.

Groups of neurons that control lordosis behavior are found in the ventromedial nuclei within the hypothalamus. An elevation of the levels of estrogen produces morphological changes in these and other neurons, increasing the size of their somata and their dendrites, and increasing the synthesis of a variety of proteins including the receptors for progesterone. Subsequent elevation of progesterone levels, as would normally occur during the estrus cycle, induces the synthesis of further proteins that are essential for lordosis behavior to occur. ❽

Summary

The development of the nervous system requires the participation of a variety of growth factors that influence neuronal determination, proliferation, migration, and differentiation. Molecular genetic approaches using *Drosophila*, as well as other creatures whose genetics is well understood, have provided insights into the mechanisms of action of some of these developmental factors, such as Notch and Delta that interact on the surface of next-door-neighbor cells. Other factors are soluble and are secreted by nearby cells or other neurons. These include neurotrophins such as NGF and BDNF, cytokines such as CNTF, as well as GDNF and steroid hormones. A great deal of current research is aimed at identifying the key growth factors required for producing different types of neurons and the different patterns of transcription factor that are activated by different combinations of these factors. This knowledge may eventually allow the development of medical therapies to convert a stem cell into a sympathetic neuron, a motor neuron, or any one of the thousands of other types of neurons that make a mature nervous system.

Adhesion Molecules and Axon Pathfinding

The electrical activity of neurons can be transformed and modulated in ways that allow a neuron to control specific behaviors, as we will discover later in this book. Electrical behavior, however, is not the only aspect of neuronal activity that is subject to regulation by other cells or external stimuli. Structural features, such as the number, size, and type of synapses that a neuron makes, can also be modified. Furthermore, the shape of a mature neuron's axonal and dendritic branches may alter over time. This is illustrated in Figure 16–1, which shows high-resolution time-lapse images of an axon in the cerebral cortex of an adult mouse brain over a period of 9 days. A subset of neurons in these mice synthesize the green fluorescent protein (GFP), allowing them to be imaged through a small window over the cortex. While most axon terminals and dendrites are stable over the entire period, new axon terminals and spines can still form rapidly, and can disappear again just as quickly. Such experiments lend support to the idea that long-term regulation of neuronal activity in the mature nervous system may be accompanied by rearrangements of neuronal structure and connections. To examine such rearrangements, we must first understand the factors that lead to specific patterns of branching during formation of the nervous system.

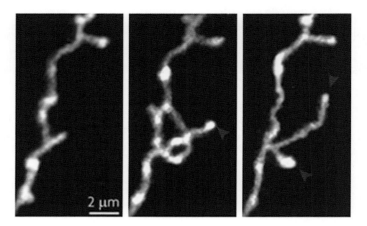

Figure 16–1. Remodeling of axonal connections. These time-lapse images of an axon with its synaptic terminals in the cerebral cortex of an adult mouse brain were made by Karel Svoboda and his colleagues. Newly formed axonal outgrowths over the 9-day period are indicated with a red arrow. Scale bar, 2 μm.

Axon Outgrowth During Development

Once an immature neuron has reached its final location in the nervous system, it must establish contacts with its appropriate synaptic partners by extending axonal and dendritic branches toward these partner neurons (see Fig. 14–1). In some cases, a neurite may have been established even during cell migration. For example, in the case of the developing granule cells of the cerebellum, discussed in Chapter 14, some neurite extension occurs before the cells migrate along the Bergmann glial cell. A trailing neurite is then left along the length of the glial fiber and eventually becomes the axon of the granule cell. In general, however, neurite extension occurs *after* cell migration.

Growth cones allow an axon to move forward. The outgrowth of neurites is guided by a specialized region of the cell known as the *growth cone*, which is found at the leading tip of a neurite. Growth cones can be investigated in cell culture, where neurite extension can be examined readily under the microscope. Figure 16–2 illustrates the major features of the growth cone region (see also the cover illustration). The *central core* is an extension of the neurite process itself and is rich in microtubules that provide the structural support for axoplasmic transport (see Chapter 2). Using video-enhanced microscopy, bidirectional transport of granules can

a

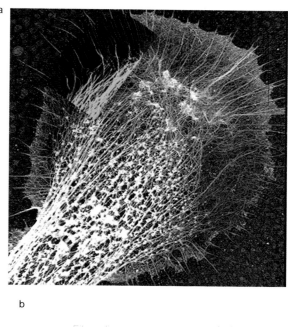

b

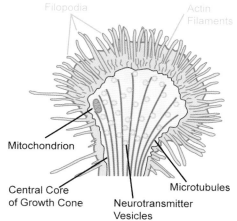

Figure 16–2. The growth cone. *a*: Photograph made by Paul Forscher of the growth cone of an *Aplysia* neuron in cell culture. Overlaid on part of the growth cone is the pattern of staining for actin (red) and microtubules. *b*: Components of a growth cone.

be observed up to the central core (Video 16–1, 16–2). ◐ In addition, the core of the growth cone is rich in mitochondria, endoplasmic reticulum, and vesicular structures.

   Surrounding the central core is a region that is generally devoid of organelles but very enriched in the contractile protein *actin*. Time-lapse

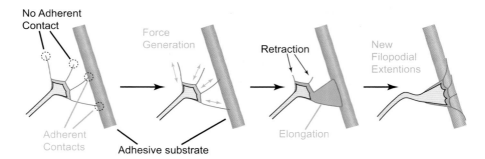

Figure 16–3. Schema showing how continual extension and retraction of microspikes (filopodia), coupled with the adhesion of microspikes to the selected substrate, may guide the direction of a growth cone.

pictures of these regions, which are known as *lamellipodia*, reveal the existence of undulating waves of movement that have been termed *ruffling*. Finally, very thin straight processes known as *microspikes* or *filopodia* are found at the extremities of the lamellipodia. These, like the lamellipodia, are rich in actin. The microspikes are in constant motion, extending from and retracting back into the lamellipodia. Growth of the neurite occurs when a microspike extends and then, instead of retracting, remains in place while the lamellipodium advances toward the end of the microspike (Fig. 16–3). New microspikes then extend from the newly advanced border of the lamellipodium.

What makes the growth cone move? The movement of the growth cone is associated with a continual cycle of polymerization and depolymerization of actin. At the very leading edge of the growth cone, actin proteins are continually assembled into the structural long filaments termed *F-actin*. This assembly of actin filaments is balanced by the ongoing depolymerization or cleavage of these filaments at the other end of the lamellipodium, at its junction with the central core of the growth cone (Fig. 16–4a). To link these cycles of synthesis and destruction, there is a continual *retrograde flow* of the actin cytoskeleton back from the leading edge toward the central core. The retrograde flow of F-actin accounts for the undulating waves of movement that can be observed with video-enhanced microscopy, and is driven by the protein *myosin 2*, a molecular motor comparable to those we encountered in Chapter 2. This continually pushes the actin filaments from the lamellipodia back toward the central core.

The rate at which a growth cone extends along a surface, or *substrate*, depends on how strongly the growth cone adheres to the substrate.

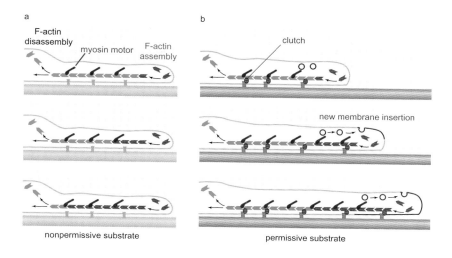

Figure 16–4. Actin assembly in the lamellipodium. *a:* Experiments by Paul Forscher and colleagues suggest that on a surface that is nonpermissive for axon elongation, F-actin filaments are not closely linked to the surface, leading to persistent retrograde flow of newly assembled F-actin. *b:* When the F-actin cytoskeleton is coupled to the substrate, polymerization of F-actin at the leading edge leads to forward extension of the growth cone. This is accompanied by insertion of new plasma membrane. See Animation 16–1. ⬤

A simple model that has been proposed for the elongation of a neurite is shown in Figure 16–4b. According to this model, the rate of growth depends on the strength of physical coupling between the F-actin cytoskeleton and the substrate. As we shall see later, there are, in the membranes of neurons, many molecules known to serve as such physical links from the cytoskeleton to the outside world. If such coupling is weak or absent, the retrograde flow of actin proceeds normally and is not coupled to the substrate. In this condition, the substrate is said to be *nonpermissive* for growth, and the growth cone remains stationary (Fig. 16–4a). In contrast, if the F-actin cytoskeleton is tightly coupled to the substrate, this coupling prevents or slows the flow of actin filaments back toward the central core. Instead, the formation of new actin filaments at the leading edge now leads to the forward extension of the growth cone. Such stronger coupling is said to occur on *permissive* substrates (Fig. 16–4b). Proteins that provide a strong physical link between the actin filament and a permissive substrate can be said to comprise a "clutch" that regulates neurite extension. The actual molecules that glue a cell to a substrate or to other cells

Table 16–1 Axonal Guidance Molecules

Guidance Molecule	Receptor
Semaphorins	Neuropilins, Plexins
Netrins	UNC-40, *C. elegans*
	DCC, vertebrates
	Frazzled (*Drosophila*)
	UNC-5 (Repulsive)
Slits (1–3)	Robo (1–3)
Ephrins	EphA and EphB receptors

are termed *substrate adhesion molecules* and *cell adhesion molecules* and will be dealt with later in this chapter.

A neurite does not elongate simply by extending its cytoskeleton. The increase in overall size of the cell requires that new plasma membrane must continually be added as the growth cone makes its way toward its target. This occurs through exocytosis of internal vesicles into plasma membrane and uses the same types of molecules we encountered in Chapter 8 for the exocytosis of neurotransmitter vesicles (Fig. 16–4b).

Growth cones have to be given directions. The physical adhesion to a substrate that allows axons to grow is only part of the story. The growth cone needs information about where to go. This is provided by *guidance molecules*, whose sole function is to provide directions to the growing axon. In contrast to cell adhesion molecules, these guidance molecules do not need to be present in sufficient abundance to provide mechanical adherence. Rather, they simply relay messages from the external environment to the growth cone. Some of these molecules and their known receptors are listed in Table 16–1; their properties will be described later in the chapter. Guidance molecules can either be soluble molecules secreted from a distant target or be present on the surfaces along which the growth cone makes its journey (Fig. 16–5). Diffusion of an attractant molecule from a distant source causes a growth cone to turn and extend toward the source, a phenomenon known as *chemoattraction* (Fig 16–5a). Conversely, in a gradient of a repulsive molecule, the growth cone experiences *chemorepulsion*, and turns away from the source (Fig 16–5b). For the case of attractant or repulsive guidance molecules on cell surfaces, a growth cone will turn onto or away from such surfaces, but only after the growing neurite has touched the surface (*contact-dependent* attraction or repulsion, Fig. 16–5c, d). ●

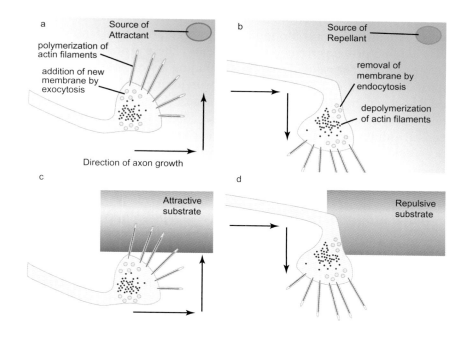

Figure 16–5. Actions of guidance molecules on the direction of movement of growth cones. *a*: Chemoattraction. *b*: Chemorepulsion. *c*: Contact-dependent attraction. *d*: Contact-dependent repulsion.

In order to make a growth cone change its direction, an attractant molecule has to accomplish two tasks. It first has to increase the polymerization of actin into new filaments that extend toward the source of the attractant. Second, it has to increase the rate of membrane addition through exocytosis on the side of the growth cone that faces the source of attractant. Conversely, a repellent has to produce a collapse of actin filaments on the side of the growth cones and increase endocytosis of membrane on the side of the growth cone that faces the source or repellant, allowing the growth cone to navigate in the opposite direction (Fig. 16–5).

As we will see later, guidance molecules act on specific receptors that trigger second messengers or activate ion channels in the growth cone. The very small differences in amount of receptor activation in the different regions of the growth cone result in retraction in one region and elongation in another. Many guidance molecules produce uneven elevations of internal calcium ions across a growth cone, such that regions with higher calcium levels undergo extension, with actin polymerization and membrane insertion, while regions on the other side have lower calcium levels, leading to retraction. The formation of the actin cytoskeleton within lamellipodia and filopodia is also regulated by a family of small GTP-binding proteins termed *Rho* proteins. These proteins are related to the ras protein that is involved in the signaling action of receptor tyrosine kinases (see Chapter 15). At least three proteins in this family, *Cdc42, Rac,* and Rho itself, are thought to regulate the actin cytoskeleton in the growth cones of neurons. Experiments in which these activated proteins have been introduced into neurons suggest that when the receptor for a guidance molecule activates Cdc42 and Rac, the subsequent biochemical events produce an attractive response. In contrast, activation of Rho triggers repulsion, by causing the local disassembly of actin.

An initial attraction can quickly change to repulsion. As with most airline travel today, axons rarely travel to their final destination along a straight line. Instead, they typically first travel to an intermediate location, at which they abruptly change their orientation, and then head toward a second or third intermediate position in the brain before reaching their eventual target. How can this be achieved with guidance molecules that act simply as attractants or repellants? The answer is that the same guidance molecule can act as either an attractant or a repellant at different times in an axon's journey. Figure 16–6 shows how an axon reaches a destination B through an intermediate location A. An initial strong attraction to source A serves to bring the axon directly to A. Once it arrives, however, this initial attraction rapidly changes to repulsion, and the axon starts to feel a new attraction to location B, which guides its subsequent movements. In this way, axons pick up new instructions at each stop on their journey.

A change from attraction to repulsion is the result of biochemical events that occur in the growth cone once it reaches its intermediate target. These events may be changes in either the levels of second messengers or the types of receptors for the guidance molecules; we shall see examples of this later in this chapter when we discuss the guidance molecules and their receptors. Before then, however, we shall first take a closer look at the adhesion molecules required for axon movement.

Cell-Substrate Adhesion Molecules

In its journey to its final destination, an axon may crawl over many types of permissive substrates, including non-neural cells and other neurons. For some types of neurons, much of their migration and axon elongation occurs not over the surface of other cells but through an *extracellular*

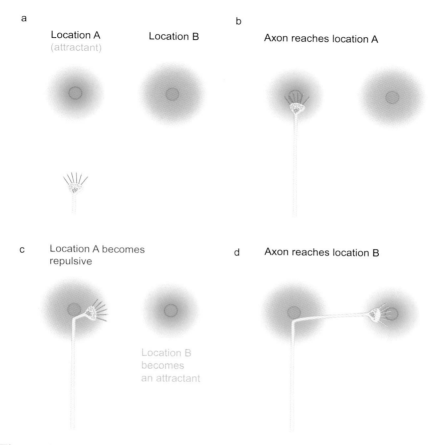

a

Location A
(attractant)

Location B

b

Axon reaches location A

c

Location A becomes
repulsive

d

Axon reaches location B

Location B
becomes
an attractant

Figure 16–6. Converting attraction to repulsion guides axons to their final destination. *a*: An axon grows toward location A. On reaching A (*b*), attraction to location A is converted to repulsion (*c*), and the axon is attracted to a new source of attractant guidance molecules in location B (*d*).

matrix, a loose latticework of glycoproteins and sugars that is relatively devoid of cells. Close to the membrane of cells that are not migrating through the space, this latticework becomes denser and forms a *basement membrane* (Fig. 16–7a). Both extracellular matrix and basement membrane contain molecules that guide axonal movement (Table 16–2). Two major components of this latticework are the proteins *fibronectin* and *laminin* (Fig. 16–7b). These can be obtained in purified form, and their effects on cell adhesion and neurite outgrowth have been measured by culturing isolated neurons on surfaces that have been coated with these proteins. Both protein complexes contain several distinct domains, each of which appears to have a specific role in binding other molecules in the

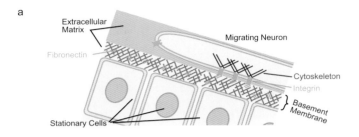

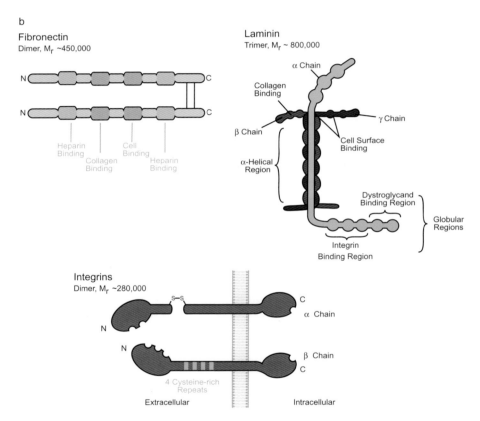

Figure 16–7. Neuron–substrate interactions. *a*: The extracellular matrix. *b*: The structure of some molecules involved in cell–substrate interactions.

extracellular matrix, or in binding to cell membranes and promoting neurite outgrowth. Thus they are a major part of the "glue" that attaches cells to the matrix.

The attachment of fibronectin and laminin to cells is mediated by receptor proteins termed *integrins*, which are located in the plasma

Table 16–2 Some Components of the Extracellular Matrix

Collagens	A family of glycoproteins rich in proline
Laminin	Elongated glycoproteins that bind to receptors on cell membranes and
Fibronectin	also to other components of extracellular matrix such as the collagens
Chondronectin	
Hyaluronic acid	Glycosaminoglycans—unbranched disaccharide polymers
Chondroitin sulfate	
Heparan sulfate	
Reelin	Large glycoprotein that regulates neuronal migration and the development of dendrites and dendritic spines
Tenascins	Glycoproteins expressed at times of active neurogenesis

membrane of many neurons and other cells (Fig. 16–7b). For example, the extracellular domain of the integrin fibronectin receptor binds to the sequence of amino acids Arg-Gly-Asp-Ser that is found in the fibronectin molecule. Similar sequences are found in many molecules that bind to other integrins in a wide variety of non-neuronal cells. The integrins are composed of two subunits. The α subunit regulates the specificity of interactions with different ligands. In contrast, the short cytoplasmic domain of the β subunit of the integrins binds directly to components of the actin cytoskeleton. This provides a direct link between the intracellular scaffold of the cells and the external latticework (Fig. 16–7a).

The integrins do more than provide a mechanical link for the extracellular matrix. The binding of a ligand induces the clustering of these receptors in the plasma membrane. This in turn allows a number of tyrosine protein kinases to associate with the integrin–cytoskeleton complex and become activated. For example, the ras/raf/MEK/MAPK pathway, covered in Chapter 15, as well as the protein kinase C pathway, is activated upon stimulation of integrins. **&**

Cell Adhesion Molecules

Although the extracellular matrix plays an important role in the migration of cells and the extension of neurites in the periphery, the central nervous system does not possess a well-defined extracellular matrix. Basement membranes are found only along the cerebral blood vessels and lining the fluid-filled cerebral ventricles, and thus the movement of neurons and their axons in the central nervous system occurs largely over other cells. Moreover, even in the peripheral nervous system, much of the growth of axons occurs over the surface of epithelial cells and other axons. In some cases, as in the growth of nerve tracts, a large number of axons initially all grow in the same direction, with most of their growth cones moving over other axons. A bundle of closely associated axons is called a *fascicle*, and the formation of such bundles is known as *fasciculation*. In contrast to the molecules described in the last section, which have been termed *substrate adhesion molecules* (SAMs), the physical association of the membranes of two cells occurs through *cell adhesion molecules* (CAMs).

Cell adhesion molecules were initially discovered by Gerald Edelman and his colleagues using suspensions of cells from chick retina. When such cells are dissociated in culture, they reaggregate readily into clumps of cells. Antibodies that can specifically prevent this reaggregation were made and then used to isolate a membrane glycoprotein that binds to these antibodies. The first protein to be isolated in this way was termed *N-CAM*, for neuronal-CAM. N-CAM is one of a family of cell adhesion molecules that has now been found to be expressed on neurons and on a wide variety of other cells. Figure 16–8 shows the similarity of N-CAM to three other CAMs, which represent only some of the known CAMs. Ng-CAM is an adhesion molecule found on specific axonal tracts. Myelin-associated glycoprotein (MAG) is a glial cell adhesion molecule. Fasciclin II ❷ is a glycoprotein expressed on a subset of grasshopper neurons. The extracellular part of these molecules has a series of domains that each contains about 50 amino acids between two cysteine residues. These, by forming -S-S- bridges, can form the domains into loops. Very similar structures (termed *immunoglobulin C2 type domains*) are found in *immunoglobulins*, molecules involved in the mounting of immune responses by lymphocytes. The CAMs are therefore considered to belong to the *immunoglobulin superfamily*.

Most of the CAMs are believed to cause cell–cell adhesion by binding to the same CAM on an adjacent cell. Thus N-CAM molecules on one cell bind directly to N-CAM molecules on its neighbor, providing so-called *homophilic* interactions between the cells. However, as shown in Figure 16–8, some of these molecules also contain regions of homology to fibronectin. These latter regions (termed *type III domains*) include the sequence Arg-Gly-Asp-, which is used in binding to the integrins, indicating that these molecules interact with cell surface proteins in more than one way. Another important family of related cell adhesion molecules is the *cadherins*. N- (for neuronal) cadherin is the best-studied example in the nervous system, although it is also found in non-neuronal cells. Like the CAMs, an N-cadherin molecule binds to another N-cadherin on an

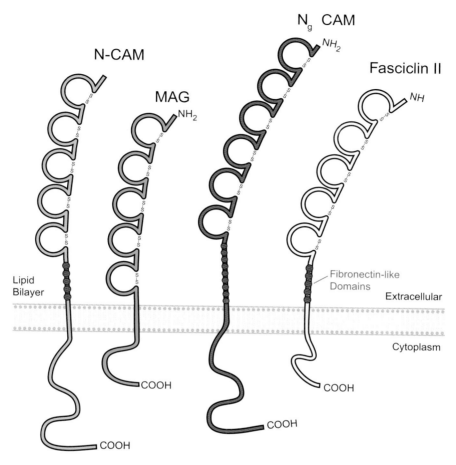

Figure 16–8. Cell adhesion molecules. Examples of the structures of four CAMs.

adjacent cell. In contrast to the CAMs, this binding requires the presence of external calcium ions.

Animals lacking N-cadherin do not survive past the embryo stage, although, perhaps surprisingly, genetic knockout of N-CAM produces relatively minor changes in neuronal development. A central test for the involvement of CAMs and cadherins in normal development has therefore been to expose developing cells to antibodies raised against these proteins. For example, antibodies that block the homophilic binding of N-CAM or N-cadherin disrupt the normal development of the retina into sharply defined cell layers. Another test for the role of N-cadherin in the extension of axons from the retina has been to place fragments of retina onto a

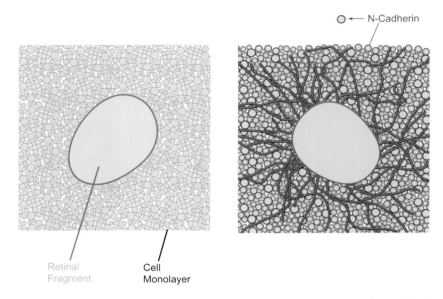

N-Cadherin

Retinal
Fragment

Cell
Monolayer

Figure 16–9. N-cadherin guides optic nerve fibers over the surface of other cells. This figure shows an experiment by M. Takeichi and his colleagues. Fragments of retina, placed over a layer of cells that lack N-cadherin, do not extend neurites from the explant. When the cells forming the layer are modified to express N-cadherin, outgrowth occurs (Matsunaga et al., 1988).

single layer of cells that do not normally express N-cadherin on their surface. In this condition, although retinal cells in the explant possess N-cadherin, no neurites extend out from the retinal fragments onto the monolayer of cells. When, however, the cells constituting the monolayer are induced to make N-cadherin, by introduction of an active gene for this adhesion protein into the cells, vigorous neurite outgrowth is observed (Fig. 16–9). ❽

Molecular Guidance Cues

To get to their eventual destination, axons and their growth cones have to get a firm grip on their road surface, and, as we have seen, this is provided by the substrate-adhesion and cell-adhesion molecules we have discussed. By itself, however, a firm grip is like having a car with good tires and gasoline but no map or GPS. As we saw earlier (Figs. 16–5, 16–6), a different set of molecules provide the *guidance cues*, which actually tell the growth cone which road to take. Four of the best-characterized classes of guidance factors are the *semaphorins*, the *netrins*, the *slits*, and the *ephrins* (Table 16–1). We shall discuss the first three later in this chapter, and cover some of the action of the ephrins in Chapter 17, when we deal with synapse formation. In addition to these four main families of guidance molecules, some of the factors that regulate earlier stages of neuronal differentiation, and which we covered in previous chapters, such as the BMPs, Wnts, Hedgehogs and other growth factors, are known to take on roles in guidance once their other work is done.

Semaphorins

A very important class of guidance molecules are the *semaphorins*. These can exist either as fixed membrane glycoproteins or as soluble proteins that are secreted from cells (Fig. 16–10a). Moreover, they can provide either attractant or repulsive cues to an advancing growth cone. All semaphorins contain a large conserved extracellular sequence of approximately 500 amino acids termed the *semaphorin domain*. In addition, semaphorins can be divided in nine classes depending on other characteristic features, such as, for example, immunoglobulin C2 domains. The membrane semaphorins have a single transmembrane domain with a very short cytoplasmic tail, which is not thought to play a role in signaling beyond anchoring the protein in the cell membrane. In common with cell adhesion molecules, semaphorins are present on subsets of axons and on non-neuronal cells along which growth cones travel. In contrast to the adhesion molecules, however, the presence of semaphorins in the membranes of dissociated cells does not cause the cells to aggregate into clumps. Rather, changes in direction of advancing growth cones are effected by biochemical changes produced by local activation of membrane receptors.

Two different types of receptors on neurons are known to bind semaphorins (Fig. 16–10b). The major receptors are the *plexins*. Technically, these are themselves semaphorins, since they possess a large extracellular semaphorin domain. In contrast to the semaphorin ligands, however, the plexins have two large cytoplasmic domains that are conserved and that participate in signal transduction once the extracellular domain binds a semaphorin ligand. Some semaphorins bind plexins directly, but others, particularly the secreted semaphorins, first bind a second type of receptor, termed a *neuropilin*, which exists in a complex

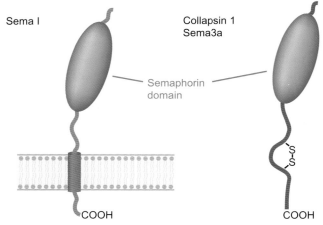

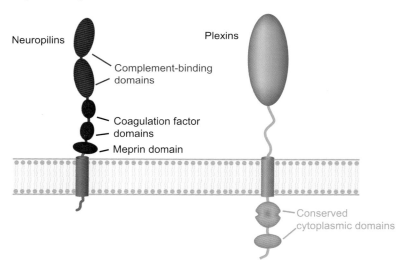

Figure 16–10. Semaphorin molecules and their receptors. *a*: G-sema I is a membrane-bound semaphorin ligand in grasshopper cells. Collapsin-1 is the original name for Sema3A, a soluble semaphorin first isolated from chick cells. *b*: Neuropilins and plexins are two structurally distinct receptors for semaphorins.

with a plexin. Neuropilins contain domains first identified in proteins involved in blood coagulation, the *complement-binding* and *coagulation factor* domains. They also contain a *meprin* domain, named after a conserved domain in meprin proteases, which is thought to be required for certain protein–protein interactions.

Semaphorin signaling in grasshopper neurons. The first semaphorin was identified by studying the navigation of a set of neurons, termed *Ti1 pioneer neurons*, in the embryonic nervous system of the grasshopper. In the early growth of a fascicle of axons, the first neuron to enter the pathway follows cues from the environment; this neuron is frequently called the *pioneer* cell. The growth cones of the subsequent axons then extend along the axon of the pioneer neuron to form the fascicle (although follower neurons may also be able to make the appropriate navigational decisions in the absence of the fibers of pioneer cells).

The cell bodies of the Ti1 neurons are present in the limb buds of the grasshopper, and during development their axons extend from the periphery into the central nervous system (Fig. 16–11a). As the axons first grow away from the cell bodies, they chart a course directly toward the central nervous system. At one point on their journey, however, they encounter a layer of epithelial cells that extends in the dorsal–ventral direction. These epithelial cells have, on their surface, a semaphorin termed *G-sema I* (G for Grasshopper) that causes the collapse of the neuronal growth cone. The growing axons are not able to cross this barrier of epithelial cells. Instead, on contacting the surface of these cells, the growing axons stop and then reorient their growth ventrally along the distal edge of the semaphorin-containing cells. It is only after the filopodia of the axons contact another type of cell, the Cx1 neurons, that the growing axons can cross the epithelial cells and proceed in the direction of the central nervous system (Fig. 16–11a).

The change in the direction of growth that occurs when the Ti1 growth cones meet the epithelial cells can be attributed to the presence of G-sema I on the epithelial cells. For example, the cells can be exposed to antibodies against G-sema I during the period of axon outgrowth, thereby preventing these molecules from interacting with the growth cones. When this happens, the ingrowing axons are able to extend directly across the stripe of epithelial cells (Fig. 16–11b).

Semaphorin 3A guides the formation and movement of axons and dendrites in opposite directions. G-sema I causes the axons of Ti1 neurons to follow a particular path because of its repulsive effect on their growth cones. Similar effects on axonal growth cones of many vertebrate neurons occur with *Sema3A*, one of the soluble semaphorins. In common with its membrane-bound relatives, Sema3A contains a large semaphorin domain (Fig. 16–10a). Instead of a transmembrane segment, however, Sema3A contains an immunoglobulin C2-type domain. This type of semaphorin was first discovered as a soluble factor that causes the collapse of the actin cytoskeleton in the growth cones of chick sensory neurons. As expected, such a collapse of the lamellipodium prevents further elongation of the

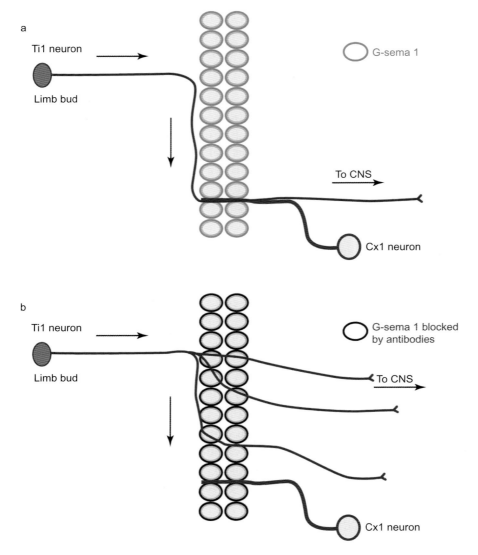

Figure 16–11. Guidance of axons by semaphorins. *a*: In the grasshopper nervous system, the Ti1 neuron in the limb bud sends its axon toward the central nervous system (CNS). On encountering a layer of epithelial cells bearing G-sema I on their surfaces, the axon is repelled and becomes diverted ventrally until it contacts the processes of the Cx1 neurons, whereupon it crosses the G-sema I barrier. *b*: Preincubation of cells with an antibody to G-sema I neutralizes its repellant effect, allowing axons to cross the epithelial cell layer.

axon; because of this biological action, the chick homolog of Sema3A was given the name *collapsin-1*.

Sema3A has a very interesting two-fold action on many growing neurons. In addition to serving as a repellant for axonal growth, it can actually suppress the formation of axons. In contrast to its effect on axons, however, it promotes the formation and growth of dendrites. As a result, it determines the *polarity* of a neuron. In the presence of a gradient of Sema3A, a newly formed neuron will extend its dendrites toward the source of Sema3A, and its axon will grow in the opposite direction. A clear example of this is found in the cerebral cortex of mammals, in which levels of Sema3A are high close to the external surface, also termed the *pial surface*, and progressively drop at locations deeper in the cortex (Fig. 16–12). The major output of the cortex comes from pyramidal neurons, which send their axons down into the white matter underlying the cortex. These neurons receive and integrate synaptic inputs on their apical dendrites, which extend toward the pial surface.

The growing pyramidal cells express the neuropilin-1 receptor for Sema3A. The effects of exposure to the gradient of Sema3A can be observed by placing the cell body of a single isolated pyramidal neuron, labeled with a fluorescent marker, on top of a slice of immature cortex. Over a period of several days, the growth cones of the dendrites of the labeled cell move toward the pial surface, while the growing axon extends in the opposite direction, toward the white matter and the underlying ventricle (Fig. 16–12). Elimination of the action of Sema3A by gene knockout, or blocking of its neuropilin receptor by applying antibodies to the slices, impairs the orientation of axons and dendrites. Genetic knockout of the neuropilin-1 receptor also prevents the migration of the cell bodies of the neurons toward the upper layers of the cortex. The endogenous gradient of Sema3A can be altered by placing non-neuronal cells that have been engineered to express high levels of Sema3A onto the slice near the white matter. When this occurs, the pyramidal cells now orient their axons away from the new source of Sema3A, and their dendrites now extend toward the Sema3-expressing cells.

The opposing effects of Sema3A on formation and growth of axons and dendrites appear to result from a "yin-yang" relationship between the second messengers cyclic GMP and cyclic AMP (see Chapter 12). Studies with isolated cells have demonstrated that local elevation of cyclic GMP in a growing neurite promotes the formation of a dendrite that is attracted to Sema3A. In contrast, local elevation of cyclic AMP promotes formation of an axon that is repelled by Sema3A. A local increase in either second messenger also suppresses the synthesis of the other messenger. The neuropilin-1 receptor activates the soluble enzyme guanylate cyclase, increasing levels

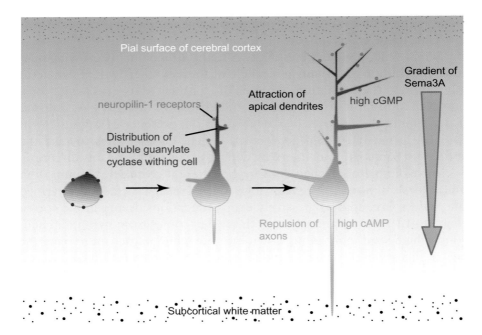

Figure 16–12. Sema3A guides the axons and dendrites of cortical pyramidal neurons in opposite directions. Work by Anirvan Ghosh and colleagues has shown that single pyramidal neurons placed on a slice of immature cerebral cortex orient their axons away from a source of Sema3A at the pial surface, while growing dendrites are oriented into the gradient (Polleux et al., 2000). Experiments by Mu-ming Poo and his coworkers demonstrated that Sema3A elevates cyclic GMP (cGMP) and decreases cyclic AMP (cAMP) levels, resulting in formation and growth of dendrites and an axon at opposite poles of a neuron (Shelly et al., 2010, 2011).

of cyclic GMP, and eventually resulting in a preferential distribution of this enzyme toward parts of the cell exposed to highest levels of Sema3A (Fig. 16–12). Accordingly, dendrites grow toward the source. Elevations in cyclic AMP can therefore only occur at parts of the cells furthest from the source, causing axons to form and grow in the opposite direction.

The Netrins—Soluble Guidance Molecules Related to Laminin

A second class of guidance molecules, which are structurally unrelated to the semaphorins, are the *netrins* (*netr* is a Sanskrit word for "one who guides"). Like peptide neurotransmitters and other secreted molecules, the netrins are first synthesized with a signal sequence, which allows them to enter the secretory pathway; the signal sequence is likely to be cleaved shortly after synthesis (see Fig. 8–6). The remainder of a netrin molecule bears a strong resemblance to part of the γ chain of the extracellular matrix protein laminin (see Fig. 16–7). Like the γ chain of laminin, the netrins contain several EGF repeats and the sequence of amino acids Arg-Gly-Asp that is a recognition signal for many of the integrin receptors (Fig. 16–13). Like the semaphorins, netrins can exert either attractive or repulsive actions on growing axons.

Netrins are found in both vertebrates and invertebrates. Indeed, the first described netrin, known as *UNC-6*, was found by genetic studies in the nematode (UNC refers to **Un**coordinated, reflecting the phenotypes of mutations in such proteins). They were first purified as chemoattractive

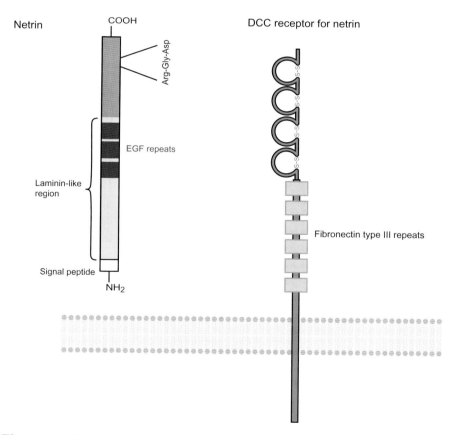

Figure 16–13. Structure of a netrin and of the DCC receptor that mediates attraction to netrins.

factors that come into play during the development of the spinal cord. Typically, netrins are present in cells near the midline of the brain, and they play a role in attracting the axons of neurons that will cross from one side of the brain to the other. Tracts of such axons are termed *commissures*. Mutant mice in which a major netrin termed *netrin-1* is present only in very low amounts fail to develop several commissures, including the *corpus callosum*, the major commissure that is composed of axons joining the two cerebral hemispheres.

Within the vertebrate embryo, neurons termed *commissural neurons* differentiate in the dorsal part of the spinal cord and then send their axons in a stereotyped pattern toward the ventral part of the spinal cord (Fig. 16–14a). At the ventral midline of the spinal cord lies the floor plate, a structure composed of epithelial cells that we encountered in the last chapter. The floor plate extends the entire length of the spinal cord and into the brain. Chemoattractant molecules, including netrin-1, are released from these floor plate cells and guide the growing axons of the

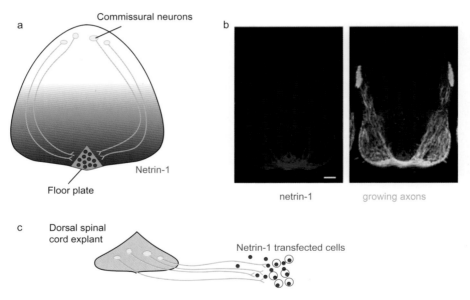

Figure 16–14. Netrins guide the axons of commissural neurons in the spinal cord. *a*: The normal course of commissural axons toward the floor plate. *b*: Staining of a section of embryonic mouse spinal cord for netrin-1 and for axonal filaments by Marc Tessier-Lavigne and his colleagues shows a gradient of netrin-1 even when axons have reached the floor plate (Kennedy et al., 2006). *c*: The axons of commissural neurons are guided from an explant of the dorsal spinal cord toward cells transfected with netrin-1.

commissural cells around the circumference of the spinal cord on their path ventrally (Fig. 16–14b). Netrin-1 is made only in the floor plate cells themselves. When applied to explants of the dorsal spinal cord, it stimulates the outgrowth of neurites. The direction of this outgrowth depends on the source of the netrin. For example, it is possible to make an artificial source of secreted netrins by transfecting a kidney cell line with the gene for netrin-1. When a clump of such transfected cells is placed next to an explant of the dorsal spinal cord, the secreted netrin-1 causes the reorientation of axon outgrowth toward the transfected cells (Fig. 16–14c).

There are at least two known receptors for the netrins. The major receptors that mediate attraction are proteins of the *DCC* family (for Deleted in Colorectal Cancer). These have four immunoglobulin repeats, as well as six repeats that resemble a region in fibronectin, termed *fibronectin type III repeats* (Fig. 16–13). In *Drosophila*, the corresponding receptor is termed *Frazzled*, and in nematodes, *Unc-40*. (The very different styles for naming proteins in flies and worms probably reflects the personalities of the investigators.). A related receptor termed *neogenin* mediates some non-neuronal effects of netrins. Another class of netrin receptor is termed the *UNC-5* family of proteins. These receptors are present on neurons that are repulsed by netrins and are required for this repulsive action. For example, for spinal cord motor neurons, whose axons grow in a direction away from the midline, netrins exert a repulsive rather than attractive influence.

How to Turn Attraction into Repulsion: Slits and Robos

Yet another family of guidance molecules is the *Slits* and their receptors, the *Robos*. In contrast to the netrins, which attract axons toward the midline, Slits were discovered by searching for factors that repel growing axons away from the midline. Slits are very large secreted proteins that contain many different motifs, including many EGF-like repeats, a laminin-like region, and four large domains rich in the amino leucine. Their Robo receptors, like DCC receptors, have both immunoglobulin and fibronectin type III repeats (Fig. 16–15a). They were discovered in mutant fruit flies that lack these receptors and that were termed *roundabout*, because axons near the midline take a roundabout path, crossing and re-crossing the midline many times instead of heading to their normal destination. Mutations that produce too much Robo signaling prevent axons from approaching the midline (Fig. 16–15b).

Slits are also present at the midline in the developing spinal cord of vertebrates. As we saw earlier, the axons of commissural neurons journey

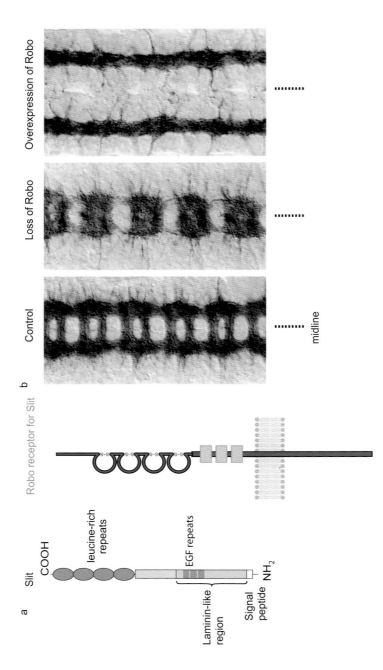

Figure 16–15. Slit/Robo signaling. *a*: Structure of a Slit and its Robo receptor. *b*: Experiments by Corey Goodman and his colleagues showing the pattern of staining of axons (in brown) within segments of the fly embryo. The normal pattern of midline crossing in the wild-type flies is altered in mutants that lack Robo (producing excessive axonal staining at the midline) or in mutants (termed *commisureless*) that have elevated Robo signaling and in which axons fail to approach or cross the midline (Kidd et al., 1998).

from the dorsal spinal cord toward the floor plate. Once they arrive, however, they cross the midline and make a right-hand turn to follow the length of the spinal cord on the opposite side, never again returning to the side where their cell bodies are located (Fig. 16–16). After reaching the floor plate, it is repulsion by Slits that keeps them moving across the midline and preventing their return. As we saw in Figure 16–6, for this to work, the repulsion can only begin once the tips of the axons have already reached the midline. This is achieved in large part by a biochemical change in Robo receptors on the axons. There are three different receptors, Robo1–3, all of which are present on the axons of the commissural neurons at one time or another. While the axons are navigating toward the floor plate, they express an isoform of one of the receptors, *Robo3.1*, that suppresses repulsion of the axons by Slits. Once they reach the midline, however, this is replaced by another isoform, *Robo3.2*. At the same time, there is an increase in Robo1 and Robo2, each of which mediates strong repulsion by the midline Slits.

Nogo, a Repellent of Axons Following Injury

A severe form of repulsion of axons occurs after damage to nerve fibers of the adult central nervous system. The devastating effects of injuries to the brain and spinal cord are due to the fact that once they have been severed, axons of neurons within the central nervous system fail to regenerate. This is because mammalian neurons are not able to extend neurites over the surfaces of oligodendroglia, the cells that make the myelin sheath of central neurons. During development, most axonal pathfinding has already occurred by the time the myelin sheaths begin to form. The repulsive effect of oligodendroglia is due to the presence in these cells of proteins that selectively repulse neuronal growth cones. A membrane protein termed *Nogo*, which bears no relationship to the other guidance factors we have discussed, appears to be the major repulsive factor in these glial cells.

Nogo is present in oligodendroglia, as well as in some neurons, but is absent in Schwann cells, the glial cells that make the myelin of peripheral nerves. For this reason, axons of the peripheral nervous system are frequently capable of regenerating. It is, however, possible to induce growing axons to cross oligodendroglia by applying antibodies to Nogo. Such antibodies neutralize the inhibitory effect of Nogo and allow axons to extend actively over the glial membranes. As might be imagined, the ability to manipulate neuronal pathfinding after injuries has enormous clinical implications for conditions such as spinal cord injuries.

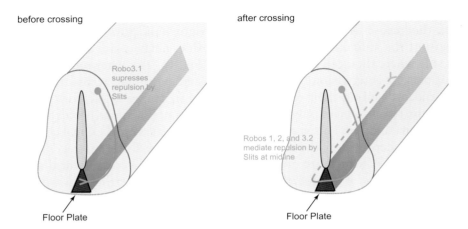

before crossing

Robo3.1
supresses
repulsion by
Slits

Floor Plate

after crossing

Robos 1, 2, and 3.2
mediate repulsion by
Slits at midline

Floor Plate

Figure 16–16. Attraction changes to repulsion. Work by Marc Tessier-Lavigne and his colleagues demonstrated that before crossing the midline, spinal cord commissural neurons express Robo3.1. After crossing, repulsion from Slits at the midline is aided by a switch to Robos 1, 2, and 3.2 on the axonal membranes (G. Chen et al., 2008; Z. Chen et al., 2008).

The Regulation of Growth Cones by Morphogens, Growth Factors, and Neurotransmitters

In addition to the "classical" guidance molecules listed in Table 16–1, it is now clear that many other secreted molecules can attract or repel axons. These include the BMPs, wnts, and hedgehog proteins covered in Chapter 15, which have an earlier role as morphogens. In fact, a gradient of Sonic Hedgehog arising from the floor plate of the spinal cord plays a role very similar to that of netrin-1 in attracting the axons of commissural neurons and contributes to the accurate navigation of their growth cones. Similarly, growth factors such as GDNF, FGF, and another growth factor termed *HGF* (Hepatocyte Growth Factor), as well as the neurotrophins, can guide axonal growth.

An early example was provided by the neurotrophin NGF. As we saw in Chapter 15, neurons from the sympathetic ganglia and sensory neurons from dorsal root ganglia are sensitive to NGF, responding to the presence of this factor with profuse outgrowth of neurites. It turns out that the direction from which NGF is applied alters the way the growth cones respond. If a gradient of NGF is established by leakage of NGF from a pipette tip in a culture dish, the axons of cultured neurons extend toward the tip of the pipette. When injections of NGF are made into the brains of newborn rats, aberrant growth of neurites from sympathetic ganglia toward the injection site occurs. Because, during the course of development, the axons of sensory and sympathetic neurons come to rely on the trophic actions of NGF only after they have established their axon pathways, it is, however, unlikely that NGF itself acts as a chemoattractant within the nervous system.

The recognition and binding of immobilized and soluble guidance molecules by receptors on the surface of a neuron have many analogies with the interaction of a neurotransmitter or growth factor with its receptor. Moreover, we have seen that changes in cyclic AMP and cyclic GMP levels can alter the direction of growth. Thus it is not surprising that application of neurotransmitters, and even the onset of electrical activity within an axon, may modify the growth of neurites. Because most of the studies on this issue have been carried out using neurons in culture, the relevance of the effects of neurotransmitters and electrical stimulation to axonal growth in the intact nervous system is not yet established. It does seem likely, however, that these influences shape the final branching patterns at times when synaptic contacts are being established. ❡

Biochemical Properties of Growing Axons

Secretion of proteases. While neurites are navigating toward their targets, their growth cones secrete *proteases*, enzymes that partially digest proteins in their extracellular environment. One way this has been demonstrated is by plating neurons in cell culture on a dish covered with a layer of protein, such as fibronectin, laminin, or gelatin, that has been modified chemically so as to fluoresce. Degradation of the proteins is detected as loss of fluorescence near the cell and the growth cone. A major role for extracellular proteolysis is to serve as a cell-to-cell signal to trigger responses that follow contact of one cell by the growth cone of another. We saw in Chapter 15 that the extracellular domain of the Notch receptor has to be cleaved before it can participate in cell-to-cell signaling. Signaling through Robo and DCC receptors is also regulated by extracellular cleavage. Two specific classes of proteases are known to regulate axonal growth in this way: matrix metalloproteases (MMPs) and ADAMs (**A** **D**isintegrin **A**nd **M**etalloproteinase domain proteins, named after *disintegrins*, components of spider venom that inhibit integrin-mediated cell adhesion). In the next chapter we shall encounter one of the actions of *Kuzbanian* (ADAM10), a protease that contributes to the extracellular cleavage of Notch (see Fig. 15–6) and whose activity has been found to be essential for normal axonal extension.

Other biochemical events required for elongation and navigation—GAP-43. As we have seen, the direction in which a neurite chooses to grow depends on how the actin cytoskeleton reacts to local changes in levels of guidance molecules. The intracellular signaling events that mediate such choices are still being unraveled and involve the activity of proteins whose function is not fully understood. These include the protein *GAP-43*, which is short for 43 kDa growth-associated protein. Many axons in animals lacking GAP-43 fail to navigate to their appropriate targets and these animals die shortly after birth. The GAP-43 protein can be modified by addition of the lipid palmitate. This allows it to attach directly to the plasma membrane, where it appears to have multiple functions, including organization of the actin cytoskeleton in the growth cone and the activation of the GTP-binding that cause extension or collapse of a local region of the growth cone. ❽

Summary

Developing neurons extend neurites, which become the axons and dendrites of the adult neuron. These neurites follow specific paths and branch in characteristic ways. The leading tip of the neurite, the growth cone, appears to sample the extracellular environment and contribute to decisions about the direction of neurite extension. Molecules of various types are essential for appropriate pathfinding by growing neurites. For example, neurites grow selectively toward or away from guidance molecules such as the semaphorins, netrins, slits, and ephrins. In addition, adhesion molecules such as fibronectin and laminin mediate specific adhesion of the neurite to the substrate over which it is growing, while CAMs and cadherins promote the adhesion of neurites of different cells to each other in specific patterns. Some of these molecules and mechanisms that regulate neuronal development and differentiation may also regulate neurite outgrowth in adult nervous systems, either during recovery from injury, as is the case for the Nogo protein, or in response to novel stimuli from the environment.

Formation, Maintenance, and Plasticity of Chemical Synapses

ollowing cellular determination and neurite elongation, developing neurons form the specific synaptic connections that are essential for brain function (Fig. 14–1). One particularly striking aspect of the formation of synapses is its extraordinary sensitivity to patterns of electrical activity in the developing pathways. We will now describe synaptogenesis during development and its guidance by the pattern of electrical stimulation to which an immature neuron is exposed. Synaptogenesis is not, however, restricted to the developing nervous system. We have hinted that reorganization of neuronal form and function occurs in the adult nervous system, and in fact, mature neurons retain most of the machinery necessary for restructuring their synaptic connections by mechanisms similar to those operating in development. We will therefore provide some examples of synaptic plasticity in adult neurons. Finally, we will describe some of the changes that occur in the properties of cells once synaptic contacts have been established.

Synaptogenesis During Development

Morphological changes during synapse formation. When growing axons approach the cell on which they will finally make synaptic contacts, changes occur in the shape of their growth cones (Fig. 17–1). The

415

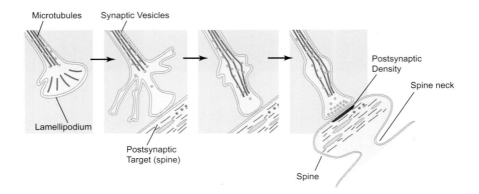

Figure 17–1. Synapse formation. Transformation of a growth cone into a presynaptic ending.

lamellipodia that are characteristic of rapid growth shrink in size, and the filopodia extend from the tip of the neurite in an irregular pattern. This change in the appearance of growth cones is associated with a slower rate of elongation. When the growth cone finally contacts a cell with which it forms a synapse, a further change takes place in its structure: The lamellipodium and filopodia disappear as neurotransmitter vesicles from the central core region advance into the tip of the neurite. The point of contact is then transformed into a full-fledged synapse with the accumulation of material in the synaptic cleft and the thickening of the postsynaptic membrane to form a *postsynaptic density*. In many cases, the postsynaptic dendrite is further fashioned into a spine, with a thick neck, that connects the new synaptic contact to the rest of the dendrite. Functional synaptic communication, however, can occur as soon as the neurite contacts the postsynaptic cell. Indeed, experiments in cell culture have demonstrated that even extending growth cones are capable of releasing neurotransmitter.

As described in Chapter 8, some neurons do not contact their targets directly but release neurotransmitter locally to influence neighboring cells without making specialized synaptic contacts. Changes in the terminals of these cells must also occur when their axons have reached their final destinations. An example of how terminal morphology can change without synapse formation is given in Figure 17–2, which shows the change in structure of a growth cone of an *Aplysia* bag cell neuron in cell culture following elevation of cyclic AMP levels. The actin-rich lamellipodium is invaded by microtubules and other organelles from the central core region to produce a club-like ending packed with secretory granules. The role of

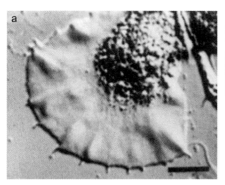
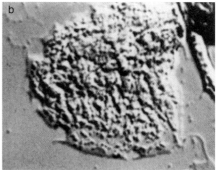

Figure 17–2. Changes in structure of a growth cone in cell culture. Photographs of the growth cone of an *Aplysia* neuron before (*a*) and after (*b*) treatment with drugs that elevate cyclic AMP levels (Forscher et al., 1987).

second messengers such as cyclic AMP in transformations of growth cones will be covered in Chapter 19.

What determines the choice of postsynaptic target? How specific is this interaction of a growing axon with its target? Can the axon of a motor neuron, for example, form a synapse on any muscle cell, or is there a strict one-to-one recognition of a specific muscle cell or a particular neuron? It appears that the very earliest stages of synapse formation are rather unselective. This lack of early specificity in connections can be seen in cell culture, where neurons make synapses with other neurons or muscle cells that are not their normal targets. It is only in time that a finely tuned pattern of specific point-to-point connections emerges in the developing brain.

The progression in time to a more and more specific pattern of synaptic connections can be explained by the finding that synapse formation occurs in two very distinct phases. There is an early phase of synapse formation that is relatively indiscriminate. This is followed by a second, more prolonged phase during which some synapses are stabilized while others are eliminated. It is this fine-tuning and pruning of contacts during the second phase that achieves a highly ordered final pattern of synaptic connections.

The two phases of synapse formation. The first phase of synapse formation is closely related to the process of neural pathfinding, considered in Chapter 16. An extending axon does not make synaptic contacts with every cell it encounters on its path. When it reaches an appropriate synaptic target, therefore, some signal must be generated that instructs the growing cell to slow its growth, contact the postsynaptic cell, and form a synaptic terminal. This initial phase of synapse formation may be termed *target selection*. As we shall see in specific examples of synapse formation in the mammalian brain, this initial phase produces a pattern of synaptic connections that is relatively "coarse-grained" compared to the final pattern seen in the adult. Another characteristic feature of this first phase is that it generally is independent of ongoing electrical activity in the growing neurons. Thus, target selection usually occurs normally, even in the presence of drugs, such as tetrodotoxin, that block neuronal action potentials (see Chapter 5).

The initial phase of synapse formation is followed by a second phase during which there is much *synapse elimination*. During this second phase, a major remodeling of the original pattern of connections occurs. The initial synaptic contacts may expand or retract and may withdraw altogether. A neuron whose process has retracted from one postsynaptic cell may subsequently form a more stable synapse on another cell. Because of this second phase, synapse formation during development is a relatively protracted process. For example, in the mammalian brain, the number of synapses formed may increase over a period of many weeks or months after birth. Thereafter, the total number of synapses declines toward that of the adult. Moreover, synapses can continue to be made and broken during adult life.

As will be described later in this chapter, the fine-tuning of synaptic connections during the second phase appears to result in large part from the competition of different axons for the same postsynaptic cell. A second, very important aspect of this fine-tuning process is that it depends on, and is entirely shaped by, the pattern of electrical activity that occurs in the synaptic pathway. One of the most thoroughly studied and clearest examples of the way activity-independent (first-phase) and activity-dependent (second-phase) mechanisms contribute to final choice of synaptic target is found in the visual system of vertebrates.

Synapse Formation in the Visual System

The retinotectal system. In many lower vertebrates, the major neuronal projection from the *retina* extends from the optic nerve to part of the midbrain known as the *optic tectum*. In higher vertebrates, the region corresponding to the tectum is the *superior colliculus*. This pathway

controls many of the rapid visual reflexes of such animals. A feature of
this pathway that has attracted much attention from developmental
biologists is the particularly clear point-to-point mapping of the input
from different parts of the retina to corresponding points on the surface
of the tectum. Such mapping can readily be demonstrated by shining
points of light on the retina and recording electrophysiological
responses in the tectum. As shown in Figure 17–3a, points of light that
fall on the ventral retina trigger responses in the dorsal tectum, whereas

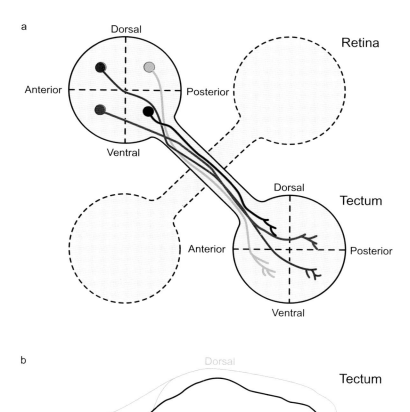

Figure 17–3. Retinotectal connections. *a*: The topology of connections.
b: Drawings of the irregular growth of retinal axons into the ventral tectum of
an adult newt (Fujisawa, 1981).

stimulation of cells in the dorsal retina produces responses in the ventral tectum. Similarly, retinal neurons that respond to light falling on the anterior retina connect to neurons in the posterior tectum, while those activated in the posterior retina project to the anterior part of the tectum.

The chemoaffinity hypothesis for the first phase of synapse formation. Although the final connections to the tectum are arranged in an exquisitely precise array, the path that the axon of a retinal cell follows to reach its final destination may not be direct. An ingrowing axon may bypass many potential postsynaptic targets in the tectum before establishing its synaptic contacts (Fig. 17–3b). The major hypothesis to account for the specificity of these connections is the *chemoaffinity hypothesis* of Roger Sperry. The simplest form of this hypothesis states that there exist specific molecules in the presynaptic and postsynaptic cells (usually thought of as being on the surface of the cells) that differ either in their chemical identities or in their relative amounts in different regions of the tectum. These differences constitute a biochemical label for each cell. The fact that cells from one region of the retina connect only to the appropriate region of the tectum is explained by the required correct matching of presynaptic and postsynaptic labels for synapse formation to occur.

Many experiments support the chemoaffinity hypothesis. Some of these involve surgical manipulation of the inputs to the tectum. For example, the optic nerve of a frog can be severed and the eye rotated through 180°. The axons of the retinal cells will in time regenerate and reinnervate the tectum. Under these conditions, cells in the retina form new synaptic connections with the same part of the tectum that they innervated before, even though the region of visual space projected by a given region of retina to tectal sites is now 180° different. Moreover, if half of the severed retina is removed, the remaining cells in the retina will re-extend axons that make synaptic contacts with the appropriate half of the tectum. Initially, these axons will not innervate the remainder of the tectum, which does not bear the appropriate label. (As will be described later, however, slower remodeling of the connections does occur, and the entire tectum is eventually innervated by the half-retina.)

Further evidence that molecules in the membrane of retinal cells determine the region of the tectum that they come to innervate comes from experiments in which the physical adhesion of retinal cells to tectal cells is tested in a dish. Retinal cells are dissociated and then incubated with the tectum. Cells from the dorsal retina are found to bind preferentially to the ventral tectum, while cells in the ventral retina adhere more strongly to dorsal tectum. Such preferential stickiness of retinal cells matches the pattern of synaptic connections that eventually is formed by these cells.

In a variant of this experiment, explants of retina are placed next to narrow, alternating stripes of cell membranes prepared from posterior and anterior tectal cells (Fig. 17–4). When a piece of the anterior retina is placed next to this striped "carpet" of membranes, the axons that grow out of the explant travel over both types of stripes. In contrast, axons

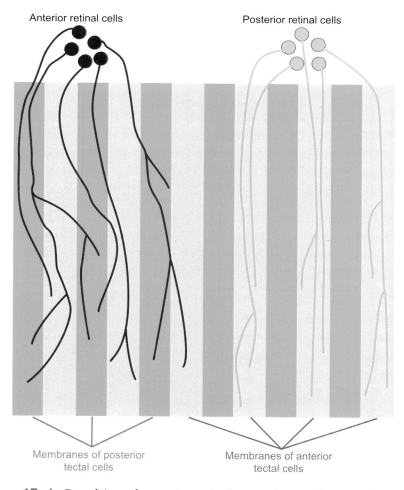

Anterior retinal cells

Posterior retinal cells

Membranes of posterior tectal cells

Membranes of anterior tectal cells

Figure 17–4. Repulsion of posterior retinal axons by membranes of posterior tectal cells. In an experiment by Boenhoeffer and his colleagues, retinal cells were allowed to extend their neurites over alternating stripes of membrane fragments from posterior and anterior tectal cells. The posterior retinal cells avoided the stripes of posterior tectal cell membranes (Walter et al., 1987).

from the posterior retina grow only along the stripes of membranes from the anterior tectal cells. This preference results from a repulsive factor in the membranes of the posterior tectal cells. Heating or protease treatment of the posterior tectal membranes neutralizes this repulsion, and the posterior retinal cells then fail to discriminate between posterior and anterior tectal membranes. The factors that repel the axons of posterior retinal cells, causing the collapse of their growth cones, have been identified as *ephrins*. Gradients of ephrins and their receptors appear to be important components of the biochemical labels of Sperry's chemoaffinity hypothesis.

Ephrins and the Eph Receptors

As stated in Chapter 16, ephrins and their receptors, termed the *Eph receptors*, function like the other guidance molecules. They are not specific to the nervous system and operate during the development of many different tissues. In the developing nervous system, like other guidance molecules, they exert either attractive or repulsive influences on axons. As we shall see later, however, there are several reasons why it may be more appropriate to consider them here, in the context of synapse formation.

Ephrins come in two forms (Fig. 17–5). The ephrin A ligands (ephrins A1–A5) are extracellular proteins tethered to the plasma membrane by a lipid anchor (a glycosyl phosphatidylinositol, or *GPI*, anchor). The B class of ephrins (ephrins B1–B3) are true integral membrane proteins that possess, in their cytoplasmic domains, a set of highly conserved tyrosines that can undergo phosphorylation. Corresponding to these two types of ligands are two classes of receptor tyrosine kinases. EphA receptors bind clusters of ephrin A ligands (there are nine different EphAs), while EphB receptors bind ephrin B ligands (there are five EphBs). Both Eph receptors and ephrin B ligands possess PDZ binding motifs at their carboxyl termini. As we saw in Chapter 12, these motifs are present in a variety of channels, receptors, and adaptor-proteins that cluster proteins containing PDZ domains at presynaptic terminals and postsynaptic densities. The extracellular, ephrin-binding region of Eph receptors possesses a large globular domain, a cysteine-rich domain and regions related to fibronectin (type III motifs). The intracellular region contains, in addition to a tyrosine kinase domain, sites for autophosphorylation and a domain termed the *SAM domain*, through which the receptor binds other signaling molecules. One of the interesting aspects of the ephrin B proteins is that, following their binding to their EphB receptors, they may themselves undergo phosphorylation on tyrosine residues and trigger signaling events

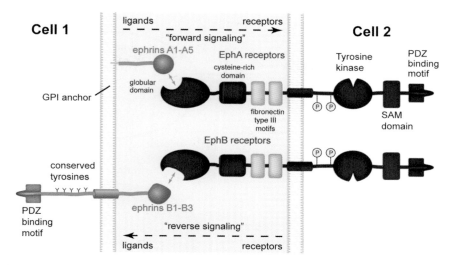

Figure 17–5. Ephrins and the Eph receptors. EphA receptors bind the lipid-anchored ephrin A ligands, while EphB receptors bind ephrin B ligands, which are integral membrane proteins. Autophosphorylation sites are shown on the receptor.

in their cytoplasmic regions. Thus these interactions constitute a bidirectional signaling system that produces changes in two interacting cells. When binding of an ephrin to an Eph receptor produces changes in the ephrin "ligand"-containing cell, this can be termed *reverse signaling* (Fig. 17–5).

Gradients of ephrins and Eph receptors in both the retina and the tectum shape the eventual pattern of synapse formation. For example, there exists a gradient of EphA receptors in the retina with highest levels on ganglion neurons in posterior (or *temporal*) retina (Fig. 17–6). There is also a corresponding gradient of several ephrin A ligands, with which axonal EphA receptors interact, in the tectum with highest ephrin A levels in the posterior tectum. As surmised from the experiments with alternating stripes of tectal membranes, the interaction between these receptors and ligands is a repulsive one. Thus the EphA-bearing axons from the posterior retina fail to enter the posterior tectum and instead terminate in the anterior tectum.

A complementary story exists for the anterior (or *nasal*) retinal neurons. Because their axons lack the EphA receptors and are not sensitive to the repulsive effects of ephrin A, they can extend toward the posterior tectum (Fig. 17–6). A second set of gradients, however, ensures that they do not synapse indiscriminately throughout the tectum. Ephrins are found

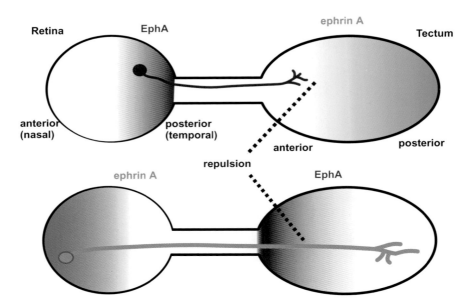

Figure 17–6. Gradients of Ephs and ephrins in the retina and the tectum. Ephrin A in the posterior tectum repels axons of EphA-containing axons from the posterior retina. EphA receptors in the anterior tectum repel the ephrin A–containing axons of anterior retinal neurons, ensuring that their terminals find a home in the posterior tectum.

on neurons in the retina itself, and their gradient is in the opposite direction from that of EphA in this location. Similarly, there is a gradient of EphA receptors in the tectum with highest levels at the anterior end. As a result, repulsion between the ephrin A on anterior retinal neurons and EphA in the anterior tectum prevents synapse formation, ensuring that the neurons make connections only in the posterior tectum. Like the EphAs, EphBs and ephrin Bs regulate axonal navigation, and in the retinotectal system they contribute to the dorsal–ventral pattern of connections. In general, axons end up terminating in regions where they find the minimal amount of repulsion.

Ephs and ephrins induce presynaptic and postsynaptic specializations. Everything we have said about ephrins thus far could equally well have been covered in the previous chapter on axonal pathfinding. It is evident, however, that ephrins do more than simply mediate attraction or repulsion: They help to fashion both the axon terminal and the site with which it eventually makes contact into a morphological synapse. Clearest evidence for this has been obtained for the EphB–ephrin B pathway. When a

non-neuronal cell is engineered to express the EphB2 receptor on its surface and is then co-cultured with neurons from the cerebral cortex, the
neurons are fooled into making functional presynaptic terminals when
they contact the non-neuronal cell (Fig. 17–7a). This occurs, however, only
if the axon has the ligand ephrin B1 or B2 on its axon. Moreover, in the
retinotectal system, ephrin B1 is present on axons of retinal ganglion cells,
and the simple addition of the extracellular domain of EphB2 receptor to
the developing tectum is sufficient to activate reverse signaling, stimulating
an increase in the number of presynaptic release sites (Fig. 17–7b).

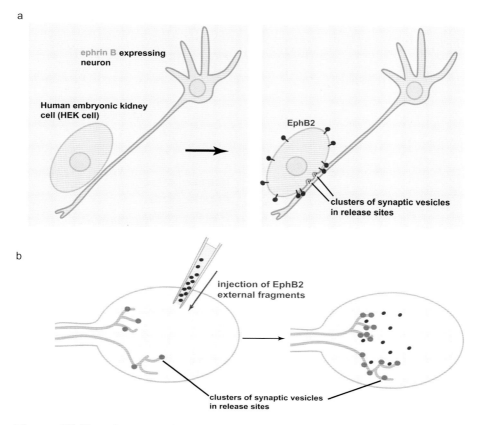

Figure 17–7. Ephrin B–EphB interactions shape presynaptic neurotransmitter
release sites. *a:* Experiments by Matt Kayser, Matthew Dalva, and their colleagues showed that cerebral cortical neurons make functional presynaptic terminals when they contact non-neuronal cells that make EphB2 (Kayser et al.,
2006). *b:* In the developing frog retinotectal system, injection of proteins containing the extracellular domain of Eph2B stimulates an increase in the number
of presynaptic release sites (Lim et al., 2008).

At least part of the way the presynaptic release site is organized following the EphB–ephrin B interaction is that the PDZ-binding domain on the presynaptic ephrin binds a PDZ domain–containing protein termed *syntenin-1*, which then triggers the clustering of synaptic vesicles (Fig. 17–8).

Contact between an ephrin B on an axon and EphB receptors in its target cell also triggers the first steps in organizing the postsynaptic side of an excitatory synapse (Fig. 17–8). These steps include changing the cytoskeleton so that postsynaptic filipodia are transformed into dendritic spines. This effect occurs through binding of the PDZ-binding domain of the EphB to a protein called *GIT1*, which activates the GTP-binding protein Rac (see Chapter 16) and remodels the actin cytoskeleton into a spine. Through their external domains, activated EphBs bind to the

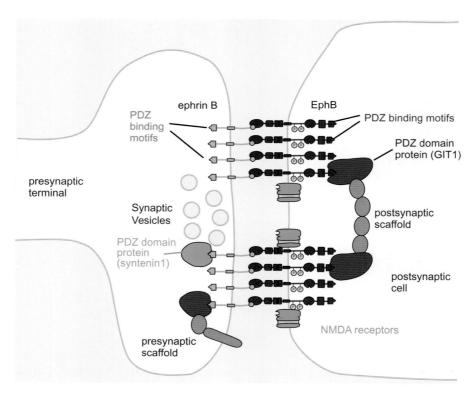

Figure 17–8. Possible contributions of ephrin B–EphB interactions in organizing synaptic contacts. EphB receptors bind NMDA receptors and PDZ binding motifs on both ephrin Bs, and their receptors may participate in formation of presynaptic clusters of synaptic vesicles and postsynaptic clusters of glutamate receptors.

NMDA class of postsynaptic glutamate receptors, regulating their activity and inducing them to cluster together. "Side-to-side" interactions between EphBs and ephrin B that are both on the postsynaptic membrane also increase the levels of AMPA postsynaptic glutamate receptors on the post-synaptic membrane. This is accomplished by lowering the rate at which they are removed from the plasma membrane by endocytosis.

Eph–Ephrin interactions can be reversed by proteases. The interaction between an ephrin and its Eph is a tight one. In the case of an attractive interaction between two cell membranes, as in the early steps of synapse formation, this may not be a problem. If, however, the interaction is a repulsive one, as in the anterior–posterior labeling of axons in the tectum, the receptor and its ligand must be separated to allow the axon to find a more appropriate target. This separation involves the physical cleavage of the ephrin ligand by an extracellular protease. In Chapter 16 we men-tioned that growing axons secrete proteases, including a metalloprotease called *ADAM10* (but first called *Kuzbanian* by people who study fruit flies). After an initial repulsive interaction between membranes, which produces biochemical changes such as changes in actin polymerization in the two sets of interacting cells, ADAM10, which is attached to the mem-brane containing the EphA, has been shown to cleave ephrin A that is bound to the receptor (Fig. 17–9). The cleavage leaves part of the ephrin molecule bound to the receptor but allows the two membranes to sepa-rate. As we shall see in the next section, many of the strong attractive interactions that initially lead to synapse formation must also eventually be reversed before the process of synaptogenesis is complete.

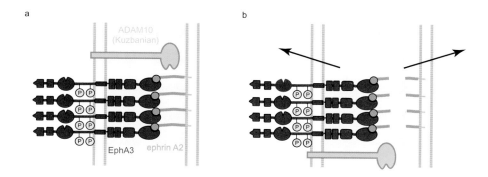

Figure 17–9. Separation of membranes by ADAM10. *a*: The protease binds to a complex of EphA3 and its ligand ephrin A2 or ephrin A5. *b*: Cleavage allows the membranes to separate (Hattori et al., 2000; Janes et al., 2005).

Rearrangement of Synaptic Connections

The second phase of synapse formation in the retinotectal system. As already mentioned, synapse formation is a two-stage process. Once an initial set of contacts is made by the incoming axons, guided by adhesive and repulsive molecules, there follows a second prolonged period of restructuring or *sorting* (sometimes termed *refinement*) of these synapses. At this time, some terminal branches and their synaptic contacts may be withdrawn from a tectal cell, while the connections from other retinal neurons may be strengthened. Although such remodeling of connections is most obvious during the initial development of the retinotectal pathway, it is clear that in some instances this continues throughout adult life.

Particularly clear examples of the ongoing restructuring of synapses occur in frogs and goldfish. In these species, the retina and tectum continue to increase in size through the addition of new cells. In the eye, new neurons are added as a ring to the circumference of the retina. In contrast, in the tectum, the new cells are added only to the posterior tectum (Fig. 17–10). Thus, to maintain the correct mapping of connections from the retina to the tectum, the entire set of synapses from the retina

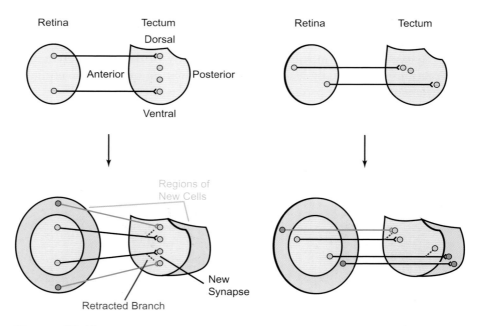

Figure 17–10. Remodeling of synaptic connections during addition of new neurons. As the size of the retina and tectum increase, connections are broken and reformed to maintain the overall map of the retina onto the tectum.

continually retracts and reconnects to a new set of tectal neurons. In the goldfish, the retina and tectum continue to increase in size, and thus to realign their connections, throughout adult life.

A shift in the pattern of synaptic connections also occurs in the experiment described earlier as one of the first tests of the chemoaffinity hypothesis. In this experiment, the optic nerve is cut and part of the retina is removed. Initially, only the area of the tectum that corresponds to the intact retina is reinnervated by the axons from the remaining retinal cells. In time, however, branches of these axons extend, to make synapses over the entire tectal area. In this way, the map of the visual field represented by the remaining part of the retina spreads out over the whole tectum.

Electrical activity of neurons determines the final pattern of synaptic contacts. Experiments with the tectum, and with many other parts of the developing nervous system, suggest strongly that the fine-tuning of synaptic connections is determined by the pattern of electrical activity in the presynaptic neurons. A given postsynaptic cell may initially be innervated by many different presynaptic cells. In many cases, however, only those inputs whose activity is correlated in time with postsynaptic activity are stabilized (Fig. 17–11). Fibers that generate weaker responses, or fail to influence the postsynaptic cell altogether, retract their synaptic contacts. Such fibers may, however, successfully stimulate and establish stable contacts with other postsynaptic cells.

The hypothesis that excitatory synapses are stabilized when they successfully trigger action potentials in the postsynaptic cell was proposed by the Canadian psychologist Donald Hebb, in 1949. It has been used

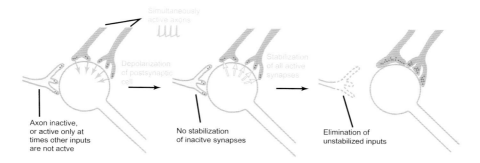

Figure 17–11. Hebb's rule. Excitatory synapses that successfully stimulate a postsynaptic neuron to fire, or are active when the postsynaptic neuron is depolarized, are selectively stabilized (Hebb, 1949).

extensively in models of both development and learning (see Chapter 19). A restatement of this hypothesis, often described as the *Hebbian rule*, is that when a postsynaptic neuron becomes depolarized, it generates a biochemical reaction or a trophic factor that stabilizes the excitatory synapses that are firing at that time. An important aspect of this hypothesis is that a given presynaptic input to a cell need not, by itself, be of sufficient strength to induce a large depolarization in its target. If that input is fired at the same time as a number of other inputs, and their combined action depolarizes the cell, all of these inputs will tend to be stabilized. If, in contrast, a given input fires *asynchronously* with most of the other inputs onto that cell, this input will tend to be eliminated. The hypothesis that synapses are stabilized or eliminated according to their patterns of activity is able to explain many different findings about synapse formation and remodeling.

Ocular dominance columns. In lower vertebrates, including frogs, all the fibers from one retina cross the midline to innervate the contralateral tectum. Thus, in the frog, there is normally no competition between inputs from the two eyes in one tectum. It is possible, however, to implant a third eye into a tadpole. The fibers from this third eye grow normally into one of the tecta, where they must compete with axons from the normal eye for synaptic space on the tectal neurons. The final pathways that are established can be measured by injection of a radioactive amino acid into one of the eyes. The radiolabel is taken up by retinal cells and transported to their terminals in the tectum. The amount of radioactivity in these terminals can then be visualized directly by placing a piece of X-ray film against slices made from the tectum (Fig. 17–12). The result is that connections have been established in a pattern of alternating columns, each of which contains inputs primarily from one eye. Under normal conditions, no such columns are observed in the tecta of frogs and other lower vertebrates.

The formation of such columns, termed *ocular dominance columns*, can be understood in terms of re-sorting of synapses on the basis of their electrical activity, a process that occurs during normal development (Fig. 17–13). Because all of the photoreceptor cells in one eye point toward one general region of visual space, the inputs in that one eye will tend to be activated approximately simultaneously. The other eye, however, covers a somewhat different visual field. Although the inputs from that eye generally will also be correlated with each other, they will tend not to be active at the same time as input from the competing eye. Thus a given small region of the tectum is innervated initially by inputs from both eyes. A small increase in the amount of input from one eye, however, tends

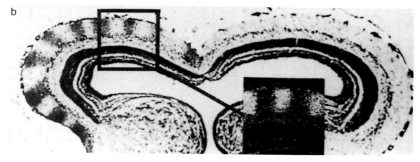

Figure 17–12. The three-eyed frog. *a*: Three-eyed frogs have been studied by Martha Constantine-Paton and her colleagues. *b*: An autoradiograph of the tectum shows the formation of stripes of inputs from the normal and the implanted eye. The inset shows an enlargement under dark-field illumination (Constantine-Paton and Law, 1978).

to stabilize all of the approximately synchronously active synaptic inputs from that eye, at the expense of inputs from the other eye. Such stabilization leads to the formation of areas or columns that are preferentially innervated by one eye or the other.

Segregation of inputs from two eyes also occurs *normally* in the visual systems of many mammals, in which each retina sends some fibers to both hemispheres of the brain. Indeed, it was the Nobel Prize–winning experiments of David Hubel and Torsten Wiesel on kittens and monkeys that first led to the idea that patterns of electrical activity shape synapse rearrangement. Sheets of cells that respond preferentially either to one eye or the other are found in the *lateral geniculate*, a visual relay station, of adult cats and monkeys. In the visual cortex, inputs from the two eyes are segregated into ocular dominance columns. Synaptic refinement similar to that occurring in the three-eyed frogs may contribute to the sorting of these inputs.

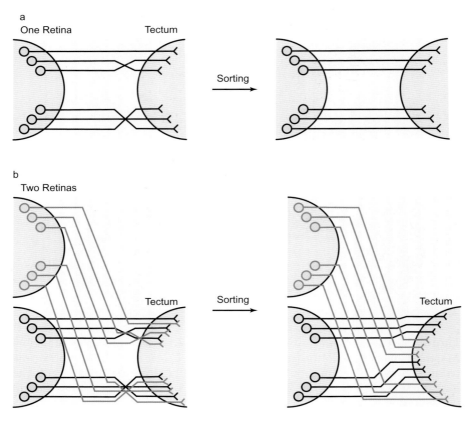

Figure 17–13. Segregation of retinal inputs to the tectum. *a*: Sorting of inputs from one eye only. *b*: Sorting in the presence of input from a second eye.

Activation of NMDA receptors and TrkB receptors by electrical activity is required for refinement of synapses. The normal formation of these segregated patterns of synaptic connections depends on the electrical activity of the presynaptic fibers. This can be demonstrated in a number of ways. For example, tetrodotoxin can be introduced to block action potentials. This treatment does not interfere with the initial innervation itself, but results in the presynaptic inputs from the two eyes becoming uniformly distributed. In other experiments the two eyes have been stimulated either synchronously or asynchronously. For example, if the two optic nerves of a cat are stimulated such that activity from the two eyes occurs at different times, then ocular dominance columns develop. If, however, the two nerves are stimulated simultaneously, such that the input from the two sets of presynaptic fibers is identical, then no segregation occurs.

We emphasized earlier in this book that many neurons are capable of generating spontaneous patterns of electrical activity in the absence of external inputs. Interestingly, such spontaneous activity plays a role in the sorting of synapses before the visual system is fully functional. Spontaneous waves of electrical activity are generated in each retina prior to eye opening, and this activity is required for the normal pattern of synapse formation in the tectum and other visual areas. (A similar phenomenon occurs in inner hair cells of the cochlea, which generate spontaneous action potentials before the onset of hearing, after which they can no longer fire action potentials and become the passive transducers described in Chapter 13). Thus, in three-eyed frogs, as well as in mammals, inputs from the two eyes can still segregate even when animals are reared in the dark or before visual inputs are functional. The ocular dominance columns formed in these circumstances are still blocked or reversed by treatment with tetrodotoxin, indicating that ongoing electrical activity is required for the segregation.

Activation of the NMDA type of glutamate receptor by ongoing electrical activity during development is required for the selective stabilization of retinotectal projections in the frog and for related phenomena in higher vertebrates. Treatment of tectal cells with the NMDA receptor antagonist APV does not prevent the stimulation of the tectal cells by retinal afferents but reverses the segregation into ocular dominance columns. As we saw in Chapter 11, the properties of the NMDA receptor make it particularly appropriate to induce biochemical changes in a cell when a set of inputs are synchronously active (see Fig. 17–11). For example, when several inputs that use glutamate as a transmitter are activated at the same time, the postsynaptic cell undergoes a depolarization. If NMDA receptors are present at these synapses, this depolarization allows calcium ions to enter the postsynaptic neurons. This in turn sends a signal back to the presynaptic terminals, selectively stabilizing those that are active.

Among its many other actions, an elevation of postsynaptic calcium leads to the secretion of the neurotrophin BDNF, a factor we encountered in Chapter 15. BDNF can be released from cells by both constitutive and regulated secretion pathways (see Fig. 8–5). BDNF released locally at the newly forming synapse can act both on presynaptic and postsynaptic TrkB receptors (Fig. 17–14). When it binds to receptors on the active presynaptic synaptic terminals it acts as a *retrograde messenger*. In Chapter 15, we covered the major signaling pathways brought into play on activation of Trk receptors. These can produce rapid local morphological changes in the organization of the presynaptic terminals. Thus, when

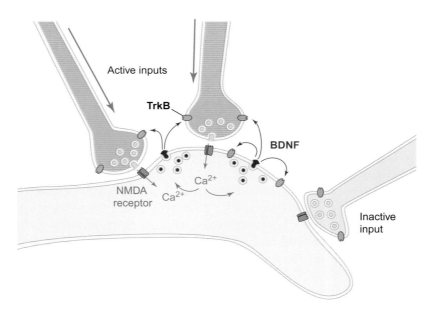

Figure 17–14. Stabilization of active presynaptic inputs by the actions of NMDA receptors and BDNF.

TrkB receptors in individual frog retinal neurons are inhibited by introduction of an excess of a mutant receptor that binds to the endogenous TrkB receptor, many of the axon terminals fail to make mature connections onto tectal cells but instead retract and degenerate. BDNF may also act at TrkB receptors on the postsynaptic membrane from which it is released (Fig. 17–14), a so-called *autocrine* action that may produce longer term biochemical events, such as the synthesis of new transmitter receptors and scaffold proteins. Finally, there is some evidence that BDNF acting at its p75NTR receptor, rather than at TrkB, contributes to the elimination of the synapses that are inactive. Although the full molecular events determining why some synapses are stabilized while others are subject to elimination are still being unraveled, they closely resemble the changes that occur in long-term potentiation and long-term depression, events that have been proposed to underlie some forms of learning and memory, and that will be discussed in more detail in Chapter 19.

Synaptic rearrangement at the neuromuscular junction. This kind of remodeling of axonal and dendritic branches and synaptic connections occurs not only in visual pathways but perhaps also in all regions of the nervous system of both vertebrates and invertebrates. Another well-studied example is found at neuromuscular junctions of vertebrates. When the axons of motor neurons first contact their skeletal muscle targets, the different branches of an axon contact several muscle fibers. Moreover, each muscle fiber is contacted by the terminals of several different motor neurons. This situation is termed *polyneuronal innervation*. Over a period of a few weeks, however, many of the motor neuron branches are withdrawn. Eventually, in the adult, each muscle fiber comes to be innervated by only one motor neuron. This elimination of synapses can be observed both morphologically and electrophysiologically. As shown in Figure 17–15a, at the neuromuscular junction of the neonate, graded stimulation of a presynaptic nerve produces graded postsynaptic

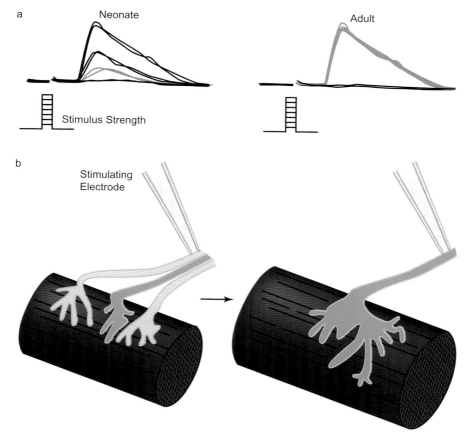

Figure 17–15. Reorganization of synapses at the neuromuscular junction. *a*: Postsynaptic potentials of varying sizes are recorded in the neonate, whereas stimulation of an adult junction gives an all-or-none postsynaptic potential. *b*: Elimination of synapses at the neuromuscular junction.

potentials in the muscle. This is because the postsynaptic response is made up of responses to several presynaptic axons, such that graded stimuli to the nerve trigger action potentials in a progressively larger proportion of these presynaptic fibers. In the adult, however, different intensities of stimuli either trigger an action potential in the axon of the one motor neuron contacting the muscle or fail to excite this one axon, resulting in an all-or-none postsynaptic response.

Although the loss of polyneuronal innervation during development of the neuromuscular junction is produced by removal of axonal branches and their associated synapses from the muscle fiber, the number of individual synaptic terminals made by branches of the one axon whose inputs are stabilized actually increases (Fig. 17–15b). This restructuring of neuritic branches and synapses depends on the electrical activity of the presynaptic terminals, and, as in the retinotectal system, the process of synapse stabilization and elimination at the neuromuscular junction is a Hebbian one. Insights into the role of electrical activity have come from studies using isolated neurons and muscles in cell culture, where the activity of the presynaptic and postsynaptic cells can be controlled precisely. When an isolated embryonic muscle cell is innervated by two different motor neurons, stimulation of either neuron alone is at first sufficient to activate large postsynaptic currents in the muscle. Provided that stimulation to either neuron is applied only infrequently, both functional synapses may exist on the muscle for a prolonged period. If, however, one of the neurons is made to fire a rapid train of action potentials (a tetanus), the strength of the *other* synapse is immediately suppressed (Fig. 17–16a). This *heterosynaptic suppression* does not occur if the tetanus is applied to both motor neurons at the same time. The effect of a tetanus in one neuron to depress transmission in the synapses of other inactive neurons can be mimicked by applying pulses of acetylcholine to the muscle from a nearby pipette (Fig. 17–16b). If such pulses are given while a presynaptic neuron is inactive, its synapses are immediately suppressed. Synapses are, however, protected from this suppression if the neuron is firing action potentials at the time of acetylcholine application.

This process of rapid suppression of inactive synapses is likely to be the first step in the complete elimination of these synaptic junctions. One of the actions of acetylcholine pulses or tetanic stimulation of the muscle cell is to produce an elevation of intracellular calcium. This is known to be required for the suppression of inactive synapses, and is believed to trigger the release of a retrograde messenger that is released from the muscle cell. According to this line of thinking, the presynaptic terminals that are active during the release of this retrograde messenger are protected from elimination.

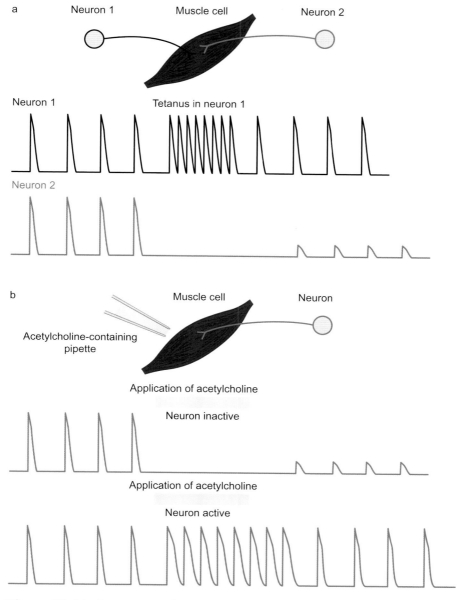

Figure 17–16. Experiments by Mu-Ming Poo and colleagues showing heterosynaptic suppression of synaptic inputs at the neuromuscular junction. *a*: A tetanus in one motor neuron causes a subsequent suppression of the inputs from a second neuron (Lo and Poo, 1992). *b*: Application of acetylcholine to a muscle cell causes suppression of synaptic inputs from a motor neuron if that neuron is inactive (*upper trace*), but does not suppress the synapse if the neuron is active at the time of exposure to the transmitter (Dan and Poo, 1992).

Developmental stabilization of active synapses has been observed in many other parts of the nervous system, and finds a parallel in schemes to account for plastic changes occurring in the adult brain. There is no reason to expect, however, that the detailed mechanisms of synaptic rearrangement are the same in all parts of the nervous system. For example, the retinotectal system requires activation of NMDA receptors, which are absent at the neuromuscular junction. Although, as a result of competition, the number of presynaptic neurons that innervate a single postsynaptic cell may decrease during development, not all such competition leads to a single "winning" presynaptic cell. The competition may be localized to relatively small regions on the postsynaptic membrane. For example, on a cell that has a complex pattern of dendritic branches, competition may lead one cell to establish its terminals on one branch, at the expense of synapses from other cells. On another branch or a different region of the dendritic tree, the terminals of a different cell may gain precedence, and the mature postsynaptic cell will come to be innervated in different regions by different axons.

Changes in Cell Properties Following Synapse Formation

The formation of a synaptic contact is followed by a sequence of changes in the properties of the postsynaptic cell. Receptors become reorganized in the postsynaptic membrane, and the types of proteins synthesized by the postsynaptic cell may alter dramatically. We have seen that at some synapses these changes can promoted by Eph–ephrin interactions. Many other pathways also contribute. Some of these effects occur because the incoming axons induce new patterns of electrical activity in the postsynaptic cell. Other changes are the result of factors that do not depend directly on stimulation of the new input.

Receptor reorganization at the neuromuscular junction. Again, the neuromuscular junction of vertebrates has provided a classic preparation for the study of many of these effects. One of the first events observed following the arrival of the growth cone of a motor neuron at the muscle is the *clustering* of acetylcholine receptors under the newly formed presynaptic terminals. Even before the arrival of the motor neuron fibers, some clusters of receptors exist on the surface of the immature muscle cells. These are said to be *prepatterned* and awaiting the arrival of the axons.

The prepatterned clusters are distributed relatively uniformly in the center of the developing muscle fibers. When the nerve arrives, it stabilizes those clusters that happen to lie under the newly formed presynaptic terminals and induces the appearance of new clusters at this location by allowing newly synthesized receptors to be inserted preferentially into the membrane under the terminal (Fig. 17–17a). The receptors under the terminals are termed *junctional* receptors, while those in the uninnervated parts of the membrane are termed *extrajunctional*. The extrajunctional receptors are synthesized and degraded at a higher rate than the receptors under the synaptic cleft, and in time the extrajunctional receptors disappear altogether, leaving a mature pattern of dense receptor clusters that are confined to the postsynaptic membrane and typically described as "pretzel-shaped" (Fig. 17–17b).

Rapsyn links acetylcholine receptors to the dystrophin glycoprotein complex. A vital link in the clustering of acetylcholine receptors clusters is *rapsyn*, a 43 kDa protein that binds the receptor and couples it to other proteins. One of these is an assembly of proteins termed the *dystrophin–glycoprotein complex* (Fig. 17–18). The very large dystrophin–glycoprotein complex contains four transmembrane proteins, including *adhalin* and *β-dystroglycan*, an extracellular protein (α-*dystroglycan*), and several intracellular proteins such as *syntrophin*. This complex is able to bind other proteins in the synaptic cleft and the muscle cytoskeleton. On the extracellular side, the complex binds laminin, while on its intracellular side, the complex binds two cytoskeletal proteins: *dystrophin*, which is present at extrajunctional sites, and the closely related protein *utrophin*, which is found under the synaptic membrane. This protein scaffold links the basal lamina to the cytoskeleton and provides a firm anchor for the acetylcholine receptors that associate with it. For a receptor to bind to rapsyn and to enter the cluster, however, the β subunit of the receptor must undergo phosphorylation on tyrosine residues.

Agrin: a stabilizer of postsynaptic receptor clusters. Following the arrival of a neuron, the formation of junctional clusters does not depend on the electrical activity of the presynaptic nerve or of the muscle itself but on *agrin*, a factor made by the growing motor neurons and released by the incoming nerve terminal. Agrin is transported to the synaptic terminals and is then secreted into the *basal lamina* of the synaptic cleft. The basal lamina is an extracellular layer of proteins that surrounds the muscle, even at sites of synaptic contact, where it forms a thin permeable layer between the presynaptic and postsynaptic membranes. We shall encounter

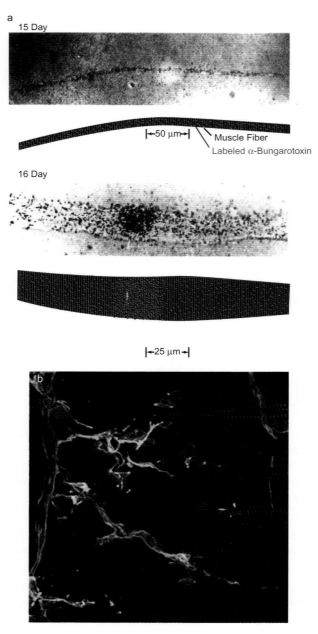

a
15 Day

|←50 µm→| Muscle Fiber

Labeled α-Bungarotoxin

16 Day

|←25 µm→|

b

Figure 17–17. Receptor clustering. *a*: Autoradiographs of neuromuscular junctions from 15- and 16-day rat embryos were made by Bevan and Steinbach (1977). The junctions were incubated with α-bungarotoxin, a ligand that binds to acetylcholine receptors. At the 16-day stage clustering is apparent. *b*: Image of a mouse neuromuscular junction with axon terminals (synaptophysin and neurofilament proteins) stained in green, clusters of α-bungarotoxin-labeled acetylcholine receptors in red, and muscle cells in blue. The image was made in the laboratory of Bob Darnell (Ruggiu et al., 2009).

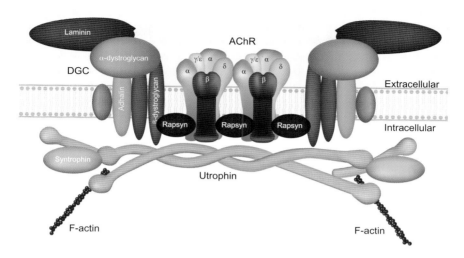

Figure 17–18. Immobilization of acetylcholine receptors at the neuromuscular junction. The protein rapsyn links the receptors to the dystrophin–glycoprotein complex. (Modified from Apel et al., 1995.)

the basal lamina again when we discuss the phenomenon of regeneration.

Agrin has many of the features of other molecules that we have encountered that regulate the growth and differentiation of neurons (Fig. 17–19). The molecule contains several EGF-like regions, a region that resembles laminin, and also has a repeated sequence that resembles the active domain of certain protease inhibitors. A slightly different variant of the agrin protein is also synthesized by muscle cells, but the muscle form does not induce clustering. It is the form that is released by the neurons, termed the z+ *agrin* isoform, that produces receptor clustering.

The formation of mature junctional clusters to the synaptic membrane depends on the binding of agrin to a two-component receptor containing an agrin-binding protein called *LRP4* and a tyrosine kinase termed *MuSK* (for **Mu**scle **S**pecific **K**inase) in the postsynaptic membrane (Fig. 17–19b). The MuSK receptor does not phosphorylate the acetylcholine receptor directly but acts by recruiting additional cytoplasmic proteins. The activation of this agrin receptor eventually leads to the phosphorylation of the acetylcholine receptors on tyrosine residues. This, in turn, appears to produce a tight and selective association of the receptors with the membrane under the synaptic cleft and with cytoskeletal

a   Agrin

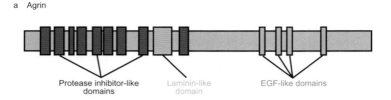

b

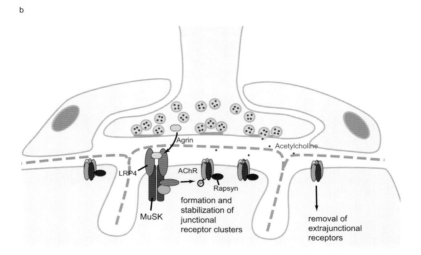

Figure 17–19. Aggregation of acetylcholine receptors at the neuromuscular junction. See Animation 17–1.◐ *a*: The structure of agrin. *b*: Scheme showing the actions of agrin acting through its LRP4/MuSK receptors to stabilize and trigger clustering of acetylcholine receptors following innervation. (Modified from Song and Balice-Gordon, 2008.)

components under this membrane, effectively immobilizing the receptors at this location.

Neuron-induced changes in receptor synthesis. After the initial clustering of preexisting acetylcholine receptors, there is an increase in synthesis of new receptors. In contrast to most types of cells, muscle cells have multiple nuclei, and some of these nuclei are strategically placed immediately under the sites of synapse formation. It is these subsynaptic nuclei that are dedicated to making mRNA for the acetylcholine receptor and other proteins of the postsynaptic membrane.

The signals that activate the synthesis of new acetylcholine receptors at the synapse are quite distinct from those that produce the clustering of preexisting receptors. Two molecules that are released from the terminals of the incoming motor neurons and have the ability to increase receptor synthesis are *calcitonin-gene-related peptide* (CGRP) and *neuregulin*. Nevertheless, preventing their action on muscle cells in development does not prevent the normal increase in receptor synthesis following innervation. Neuregulin does, however, have an important role in development of neuromuscular junction. Several different growth factor–like neuregulins can be generated by alternative splicing of messenger RNA from the neuregulin gene, and these were originally termed *glial growth factors*. By interacting with its receptors *erbB2* and *erbB3*, which are found on both neurons and glial cells, neuregulin ensures the survival of the surrounding Schwann cells that are essential for normal axonal growth and differentiation.

Not only do the physical location and the amount of the receptor change during maturation of the nerve-muscle contact, but the nature of the acetylcholine receptors themselves alters at this time. Patch clamp recordings have shown that the properties of the embryonic receptor channel are different from those of the adult. As we saw in Chapter 11, the fetal muscle receptor is made up of α, β, γ, and δ subunits. As the neuromuscular junction matures, the synthesis of the γ subunit ceases and synthesis of a new ε subunit begins. The change in channel properties can be attributed to this switch from the γ to the ε subunit.

Effects of electrical activity on the development of the neuron-muscle synapse. We have seen that the clustering and the synthesis of new acetylcholine receptors do not depend directly on electrical activity in the nerve or muscle but on the constitutive secretion of factors in the basal lamina. Earlier in this chapter, however, we learned that electrical activity in one motor neuron can lead to the elimination of synapses from other neurons contacting the same muscle cell. The removal of the extrajunctional receptors, which are not activated by the winning presynaptic terminal, depends directly on the stimulation that the presynaptic neuron provides. Thus, blocking the occurrence of action potentials in the motor neurons prevents the removal of these receptors. Even in the adult, block of the motor neurons produces an increase in the number of extrajunctional acetylcholine receptors, leading to supersensitivity of the muscle to acetylcholine. Direct electrical stimulation of the muscle itself can largely prevent these changes, indicating that continued activity in the muscle is required for its normal characteristics.

Two other proteins whose distribution and levels are altered by ongoing electrical activity are acetylcholinesterase, the enzyme that terminates the actions of acetylcholine, and the cell adhesion molecule N-CAM. Like the clusters of acetylcholine receptors, both of these molecules are initially distributed uniformly across the muscle. After innervation by the nerve terminal, however, both are lost from the extrajunctional regions but come to be clustered at high levels at the postsynaptic junction. Normal aggregation of these proteins at the synapse appears to require ongoing electrical activity in the muscle and is compromised by treatments that paralyze the muscle. Conversely, direct electrical stimulation of muscles promotes their aggregation at the synapses. In addition, when the activity of the presynaptic axons is eliminated, either by denervation or by pharmacological agents, there is a change in the properties of the voltage-dependent sodium channels in the muscle, which revert to a form that is insensitive to the blocking agent tetrodotoxin. Such tetrodotoxin-insensitive sodium channels are normally found only early in development.

Finally, the temporal pattern of action potentials in the presynaptic nerve actually alters the characteristics of the muscle, a subject of great interest to athletes and bodybuilders. Mammals have two forms of skeletal muscle, fast and slow. Neurons that innervate fast muscle fire rapid bursts of action potentials at irregular intervals, while those that innervate slow muscles fire more continually at a lower rate. When the nerve input to a fast muscle is stimulated with a pattern normally experienced by slow muscle, the fast muscle begins to question its identity and starts to synthesize new contractive proteins characteristic of slow muscle. ❷

Synaptic Plasticity in the Adult Nervous System

Sprouting in adult nerves. We have given an account of the extension of neurite branches and the making and breaking of synapses during the formation of the nervous system. It is clear that the adult nervous system retains nearly all of the machinery required for such synaptic plasticity. This can be demonstrated at the adult neuromuscular junction, where the stability of the synaptic connections depends on ongoing electrical activity in the muscles. Synaptic transmission can be blocked by application of tetrodotoxin to the motor axons, by local injection of an agent such as α-bungarotoxin, which blocks the response of muscle cells to acetylcholine, or even by making the neurons express new potassium channels that suppress their excitability. When this occurs, new branches are formed at the synaptic terminals. They extend over the muscle fiber, forming new areas of synaptic contact.

Further evidence that adult axons are capable of sprouting new processes and forming new synapses comes from studies of recovery from injury of motor nerves at the neuromuscular junction. Following partial denervation of a muscle by cutting some of its incoming axons, the remaining intact motor neurons form new branches that extend toward the denervated region and establish new synapses (Fig. 17–20). These new *sprouts* may extend either from the terminals of motor neurons (terminal sprouts) or from the axons at the nodes of Ranvier (nodal sprouts). The sprouts that establish synapses on the denervated muscles become stabilized, perhaps through the action of some trophic factor from the muscle cells (see later discussion). Sprouts that fail to reach a target, however, are eventually retracted.

In addition to being able to generate sprouts near a terminal region, mature neurons are capable of regenerating full axons. In the case of the neuromuscular junction, when the axon of a motor neuron is severed, the distal part of the axon degenerates. The remaining part of the axon, which

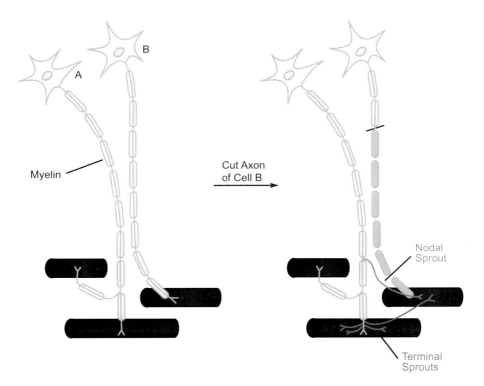

Figure 17–20. Axonal sprouting at the adult neuromuscular junction. (Modified from Brown, 1984.)

is attached to the soma, forms a new growth cone at its distal end and grows back to the denervated muscle. When the growth cone reaches the muscle, new synapses are formed. If the original sites of synaptic contact have come to be filled by branch sprouts from neighboring terminals whose axons were not severed, then these sprouted branches may retract as a result of the reinnervation. In many respects, therefore, this restructuring of synaptic branches in the adult nervous system resembles what occurs during development.

Proteins in the basal lamina bind presynaptic calcium channels, providing a memory of earlier synaptic connections. The process by which a neuromuscular synapse is established by a *regenerating* axon differs slightly from synapse formation by the first axons of embryonic motor neurons. In particular, the first synapses of the embryonic axons may form at random locations on the muscle. Regenerating axons, on the other hand, form synapses specifically at those sites where a synapse had previously existed. This is because the sites of former synaptic contact are marked by specific molecules that induce the growth cone to form a presynaptic structure. Interestingly, these molecules are not located on the surface of the muscle membrane. Instead, they are in the basal lamina, the extracellular layer of proteins that is secreted by the muscle cells and surrounds the muscle (Fig. 17–21).

The existence in the basal lamina of molecules that mark the sites of former synaptic contact can be demonstrated by severing both the motor axons *and* the muscle fibers, causing them both to degenerate (Fig. 17–21). In this condition, only the basal lamina remains. The sites of former synaptic contact can be identified both morphologically and

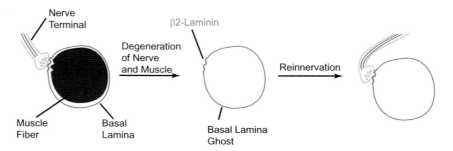

Figure 17–21. Basal lamina marks sites of synapse formation at the neuromuscular junction. This experiment, by Marshall et al. (1977), demonstrated that a regenerating motor axon specifically forms synapses at sites on basal lamina that had previously been occupied by a synapse.

by the presence of the enzyme acetylcholinesterase. When the regenerating motor axons reach the basal lamina, they form apparently normal synaptic terminals specifically at these sites. This is because, some time after an initial synapse is established, muscle cells secrete marker proteins into the basal lamina at the synaptic cleft. Antibodies have been generated against various proteins of the basal lamina, and some of these have been found to be localized selectively to the synaptic part of the basal lamina. One, called β2-*laminin* (originally termed *S-laminin*, for synaptic laminin), turns out to be an isoform of the β chain of the laminin protein we discussed as a normal component of the extracellular matrix, in Chapter 16.

Although the laminin trimer (see Fig. 16–7b) is present throughout the basal lamina, the muscle cells substitute the β2-laminin chain for the normal β1 isoform into the synaptic space once a stable synaptic contact is established. β2-Laminin differs from the normal β1 chain in an important way: The β2-laminin molecule *selectively* binds the membrane of motor neurons. Interestingly, a sequence of only three amino acids (Leu-Arg-Glu) within β2-laminin appears to be all that is required for this selective adhesion. The key presynaptic proteins that are recognized by β2-laminin are the $Ca_V2.1$ and $Ca_V2.2$ voltage-dependent calcium channels that trigger the release of neurotransmitter from the terminal. Binding of β2-laminin to a sequence of amino acids located on the extracellular face of these channels causes the channels to cluster together (Fig. 17–22). These clusters of calcium channels then recruit some of the scaffold proteins of the presynaptic terminal, including Bassoon (see Chapter 8), to form the dense web that links synaptic vesicles to the active zone. Evidence that clustering of the calcium channel by β2-laminin is required for formation and maintenance of the active zones has been provided by injecting animals with short peptides corresponding to the part of the channel that interacts with β2-laminin. These peptides prevent the β2-laminin from finding the real channels on the presynaptic membrane and prevent the normal construction of presynaptic active zones (Fig. 17–22).

During regeneration of a nerve, it is the physical binding of calcium channels by β2-laminin that prevents the ingrowing axons from finding new sites of synaptic contact once they have contacted a previously innervated site. Another important function of β2-laminin is to cause the repulsion of the processes of Schwann cells, which would impair access of the growing nerve onto the synaptic site. In addition to β2-laminin, there are several other protein markers of the synaptic basal lamina. These include the neuron-specific *z+ agrin* isoform, which produces receptor clustering

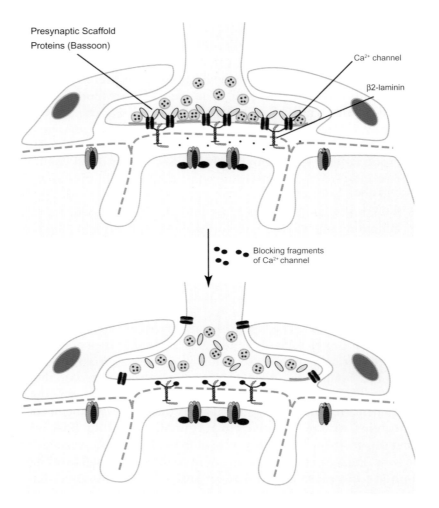

Figure 17–22. β2-laminin in the synaptic basal lamina binds and clusters presynaptic calcium channels, triggering the organization of presynaptic active zones. An experiment by Hiroshi Nishimune and colleagues demonstrated that presynaptic organization could be disrupted by a peptide that blocks the interaction of β2-laminin with the channel (Nishimune et al., 2004).

and continues to provide signaling to the muscles even after a presynaptic terminal has degenerated.

Loss of presynaptic terminals after axotomy or interruption of axoplasmic transport. We have seen that cutting an axon, or interrupting transport in an axon, produces a restructuring of the synaptic contacts made *by* that axon and others that innervate the same target. These same manipulations can also produce changes in the presynaptic contacts *onto* the cell whose axon has been cut. For example, if the axons of neurons of the sympathetic ganglion are severed or exposed to colchicine, the synapses they receive from other neurons retract (Fig. 17–23). When transport is restored or the axons of the sympathetic ganglion cells are allowed to regrow and form new contacts, the branches of the presynaptic axons also extend to reestablish their full complement of synapses. Thus the normal maintenance of synaptic endings may depend on factors available only from an intact postsynaptic cell. For sympathetic neurons, NGF plays a role here. For example, application of NGF to the sympathetic ganglion following axotomy prevents the loss of presynaptic terminals, illustrated in Figure 17–23. In addition, treatment of animals with an antiserum to NGF, which would be expected to bind to endogenous NGF and thereby prevent its uptake by cells, induces loss of synapses on sympathetic neurons even though their axons remain intact.

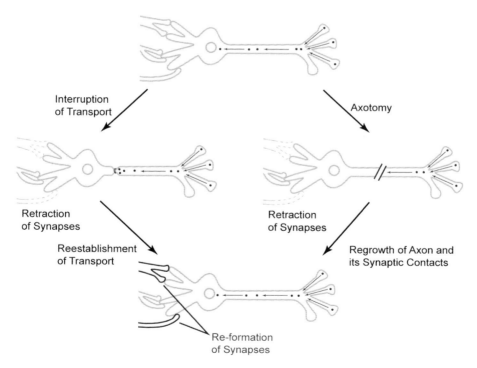

Figure 17–23. Loss of presynaptic terminals after interruption of axoplasmic transport or axotomy. Maintenance of presynaptic inputs may depend on a postsynaptic factor that is transported from the terminals back toward the soma.

Even in Adults, Transcription and Translation Are Activated by Neuronal Activity

Up to this point, this chapter, together with the previous three chapters, has focused on mechanisms that establish the synaptic properties of neurons during their development or during recovery from an injury. It has become very clear, however, that the majority of these mechanisms also continue to be responsible for the maintenance and remodeling of synapses in the adult. For example, the Notch/Delta signaling pathway, which participates in the very earliest steps in the birth of a neuron, continues to be present and active in mature neurons. We have provided several examples of profound plastic changes during development that are dependent on the ongoing electrical activity of neurons. Patterns of electrical activity in the adult nervous system continue to have a key effect on neurons, because they alter the rate at which genes are transcribed into messenger RNAs and the rate at which messenger RNAs are translated into proteins.

Immediate-early genes. When any external stimulus switches a gene on or off, it acts through transcription factors. These cellular switches control the synthesis of new sets of proteins, thereby effecting a change in the properties of a cell. In Chapter 14, we alluded to some of the transcription factors that contribute to neuronal differentiation. Some additional transcription factors are, however, of particular interest to the neurobiologist because they can be synthesized very rapidly in response to neuronal activity, even in adults. Two such transcription factors are the *fos* and *jun* proteins.

The *fos* and *jun* genes were first discovered as *proto-oncogenes*. An *oncogene* is an aberrant gene that causes uncontrolled growth in a population of cells, while the term *proto-oncogene* is used to describe the normal cellular counterpart of the oncogene. The fos and jun transcription factors belong to a large family of closely related proteins. In many non-neuronal cells, exposure to growth factors that stimulate cell division causes a very rapid synthesis of some of these proteins (Fig. 17–24a), which then move to the cell nucleus. For this reason, the genes for these transcription factors are sometimes termed *immediate-early genes*, to reflect the fact that they are the first genes to be activated in response to a stimulus to the cell. The fos and jun proteins both contain a helical region in which the amino acid leucine is found at every seventh position. This relatively common structural motif is known as a *leucine zipper*. It tends to be involved in protein–protein interactions, and its presence in the fos and jun proteins allows them to form a dimer. In fact, the jun protein also exists as a dimer with other related transcription factors. The fos-jun dimer then binds to certain sequences on DNA, termed *AP-1 sites*, and thereby regulates the ability of nearby genes to be transcribed into messenger RNA. The fos-jun complex and its relatives can therefore be thought of as messengers linking growth factor receptors to the synthesis of specific proteins.

The *fos* gene was the first gene found to be regulated by neuronal activity. Transcription of the *fos* gene into RNA can be detected as early as 1 minute following the onset of stimulation. Bursts of action potentials or exposure to neurotransmitters or growth factors can rapidly induce *fos* expression as much as 100-fold. In fact, the location of neurons that begin to synthesize fos in response to sensory stimulation has been used as an anatomical tool to determine the neuronal pathways activated by the stimulus (Fig 17–24b).

Because fos is found in a very wide range of cells, its rapid synthesis in response to stimulation may not reflect a process that is specific to neurons. Other immediate-early genes, however, are not transcription factors but encode proteins that have a direct effect on synaptic transmission and neuronal excitability. One is the gene for BDNF, whose neuronal effects we have already considered. Another is the gene for *Arc* (**A**ctivity-**r**egulated **c**ytoskeletal protein), a protein that is rapidly induced when animals are acquiring new memories. Arc is located at the postsynaptic densities of excitatory synapses in the cerebral cortex and hippocampus, and it is thought to play a role in determining the appropriate number of glutamate receptors needed in the plasma membrane. Yet another is the gene for *Homer1*, a synaptic scaffolding protein. In fact, it appears that quite a number of neuron-specific genes, for example, those for specific potassium channels that determine neuronal firing rates, fall into the immediate-early gene category.

Calcium is an important signal for the transcription of immediate-early genes. For stimulation of a neuron to induce rapidly the transcription of specific genes, a signal must be generated that links events at the plasma membrane, such as calcium entry through calcium channels or the activation of a neurotransmitter receptor, to the nucleus. Very often, this signal is calcium itself (Fig. 17–24a). The genes for fos, Arc, and many other immediate-early genes have a sequence of DNA in them that is termed a *CRE* for (**C**yclic AMP/calcium **R**esponse **E**lement). Neuronal stimulation that elevates calcium activates *calcium/calmodulin-dependent protein kinase type IV* (Ca^{2+}/Cam kinase IV), which is related to Ca^{2+}/Cam kinase II, discussed in Chapter 8. This enzyme phosphorylates a transcription

a

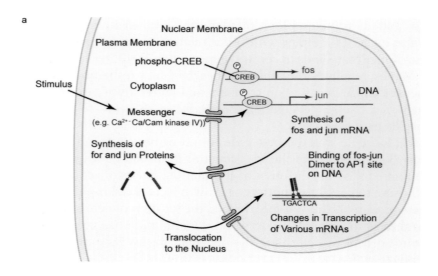

b

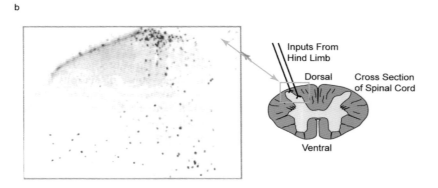

Figure 17–24. Immediate early genes *fos* and *jun*. *a*: Formation and nuclear translocation of the fos and jun proteins after stimulation of a cell (Curran and Morgan, 1995). *b*: Steven Hunt and his colleagues stained sections of spinal cord with antibodies to c-fos. This section came from an animal that had received stimulation of a hind limb. Dark spots show the location of neuronal nuclei in which c-*fos* has been induced (Hunt et al., 1987).

factor termed *CREB* (Cyclic-AMP Response Element Binding protein), allowing it to bind DNA at the CRE. In combination with a complex of other proteins, this recruits RNA polymerase II to the gene, beginning the business of transcribing DNA into RNA. We shall encounter CREB again in Chapter 19.

Activity can also initiate RNA synthesis by activating other transcription factors such as *CREM* (**C**yclic AMP **R**esponse **E**lement **M**odulatory Protein), *CaRF* (**Ca**lcium-**R**esponse **F**actor), *SRF* (**S**erum **R**esponse **F**actor), and *MEF2* (**M**yocyte **E**nhancer **F**actor **2**), each of which binds to specific sequences of DNA. Another, *MeCP2* (**Me**thyl **C**p**G** binding protein), a protein that binds to DNA to which methyl groups have been added, is of interest because human mutations in this protein produce *Rett syndrome*, a condition with severe cognitive disability that may result from abnormal synaptic maturation.

FMRP and local activity-dependent translation of proteins. When a change in neuronal firing alters the activity of transcription factors in the nucleus, the resultant increases in mRNA and proteins encoded by these mRNAs are available for the entire cell to use. We have seen, however, that changes in neuronal activity may selectively produce changes in the form and structure of only a subset of synapses. How can an increase in mRNA at the cell body lead to increases in protein synthesis only at restricted locations such as subsets of active dendrites? The answer is that neurons are capable of *local translation*. Ribosomes are located at specific postsynaptic sites and can even be tethered to the postsynaptic density. Proteins made by translating mRNA at these locations will be available only for that synapse.

Local translation of proteins in dendrites requires that RNA be transported away from the cell body toward distal synapses in the dendrites, and then to await a signal before being translated. Insights into this process have come from studies of *fragile X syndrome*, an inherited disorder that is one of the most common causes of inherited autism and intellectual disability. It results from the inability to make *FMRP* (**F**ragile **X** **M**ental **R**etardation **P**rotein). This is an mRNA-binding protein that can associate with any one of ~ 800 different mRNAs, which represents about 4% of the total number of mRNAs translated by a neuron. It can be transported with its mRNAs to local sites in the dendrites. In the absence of stimulation, FMRP suppresses the translation of the mRNAs to which it is bound (Fig. 17–25). When a local region of the neuron becomes depolarized or receives an appropriate neurotransmitter signal, however, FMRP liberates its target mRNAs, causing an increase in their rate of translation. For example, activation of metabotropic glutamate receptors can increase local protein synthesis, most likely by triggering the dephosphorylation of FMRP.

Genetic loss of FMRP produces a number of cellular effects, including an increase in the overall rate of neuronal protein translation. Although synaptic transmission generally appears intact in animals lacking FMRP,

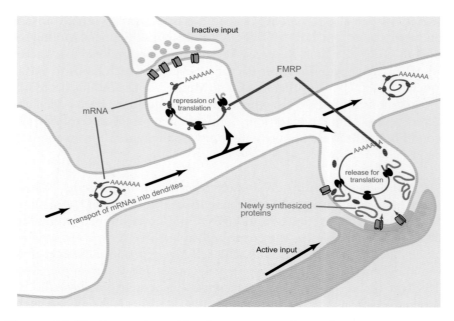

Figure 17–25. Repression of local protein translation by FMRP at inactive synapses and activity-dependent synthesis of new proteins at activated postsynaptic sites.

many brain regions in both humans and mice have been found to contain increased numbers of dendritic spines that appear "immature" in morphology, with an excess of filipodia. In addition to its role in regulating translation, FMRP also appears to have other functions. For example, it is found in axons and in presynaptic terminals, where it is not thought to regulate translation, and it can bind directly to certain potassium channels, directly influencing neuronal excitability.

Summary

When the developing axon reaches its appropriate postsynaptic target, it stops elongating. A series of characteristic morphological and biochemical changes, culminating in synapse formation, then occur. Some of the key events covered in this chapter are summarized in Table 17–1. Among the cues used by an excitatory neuron in choosing its correct postsynaptic partner are chemical labels such as Ephrins and Eph receptors. Not all synapses that form during development persist in the adult animal. Certain synapses are selectively stabilized, whereas others are lost. In many cases

Table 17–1 Some of the Steps in the Formation of a Chemical Synapse

1. Contact of the growth cone with an appropriate target
2. Increase in the release of neurotransmitter
3. Increase in the adhesion of the presynaptic terminal to the target
4. Elimination of other competing synapses by heterosynaptic suppression
5. Clustering of receptors at the postsynaptic membrane
6. Synthesis and insertion of new receptors at the postsynaptic membrane
7. Elimination of extrajunctional receptors

such rearrangements follow a Hebbian rule, in which excitatory synapses are stabilized when they trigger postsynaptic action potentials. Such reorganization of connections is therefore regulated by patterns of electrical activity and involves the coordinated activity of metalloproteases, NMDA receptors, and the secretion of factors such BDNF. A wealth of information on factors that regulate the clustering of postsynaptic receptors and the way that postsynaptic sites come to be linked to presynaptic terminals has also come from studies of molecules such as agrin, rapsyn, and β2-laminin at the neuromuscular junction.

Synapses may also be broken and reformed continually in the adult animal, either following injury or in response to normal physiological patterns of activity. Stimulation of neurons in adults engenders many of the same biochemical responses that occur in development. These include the activation of the transcription of immediate early genes, such as those encoding fos, Arc, and BDNF, and the stimulation of protein synthesis by mechanisms such as loss of repression by the mRNA-binding protein FMRP. Neurobiologists are still putting together a jigsaw puzzle in which each of these molecules represents a piece. When the picture is complete, it should be possible to view the path by which an undifferentiated cell becomes a mature neuron, the involvement of electrical activity in determining the characteristics of the mature cell, and the role of various molecules in the plastic properties of the adult neuron.

18

Intrinsic Neuronal Properties, Neural Networks, and Behavior

As cells go, neurons are not loners. Every function of the nervous system, from regulation of autonomic activities, such as heartbeat, to the control of complex animal behaviors, such as dating and mating, reflects the coordinated action of a *network* of interacting neurons. A major challenge of neurobiology is to understand the nature of computations that neural networks carry out. To even begin tackling such questions requires two rather different types of knowledge. First of all one has to know how many neurons there are in the network and exactly how they are connected to each other. Secondly, one must understand the biochemical and electrical properties of the individual neurons. In this chapter, we describe a number of rather simple neural networks whose biological roles are known. These representative examples have been chosen to illustrate how the intrinsic cellular properties of neurons are as important as the "wiring diagram" in the function of the network as a whole.

Models of Neural Networks

Technologies are emerging that will soon provide us with three-dimensional maps showing how every neuron in a nervous system is physically connected to other neurons, with detailed pictures of each synaptic connection. Such a static map has been termed a *connectome*. For many invertebrates, significant parts of the connectome have are already been

457

established. Figure 18–1a provides a map of the connections between the 302 neurons that comprise the entire nervous system of the nematode *C. elegans*. We came across this creature in Chapter 14 when discussing cell death, because its genetic analysis has provided major insights into development. Its neurons can be categorized into sensory neurons, interneurons, and motor neurons. These control a repertoire of behaviors that include feeding, mating, escaping from noxious stimulation, and moving toward environments that are sensually attractive to the animal. It is also

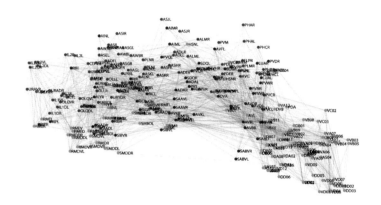

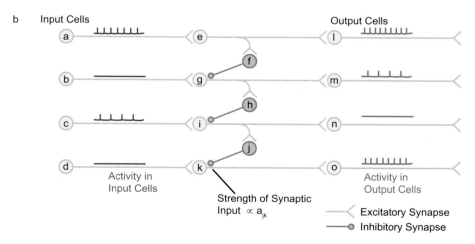

Figure 18–1. Neural networks. *a*: The connectome of *C. elegans*. Each neuron has its own label. Sensory neurons are shown in red, interneurons in blue, and motor neurons in green (Varshney et al., 2011). *b*: Model neural network used in computations with synaptic weights assigned to each connection.

capable of simple forms of learning. The study of mutations that influence these behaviors has revealed that neurons with similar biological functions can express very different types of ion channels and receptors, providing each neuron with a distinct biochemical personality.

It is evident that a network such as that of *C. elegans*, comprising many neurons, can generate patterns of activity that cannot be predicted by the study of a single cell in isolation. The properties of a network that can be attributed to interactions between cells are referred to as its *emergent properties*. Long before any connectome was established, many attempts had been made to understand such emergent properties by analyzing simple mathematical or computer models of interacting units. Figure 18–1b illustrates a typical model network with a set of input units, some internal units, and a set of output units. The lessons learned from such computer models is that, even if all of the units in the model are identical, the network can be made to "learn" to categorize different patterns of inputs, provided the strength of a "synaptic" connection between one neuron and another, for example, neuron i and neuron j in the figure, which is set by a parameter a_{ij} known as the *synaptic weight*, is allowed to vary according to a fixed set of rules. ❷

We shall now give an account of several *real* neural networks. These can be seen to function in ways that differ from those of computer simulations with many identical units. Our goal is not to give an exhaustive review, but rather to provide selected examples of how the intrinsic properties of neurons and their synaptic interactions shape the behavior of a network.

Networks Generating Rhythmic Movements

Central pattern generators. Many animal behaviors, such as walking or swimming, require the rhythmic contraction of muscles. We have already seen that a single neuron is capable of generating rhythmic bursts in the absence of external stimulation. However, most rhythmic behaviors require that opposing groups of muscles be contracted and relaxed in a coordinated manner. This coordination can be carried out only by a network in which different neurons innervate different muscles.

In theory, there are many ways to build a rhythmic network. For example, the generation and coordination of rhythmic movements could occur through a chain of reflexes in which receptors in the muscles signal the state of extension or contraction of each muscle to the remainder of the network, and this information would be essential for the network to function rhythmically. If this were the case, the muscles themselves would be integral components of the network. This, however, does not appear to be the case for most rhythmic networks that have been examined in detail. Rather, the pattern of outputs to different muscles is generated by a *central pattern generator*, a network of neurons that, even in the absence of direct feedback from the muscles themselves, is capable of generating the appropriate patterns of rhythmic activity. (It is important to remember, however, that while sensory feedback is often not needed for the basic rhythmic movements, it *is* required to shape these movements to the needs of the animal in the real world.)

Rhythmic movements can be generated by networks with reciprocal inhibition. The very simplest circuit that can generate alternating contraction and relaxation in two different muscles consists of only two neurons. Each neuron makes an inhibitory synapse onto the other (Fig. 18–2a). For such a circuit to generate rhythmic output, it is not necessary that these neurons be endogenously active in the absence of other synaptic inputs. It is, however, necessary for the neurons to display *postinhibitory rebound*. This simply means that after the membrane potential of the cell has been hyperpolarized for a short period of time, the cell becomes more excitable than usual. When the membrane is then allowed to return toward its normal resting potential, one or more action potentials may result. This is a relatively common phenomenon in neurons, and when it follows an experimentally applied hyperpolarizing current pulse as in Figure 18–2b, the action potential is often termed an *anode break spike*. In some cases, the explanation for postinhibitory rebound is that an inward current, such as a voltage-dependent sodium current or a T-type calcium current, is partly inactivated at the resting potential. Transient hyperpolarization, for example, by an inhibitory input, removes some of this inactivation so that the threshold for an action potential becomes more negative. As the cell depolarizes toward the resting potential, the increased inward current triggers an action potential before inactivation again develops. Another current that produces anode break spikes is a nonselective cation current called the *H-current*, sometimes also called a *pacemaker* current, because it is commonly found in neurons that fire repetitively. This depolarizing current is only activated after a neuron is hyperpolarized. The channels that produce the H-current are termed HCN channels (for Hyperpolarization-activated Cyclic Nucleotide-gated channels).

Networks based on the simple two-neuron circuit, sometimes called a *half-center oscillator*, do indeed exist, and contribute to locomotion in some species. Figure 18–2c illustrates the activity of two neurons in *Clione*, a small marine mollusc. This animal swims in the sea by moving a pair of wing-like structures (termed *parapodia*) that are alternately

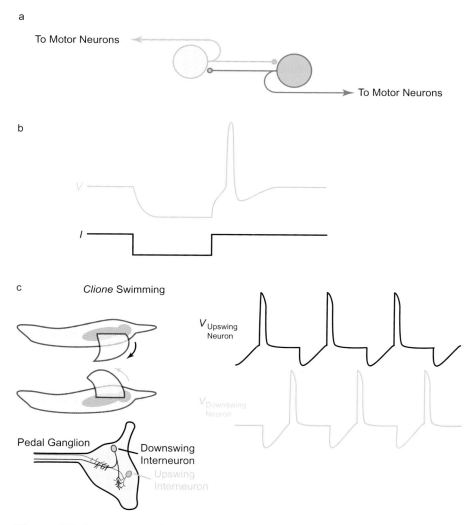

Figure 18–2. Reciprocal inhibition. *a*: Basic network that can generate alternating contraction and relaxation in two muscles. *b*: Anode break spike following a hyperpolarizing current pulse. *c*: Interneuron activity during swimming in *Clione* was studied by Satterlie (1985).

flexed in a dorsal and ventral direction. A major component of the central pattern generator for swimming appears to comprise four swim interneurons (Fig. 18–2c). One upswing neuron and one downswing neuron are found on each side of the nervous system. An action potential in an upswing neuron generates an inhibitory postsynaptic potential (IPSP) in a downswing neuron. Because of postinhibitory rebound, the downswing neuron generates an action potential at the end of the IPSP. This in turn triggers an IPSP in the upswing neuron. Thus a sustained ping pong–like activity reverberates in this two-neuron network, and is conveyed to the motor neurons that innervate the muscles in the parapodia, producing rhythmic swimming movements.

There is more than one way to design an oscillating network, even with only two neurons. For example, if individual neurons do not display postinhibitory rebound, but the firing of the individual neurons is subject to accommodation (see Fig. 3–13), a circuit such as that in Figure 18–2a can produce alternating bursts of action potentials, provided that a source of maintained excitation is provided to the cells. Moreover, in most cases it is not the firing of single action potentials that alternates between the two cells but prolonged bursts of action potentials that occur out of phase in the two cells. Central pattern generators in most nervous systems are substantially more elaborate, and hence more versatile, than simply two groups of mutually inhibitory neurons. Reciprocal inhibition is, however, a recurring theme even in more complicated pattern generators. One simple system that has been widely exploited in studies of cellular mechanisms of rhythm generation is the crustacean stomatogastric ganglion. We shall now summarize briefly some of the lessons that have been learned from this system. ❽

Rhythmic Neuronal Activity in Crustaceans

Although despised by gourmets, the stomachs of spiny lobsters and of crabs have provided pleasure to many neurobiologists. The stomach of such crustaceans is, like all of Gaul, divided into three parts: the cardiac sac, the gastric mill, and the pylorus (Fig. 18–3a). Food enters the stomach through the esophagus and is digested as it moves progressively through these three regions. Rhythmic contraction of muscles in all three regions contributes both to the physical disruption of the food and to movement through the stomach in a manner that resembles the chewing and swallowing of food by humans.

The stomatogastric ganglion. Muscles in the stomach are controlled by neurons in the *stomatogastric ganglion*, which contains 30 neurons. The three stomach regions are controlled by different sets of neurons. We shall consider only the central pattern generator for the pylorus, which consists of only 14 neurons. Synaptic connections between these are illustrated schematically in Figure 18–3b. The eight identical PY cells (lumped into a single neuron for simplicity in the figure), the two PD cells, and the VD, LP,

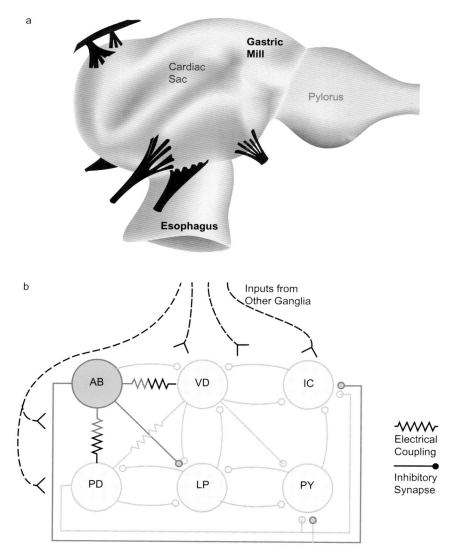

Figure 18–3. The lobster stomach. *a*: Diagram of the stomach (from Dickinson and Marder, 1989). *b*: Neuronal interactions in the pyloric network.

and IC neurons are all motor neurons that directly innervate the pyloric muscles. Cell AB is an interneuron that only makes connections within the ganglion. Note that all the chemical synapses are inhibitory and that reciprocal inhibition between pairs of neurons is a dominant theme in the network. In addition, some pairs of neurons are coupled by electrical synapses.

As shown in Figure 18–4, the rhythmic output of the pyloric neurons may be recorded both by intracellular microelectrodes in individual neurons and by extracellular electrodes placed on the nerves containing axons from motor neurons such as LP and IC to different sets of muscles. Three different phases of bursting are recorded in the different nerves.

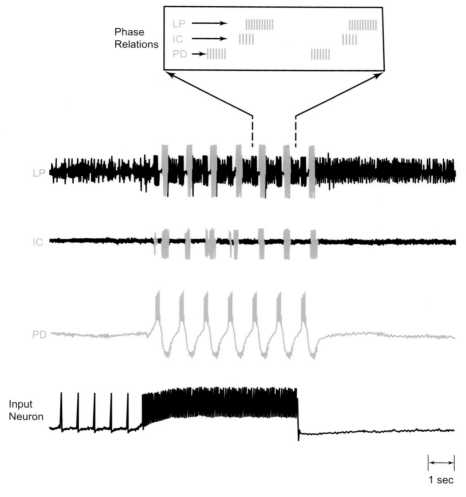

Figure 18–4. The pyloric rhythm. Recordings (made by Nusbaum and Marder, 1989) of rhythmic firing in three neurons (LP, IC, and PD) of the pyloric circuit during stimulation of a neuron with inputs to the ganglion. Because action potentials in the LP and IC neurons were detected with extracellular electrodes placed on nerves leading to the pyloric muscles, the activity of other neurons can also be seen in these recordings. The inset at the top shows phase relations among the three cells through two cycles.

Neurons in the rhythm-generating circuit need to be told how to behave. The first important lesson about the rhythm generated by the pyloric circuit is that it depends on the presence of inputs from other parts of the nervous system. Traces such as those in Figure 18–4 are recorded only when the inputs from two other ganglia are intact. When the nerve from these ganglia to the stomatogastric ganglion is blocked, rhythmic bursting ceases. This is not because the incoming inputs generate any rhythmic activity themselves. Instead, the neurotransmitters used by inputs from these other ganglia appear to act as local hormones (see Chapter 10). When the inputs are activated and release their hormones, the pyloric neurons *acquire* the specific electrical properties needed to generate rhythmic bursts. The rhythm of the AB bursting neuron is generated in full by a cycle of opening and closing of different ion channels in its membrane. These ion channels only come into play, however, when activated by biochemical pathways triggered by the local hormones. Because this pyloric neuron requires the presence of modulatory substances from the other ganglia to generate the bursts, this neuron can be termed a *conditional burster*.

The other neurons of the pyloric network are not endogenously bursting neurons. In response to a brief depolarization, however, they can generate a single long burst of action potentials. This burst occurs because of a sustained, regenerative depolarization that outlasts the brief stimulus. This regenerative depolarization is sometimes termed a *plateau potential* or *driver potential* (Fig. 18–5). Again, this plateau depolarization occurs only when inputs from other ganglia have been activated.

Rhythm-generating neurons need to be kept busy; homeostatic control of excitability. If the nerve that provides input to the stomatogastric

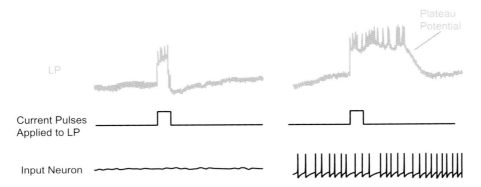

Figure 18–5. Induction of a plateau potential in the LP neuron by the activity of an input neuron (Moulins and Nagy, 1985).

ganglion is cut, the ganglion becomes silent. If, however, the ganglion is then kept alive for several days, a pyloric rhythm again resumes, now without needing the external inputs. Transcription and translation mechanisms similar to those we discussed in Chapter 17 adjust the levels of ion channels and the strength of synaptic connections to allow the circuit to become autonomous. This finding is an example of *homeostasis,* a ubiquitous biological phenomenon first described by Claude Bernard in 1865. Applied to neurons, it implies that each neuron has a target pattern of firing with which it is comfortable. Anything that causes the firing pattern to deviate from the comfort zone triggers biochemical reactions in an attempt to restore the target pattern.

A neuron can find many different ways to become comfortable with itself. Despite the fact that all stomatogastric ganglia produce similar rhythmic bursting patterns, it has been found that levels of expression of channels in PD neurons, and in other neurons of the circuit, are quite variable from animal to animal. In part, this is because it is not the absolute level of any channel that determines the firing pattern of a neuron but the ratio of that channel to other channels that oppose its activation. For example, it has been found that injection of mRNA for the Shal potassium channel into PD neurons, which greatly increases the A-current (see Chapter 7), does not change its bursting firing pattern. This clear example of homeostasis occurs because the increase in potassium current is rapidly compensated by an increase in the level of HCN channels, which generate a depolarizing H-current to oppose the increased A-current.

A demonstration that a very specific set of ionic conductances is not required for reciprocal inhibition to produce alternating rhythmic output has been provided by isolating each of the neurons of the stomatogastric ganglion and coupling them to a *dynamic clamp*. This is simply a very smart form of the voltage clamp we covered in Chapter 4. Instead of holding the voltage of the cells at a fixed value, the dynamic clamp smoothly changes the current across the membrane in a way that exactly matches that produced by a postsynaptic current and it allows the cell to fire an action potential. It then rapidly calculates the way that a second cell would respond to transmitter released from this action potential and again stimulates or inhibits the real cell. This versatile type of voltage clamp can also change current flow so as to simulate the addition or subtraction of intrinsic conductances from the cell being recorded. Isolated LP neurons do not generate rhythmic bursts. When subjected to this dynamic clamp, however, they could be made to generate typical rhythmic bursts, provided an appropriate set of synaptic connections was simulated and the neurons had a sufficient amount of the pacemaker H-current

(Fig. 18–6). Different neurons with differing intrinsic levels of excitability and firing patterns can be recruited to produce similar rhythms with slight (but very important) adjustments to the intrinsic conductances in the model cell and the strength of the "synaptic" connections.

The fact that the right sort of output can be obtained using different values for levels of ion channels could be interpreted as meaning that these channels are not important for the overall output. Nothing could be further from the truth, however. Without the right mix of ion channels in the cells, the appropriate patterns simply could not be generated. It is just that cellular mechanisms adjust levels of currents and the intrinsic excitability of neurons to maintain appropriate firing patterns in response to a wide range of external stimuli. Exactly what these mechanisms are is still being established.

Modulatory neurotransmitters may "design" different networks. If a stomatogastric ganglion can generate acceptable patterns of muscle contractions even after it has lost its input, why are these inputs normally

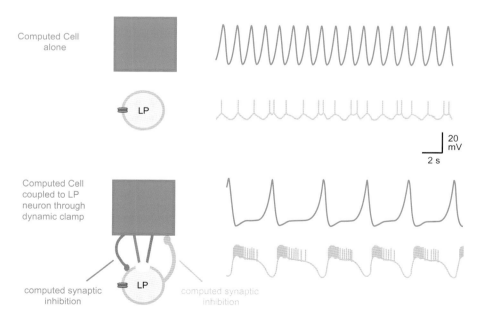

Figure 18–6. The dynamic clamp. A numerical simulation of a model neuron generates fast oscillation, while an isolated LP neuron exhibits an irregular pattern of action potentials (*top traces*). When the two are coupled through the dynamic clamp, they generate a slow, alternating rhythm that matches a normal LP pattern in the intact ganglion (*bottom traces*) (Grashow et al., 2010).

required? The answer can easily be discovered if one engages in a period of dispassionate observation in a restaurant. Chewing and swallowing in humans are not simple processes but can take many different dynamic patterns, depending on the nature of the food and the psychological state of the diner. So it is with lobsters and crabs. The rhythmic output of the pyloric circuit does not always follow the very stereotyped pattern described earlier. When recorded in intact animals, the phase, timing, and amount of activity in different motor neurons can vary with time and with the pattern of behavior of the animal. In addition to simply maintaining the rhythm, inputs from other ganglia serve to fashion these different patterns of activity. This occurs because different transmitter substances act to effectively "rewire" the network into different configurations. There is a wealth of neurotransmitters that provide input to the stomatogastric ganglion: acetylcholine, GABA, serotonin, dopamine, histamine, octopamine, and many neuropeptides. Because many of these act as local hormones, simple application of the transmitters to the external medium surrounding an isolated ganglion mimics the action of continuous firing in an input pathway. Here we shall compare the actions of two amines, serotonin and dopamine, when they are applied to the ganglion.

Figure 18–7a illustrates the effects of serotonin and dopamine on the intrinsic excitability of individual pyloric neurons when they are isolated from their synaptic inputs. Both of these agents induce endogenous bursting in interneuron AB. In contrast, their effects on the other neurons are quite different. The ability of some of the neurons to fire is strongly inhibited by serotonin, while the same neurons may be excited to fire repetitively by dopamine. This inhibition or excitation of different neurons in the full network leads to a functional reorganization of the circuit. When either serotonin or dopamine is added to a stomatogastric ganglion that has been isolated from the neural inputs that would normally allow it to burst, they are able to reinstate the rhythm (Fig. 18–7b). The different amines, however, generate different rhythms. In fact, the two rhythms give the impression of being generated by very *different networks*, which in a sense they are. This can be understood by a closer examination of the circuit. Neurons that are inhibited by serotonin are removed from the active circuit. Thus, in the presence of serotonin alone, the circuit is effectively driven by the endogenous activity of the AB-PD set of neurons (Fig. 18–7b). In the presence of dopamine, on the other hand, neurons that are excited by dopamine remain in the circuit, and the endogenous bursting of AB coupled to reciprocal inhibition between the LP and PY cells shapes the output of the circuit. The pattern and timing of impulses from motor neurons to muscles is therefore different for serotonin and for dopamine.

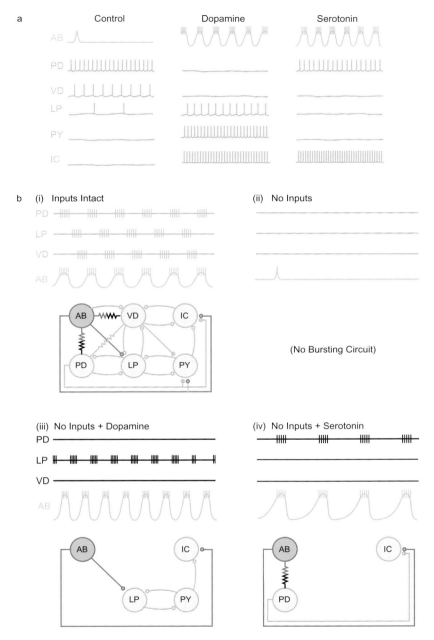

Figure 18–7. "Rewiring" the pyloric network. *a*: Actions of dopamine and serotonin on individual neurons in the pyloric circuit when synaptic connections with other cells have been eliminated. *b*: Actions of serotonin and dopamine on the intact pyloric circuit. Patterns of activity, together with the effective circuit diagram, are shown for four conditions: (i) "normal" rhythm with external inputs intact, (ii) no external inputs, (iii) no external inputs but with dopamine added to the ganglion, and (iv) no external inputs but with serotonin added (Harris-Warrick and Flamm, 1986).

Other amines and peptides have also been found to induce character-istic configurations of the circuit that differ from those produced by sero-tonin and dopamine. Each of these, in turn, differs from those observed when the combined spectrum of modulatory inputs from neurons in the other ganglia is allowed to tinker with the active pyloric circuit. Moreover, some modulators take neurons out of the pyloric but recruit them into additional circuits that we have not described but that regulate the con-tractions of the other parts of the lobster stomach, depicted in Figure 18–3a. The key lesson here is that these modulators act in large part by altering the intrinsic excitability of neurons, producing an effect that, at first glance, might be thought to require a physical rearrangement of synaptic connections.

Command Systems of Neurons

Most animal behaviors do not persist day and night. Mechanisms must exist that allow activities such as walking, eating, swimming, and mating to be turned on and off. Furthermore, even relatively simple animal behav-iors may require the coordinated activation or suppression of a number of apparently independent networks. These tasks are relegated to what are frequently termed *command systems* of neurons.

In some nervous systems, a single *command neuron* can exert control over relatively complex coordinated responses. The definition of a com-mand neuron is that its activity should be both necessary and sufficient to trigger an entire coordinated behavior. For example, a flying cricket avoids high-pitched ultrasound, similar to that emitted by a bat, by contracting a set of muscles that causes the animal to fly away from the direction of the sound. Stimulation of a single identified neuron can trigger this behav-ior in a flying cricket. Moreover, when the neuron is hyperpolarized, the animal fails to respond to the sounds. In most animals, however, impor-tant behavioral decisions are not entrusted to a single neuron. Rather, command *systems* of neurons weigh the pros and cons of a given course of action before committing the animal to a specific choice.

It is useful to make a distinction between two different ways that neuronal command systems influence other neuronal networks such as central pattern generators. Ongoing activity in the pattern generator may require another set of neurons to be continually firing, as we saw in the case of the input neuron in Figure 18–4. If the input neuron slows or stops firing, the rhythmic output stops. Such neurons can be said to *gate* the activity of the rhythm-generating network. At a higher level in the nervous system, however, stimulation of command neurons by sensory inputs or by other neurons in the brain may produce only a relatively brief period of firing that is sufficient to trigger a more prolonged pattern of activity in the central pattern generator. The role of such *trigger neurons*, as well as *gating neurons*, in the control of locomotion has been nicely demonstrated in the nervous system of the leech, and the principles are relevant to a wide variety of circuits that coordinate animal behaviors. ❽

As in the case of rhythmic networks, a thorough analysis of how networks control animal behaviors would be out of place in this book. We shall, however, now describe two systems of neurons that preside over locomotor and reproductive behaviors, with an emphasis on the cellular properties of neurons in these command systems.

Locomotion in the lamprey, a "simpler" vertebrate. It is not by chance that, in this chapter, many examples of how the cellular properties of individual neurons shape the behavior of a network come from work with invertebrates. It is the relative ease with which individual neurons can be identified from animal to animal, and the smaller numbers of cells in the networks, that has made invertebrate systems experimentally tractable. It is clear that, as in invertebrates, locomotion in mammals is produced by local central pattern generators. There are, however, several different classes of interneurons in the mammalian spinal cord, each specified by a different combination of transcription factors (see Fig. 15–3), that may all contribute to pattern generation. In the past, the complexity of the mammalian brain precluded the detailed sort of cellular analysis possible with invertebrates. Nevertheless, lower vertebrates such as the lamprey have several orders of magnitude fewer neurons than mammals; using such animals, it has been possible to confirm that modulation of membrane properties does indeed determine how a network functions.

The eel-like lamprey swims by producing alternating contractions of the left and right sides of its body. The frequency of these contractions varies from one every 5 seconds, for slow movements, to ten per second for rapid swimming. Neurons that make up the pattern-generating circuit are found in each spinal segment. The principal components of these pattern-generating circuits are excitatory interneurons (E in Fig. 18–8a) and inhibitory interneurons (I in the same figure). In contrast to the stomatogastric ganglion, the rhythmic bursts are generated by reciprocal excitation between the E neurons, while reciprocal inhibition between inhibitory interneurons (I) is not used to generate bursts but to ensure that the two sides of the spinal cord burst out of phase with each other. The pattern is shaped further by feedback from the muscles in the form of excitatory and inhibitory inputs from stretch receptor neurons (SR-E and SR-I in Fig. 18–8a).

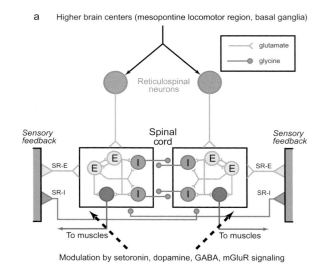

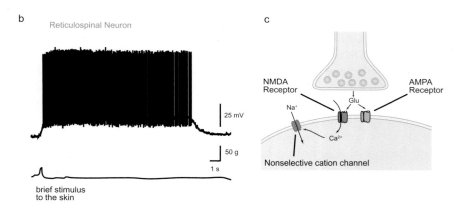

Figure 18–8. How the lamprey swims. *a*: Diagram of the pattern-generating circuit in one spinal segment and its input from the reticulospinal neurons (modified from Grillner and Jessell, 2009). *b*: Prolonged discharge of reticulo-spinal neurons induced by brief stimulation of the skin (Di Prisco et al., 2000). *c*: Following synaptic stimulation, calcium entry through NMDA receptors can activate a nonselective cation current that produces a prolonged depolarization of reticulospinal neurons.

The overall level of activity of the locomotor system is determined by neurons in the brainstem. The activity of these *reticulospinal neurons* (Fig. 18–8a) both initiates the onset of swimming and controls the overall activity of the network. In response to a brief stimulus to the skin, which is a biological stimulus to swim away from the site of touch as quickly as possible, the reticulospinal neurons generate a prolonged burst of action potentials that lasts for many seconds (Fig 18–8b). The ability to burst this way is due to the intrinsic excitability of these neurons, and in particular to the fact that their plasma membrane contains many nonselective cation channels, most likely calcium-activated TRP channels similar to the TRP channels covered in Chapters 7 and 13. Under physiological conditions, the primary effect of activating such channels is to depolarize cells by allowing an influx of sodium ions. The brief touch to the skin eventually leads to the firing of an input pathway that releases glutamate onto receptors on the reticulospinal neurons (Fig. 18–8c). These receptors include NMDA receptors, which are permeable to calcium. Thus synaptic stimulation produces an elevation in internal calcium that activates sodium influx through the TRP cation channels, causing a prolonged depolarization of the reticulospinal neurons.

The prolonged depolarization of the reticulospinal neurons can be considered another example of a plateau potential (compare Fig. 18–8b with Fig. 18–5). Plateau potentials are also important in maintaining bursts of action potentials within the excitatory E interneurons that comprise the central pattern generator itself. The prolonged bursts in the reticulospinal neurons release glutamate, which directly activates NMDA and AMPA receptors on the E interneurons to produce plateau potentials in these cells. These potentials result from the activity of voltage-dependent ion channels in the plasma membrane, including voltage-dependent calcium channels that contribute to the sustained depolarizing phase of the plateau potentials. Termination of each plateau phase occurs due to calcium and sodium entry during the depolarization, which leads to the delayed activation of calcium-dependent and sodium-dependent potassium channels. These hyperpolarize the membrane back to rest so that the cycle may begin anew. ❷

Many of the other lessons we have learned about the role of intrinsic neuronal electrical properties from studying invertebrate networks also apply to vertebrates. For the lamprey, plateau potentials induced in the pattern-generating cells are particularly critical for slow movements, because they allow the network to maintain long-lasting stable bursts of action potentials. The amount of postinhibitory rebound and the rate of accommodation (see Fig. 3–13) in the pattern-generating neurons are key to generating specific outputs from the circuit. Moreover, as in the stomatogastric ganglion, such parameters are altered by inputs that use modulatory transmitters such as serotonin, dopamine, and GABA, as well as neuropeptides and glutamate acting through metabotropic receptors. These can reshape the network to generate a wide variety of frequencies and patterns of locomotion.

How to trigger a prolonged coordinated set of behaviors: the bag cell neurons. In both mammals and lower species, many complex behaviors, such as feeding, drinking, and reproductive behaviors, are regulated by neurons that release neuropeptides. We shall now turn to another command system of neurons, this time in the marine snail *Aplysia*. Two clusters of 200–400 cells each, located in the abdominal ganglion of this animal, control a sequence of very prolonged reproductive behaviors that lead to egg laying (Fig. 18–9a). The cellular and molecular properties of the neurons in the command pathway for egg laying, termed the *bag cell neurons*, have been studied in substantial detail.

The bag cell neurons do not normally display any spontaneous electrical activity. In response to transient stimulation of an input from another ganglion, however, the cells depolarize and fire a long-lasting *discharge* of action potentials (Fig. 18–9b). Although the stimulus lasts only a few seconds, the evoked discharge usually persists for about 30 minutes. At the start of the discharge the neurons fire briskly for about 1 minute, after which they settle down to a slower period of firing, during which the action potentials become enhanced in height and width. When a discharge occurs in an intact animal, it is followed by a stereotyped sequence of behaviors. If the animal is feeding, it abandons its food. It then seeks out a vertical substrate such as the side of a rock, and begins a characteristic sequence of head movements before depositing its eggs on the rock. These behaviors occur because of the action of neuropeptides released from the bag cell neurons during the discharge.

In Chapter 8, we discussed the structure of the precursor protein from which the neuroactive peptides are cleaved in the bag cell neurons (Fig. 8–6b). The major peptide released by the bag cell neurons is *egg-laying hormone* (ELH), which, when injected into animals, induces egg laying and its associated behaviors. During a discharge, ELH and the other smaller peptides cleaved from the precursor are released locally onto other neurons in the abdominal ganglion, as well as into the bloodstream. The released ELH then induces a change in the intrinsic electrical properties of multiple neurons, including those in central pattern generators for locomotion and feeding (Fig. 18–9b).

The onset of this discharge is produced by coordinated changes in the activity of at least three different ion channels, each of which is regulated by protein kinases. As in the reticulospinal neurons, the drive for depolarization is produced by activation of calcium-activated TRP-like nonselective cation channels. The prolonged nature of the depolarization, however, occurs because of the cyclic AMP–dependent protein kinase and a tyrosine phosphatase, which are both activated on stimulation and act on the

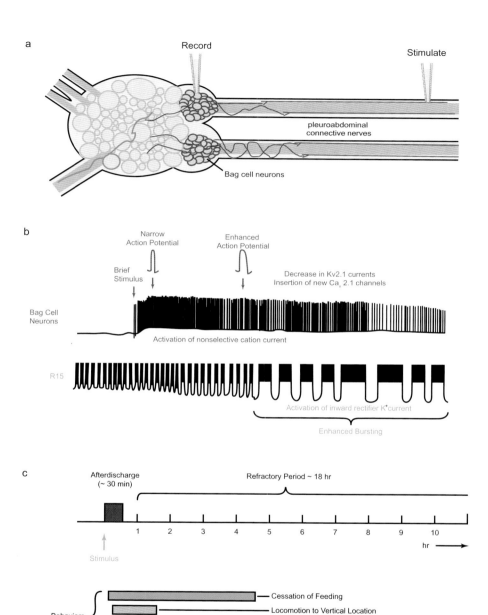

Figure 18–9. The bag cell neurons. *a*: Diagram of the location of bag cell neurons in the abdominal ganglion of *Aplysia*. *b*: A discharge in bag cell neurons and the effects of released ELH on the pattern of endogenous bursting in another neuron, termed *R15*, in the ganglion. *c*: Time scale of changes in excitability of the bag cell neurons and in associated behaviors (Conn and Kaczmarek, 1990).

channel to produce a long-lasting increase in open probability (Fig. 18–10a, b)

As the neurons begin to fire, the width of the action potentials increases. This can be attributed to the action of the cyclic AMP–dependent protein kinase on delayed rectifier potassium channels (Fig. 18–10c, d). Some of the peptides released by the bag cell neurons themselves act

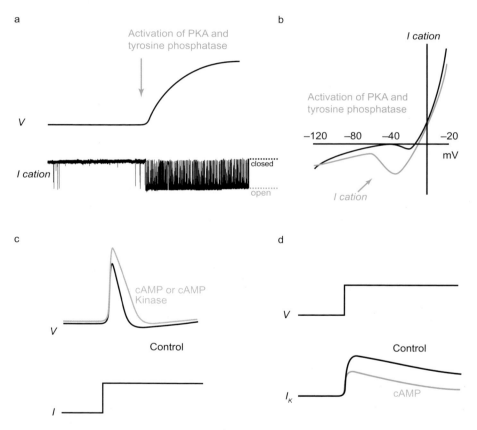

Figure 18–10. Prolonged depolarization and action potential broadening in bag cell neurons. *a*: Activation of the cyclic AMP–dependent protein kinase or a tyrosine phosphatase increases the opening probability of a nonselective cation channel (*bottom*), producing a depolarization of the cell membrane (*top*). *b*: Voltage dependence of the nonselective cation current. *c*: Amplitude and duration of action potentials, evoked by a depolarizing current pulse, increase when a cell is injected with cyclic AMP analogs or the catalytic subunit of the cyclic AMP–dependent protein kinase. *d*: In voltage clamp recordings, spike modulation is accompanied by changes in the amplitude of the delayed rectifying potassium current

through *autoreceptors* to elevate cyclic AMP levels in the neurons from which they are released. Treatment with these peptides or with cyclic AMP analogs, or direct injection of the catalytic subunit of the cyclic AMP–dependent protein kinase, broadens the action potential, and the increase in width can be prevented by injection of inhibitors of this kinase.

Finally, new voltage-dependent calcium channels are inserted from a pool of intracellular vesicles into the plasma membrane, leading to the formation of new sites of calcium entry. This results from the action of protein kinase C, another enzyme activated on stimulation. Treatment of isolated neurons with activators of protein kinase C or direct intracellular injection of protein kinase C causes an increase in action potential amplitude similar to that observed at the onset of discharge (Fig. 18–11a). The increase in action potential amplitude is produced by an increase in the calcium current, which is a major contributor to the rising phase of the spike (Fig. 18–11b). When the microscopic mechanism of this change is examined by single channel analysis, it is found that the increase in calcium current involves the recruitment of a novel calcium channel to the plasma membrane. In control neurons, the calcium current is carried by a class of voltage-dependent calcium channels with a single channel conductance of about 12 pS. After activation of protein kinase C these channels are still present. In addition, there is a new 24-pS calcium channel that is never seen in control cells (Fig. 18–11b, c). Imaging experiments have tracked the movement of these $Ca_V2.1$ voltage-dependent calcium channels to plasma membrane at the distal edge of the terminals of isolated neurons. Experiments using the fura-2 technique to measure the distribution and levels of intracellular calcium (see Chapter 9) reveal that newly inserted channels produce new sites of calcium entry following activation of protein kinase C (Fig. 18–11d). Thus the combined action of the cyclic AMP–dependent protein kinase and protein kinase C is to produce a rapid morphological and functional rearrangement of the terminals to enhance neuropeptide release.

At the end of the 30-minute discharge, there is yet another change in the intrinsic excitability of the neurons, and it is not possible to stimulate another long-lasting discharge. Recovery from this period of inhibition, sometimes termed the *refractory period* (not to be confused with the action potential refractory period described in Chapter 3), occurs gradually over about 18 hours. The decrease in excitability during the refractory period is associated with a persistent increase in BK calcium-activated potassium channel activity, and stimulation at this time also fails to activate the nonselective cation channel required for the prolonged depolarization. As the sequence of behaviors triggered by a discharge can last for several hours, the bag cell neurons are in the refractory state during these

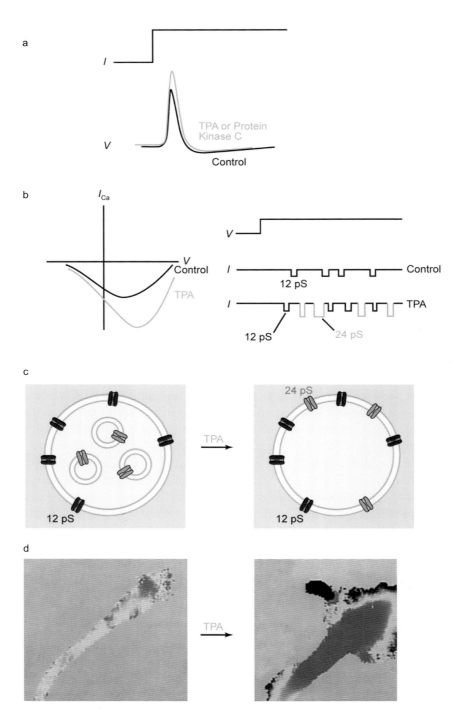

a

I

V

TPA or Protein
Kinase C

Control

b

I_{Ca}

V

Control

TPA

V

I   Control
12 pS

I   TPA
12 pS   24 pS

c

24 pS

12 pS   12 pS

TPA

d

TPA

Figure 18–11.

behaviors (Fig. 18–9c). Thus, the prolonged refractory period may prevent the reinitiation of the behavioral sequence once it is underway, and also serves to limit the frequency with which the behaviors can be evoked. Thus, by undergoing a sequence of changes in their endogenous properties, the bag cell neurons act as a sophisticated master switch for the sequence of behaviors leading to egg laying.

Clocks in the Brain: The Generation of Circadian Rhythms

Neurons that change their intrinsic electrical properties to secrete neuropeptides that trigger or promote coordinated sets of behaviors are found in many regions of the vertebrate nervous system. Among these are neurons of the *suprachiasmatic nucleus*, a small nucleus of the hypothalamus. These cells are the master controllers of *circadian rhythms* both in the brain and in peripheral organs. Their activity influences the sleep–wake cycle, locomotion, and all other physiological and hormonal behaviors coupled to the 24-hour rhythm. During the night, they are silent or fire infrequently. During the day, however, their intrinsic excitability is altered so that they generate continual spontaneous firing. This change in their intrinsic firing rate is brought about by a cycle of transcription of new proteins that are synthesized and destroyed on a daily basis. ❽

For Gathering Information, Timing Is Everything

For some neural networks, the exact timing of action potentials is not critical. For example, in the bag cell neurons and suprachiasmatic neurons just described, the occurrence of a train of action potentials triggers a

Figure 18–11. Enhancement of calcium current in bag cell neurons by protein kinase C. *a*: Protein kinase C injection or treatment with a phorbol ester (TPA) that activates this enzyme increases the amplitude but not the duration of the action potential. *b*: TPA increases voltage-dependent calcium current (*left*). This is due to the recruitment of a novel calcium channel, which is not seen in the absence of TPA (*right*). *c*: The recruitment of the new calcium channels from intracellular vesicles is illustrated (see Zhang et al., 2008). *d*: Images of intracellular calcium levels measured using the Fura-2 indicator in the same growth cone terminal of an isolated bag cell neuron before and after activation of protein kinase C, which causes the insertion of new calcium channels and an increase in size of the terminal (Knox et al., 1992). Blue, yellow, and red indicate low, intermediate, and high calcium levels, respectively.

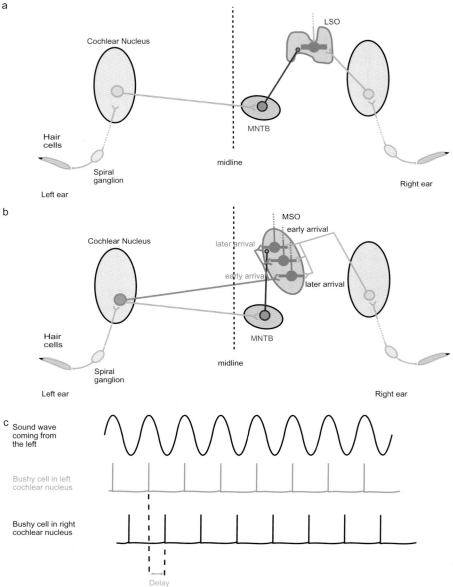

Figure 18.12 Phase-locking neurons. *a*: Connections to the lateral superior olive (LSO) are from excitatory bushy cells on one side and inhibitory medial nucleus of the trapezoid body (MNTB) neurons from the other. *b*: Pathways connecting auditory stimulation to the medial superior olive (MSO). *c*: Many auditory neurons lock their action potentials to a specific phase of a sinusoidal sound stimulus. The delay between two phase-locked action potentials in different neurons can be used to compute parameters such as the direction of a sound source.

behavior, but individual action potentials probably hold no significance. In other networks, however, the precision with which an action potential fires is of utmost importance. One of the clearest examples of this occurs in neurons in the brainstem that calculate differences in the sounds that arrive at the two ears.

Action potential timing in auditory brainstem networks. In Chapter 13, we emphasized that while most sensory systems use relatively slow second messenger pathways for transduction, the mechanics of hair cells in the cochlea allows them to respond to sounds very rapidly. The neurons that receive this information must also respond equally rapidly. Transmitter release from the cochlear hair cells activates neurons in the *spiral ganglion*, located in the cochlea itself. These, in turn, send their axons into the brain, where they terminate on a variety of neurons in the *cochlear nucleus* (Fig 18–12a, b). Among these is a class of neurons termed *bushy cells*, which have the ability to *phase-lock* their action potentials to a sound stimulus. When a sinusoidal sound with a frequency of several hundred Hz is presented to the cochlea, these neurons follow the stimulus, firing their action potentials at a very precise point during each sine wave (Fig. 18–12c). Our ability to hear accurately requires two things of these cells. First, they must be capable of firing at extremely high rates (up to 500–1000 Hz). Second, the timing of each action potential must be accurate to within a few microseconds. This accuracy of timing is required even at higher frequencies of sound, when the neurons fire action potentials during only some of the individual waves.

Probably much of what we hear and interpret, such as speech and music, requires this degree of accuracy. A particularly clear example, however, is our ability to determine the source of a sound in space. This is accomplished by neuronal networks that compare the difference in time of arrival of sounds at the two ears, as well as their relative intensities. Phase-locking neurons are essential for this. For example, one component of the network is the *lateral superior olive* (LSO) (Fig 18–12a). The LSO receives an excitatory input from bushy cells on the same side of the brain. The input from the contralateral side, however, is an inhibitory one. Rather than sending their axons directly to LSO neurons on the opposite side, the bushy cells make a connection onto neurons in the *medial nucleus of the trapezoid body* (MNTB). These neurons, which use the inhibitory neurotransmitter glycine, then continue the signal to the LSO. This integration of the excitatory inputs from one side and inhibitory inputs from the other computes the direction of a sound stimulus, using both the intensity and timing of sounds as cues.

Another comparison of inputs from left and right occurs in another auditory nucleus, the *medial superior olive* (MSO), which receives excitatory input from bushy cells on both sides of the brain, that is, from both ears (Fig. 18–12b). A neuron in the MSO may fire when, and only when, the timing of excitatory inputs it receives from each ear is exactly balanced. Thus some MSO neurons will fire only when a sound stimulus occurs directly in front of a subject. The time delay between the firing of the bushy cell and the postsynaptic potential in the dendrites of MSO neurons is, however, carefully regulated so that the delay is slightly different for each MSO neuron. Thus each MSO neuron receives inputs a little earlier or later than its neighbor. For some MSO neurons, the timing of inputs is balanced only when a sound arrives earlier at the right ear, while for others, simultaneous activation of its two inputs occurs when a sound originates on the left. In fact, there appears to be an orderly anatomical arrangement of "delay lines" such that the position of a sound stimulus can be mapped to specific locations in the MSO (Fig. 18–12b). In lower vertebrates these delays are caused simply by the different lengths of the axons from the cochlear nucleus to different parts of their MSO homolog. In mammals, however, precisely timed inputs from the MNTB appear to be required to produce the progressive delay in the transfer of information from the bushy cells to the MSO neurons.

What are the cellular specializations that allow these networks to function? Remember that these networks have to process microsecond timing differences that are a fraction of the time it takes sound to travel from one side of the head to the other, shorter than the duration of an action potential. We encountered one of these specializations in Chapter 9. The synaptic endings onto the bushy cells (termed *the end bulbs of Held*) and those onto the MNTB neurons (termed the *calyces of Held*—see Fig. 9–5) are the largest synapses in the central nervous system. These are able to generate enormous excitatory postsynaptic currents in the postsynaptic bushy cells and MNTB neurons, providing the rapidity and accuracy of transmission that is required. A specific form of glutamate receptor of the AMPA subclass, which activates and inactivates very rapidly, is used at these excitatory synapses.

The intrinsic excitability of each of the neurons in these networks also has to be optimized for accuracy and for speed. Potassium channels within the bushy cells and MNTB neurons are essential for rapid responses. One set of potassium channels is open near the resting potential to ensure that the membrane time constant (see Chapter 3) is very small, allowing the cells to respond very rapidly to stimulation. These channels also ensure that only a single action potential is evoked by each synaptic stimulus. A second set of channels allows the cells to fire at very high rates with little or no relative refractory period (see Fig. 3–6). When these channels, delayed rectifier potassium channels termed *Kv3.1*, are eliminated by genetic manipulation, the cells are unable to follow high-frequency stimuli (Fig. 18–13a).

Perhaps as important as accuracy is flexibility. For example, the reader is most likely studying this book in a very quiet environment, such as a bedroom, before going through noisy traffic to a rock concert. Each of these environments produces a very different pattern and intensity of impulses propagating through the network. Accurate processing of time differences is required, however, in all of these conditions. It appears that different sound environments alter the intrinsic excitability of the neurons by altering the phosphorylation state of channels. For example, moving from a quiet place to a noisier environment comparable to that in a trendy restaurant causes the dephosphorylation of the Kv3.1 channels (Fig. 18–13b). This increases Kv3.1 potassium current and allows the cells to fire at higher rates. Thus the electrical and morphological characteristics of the bushy cells and MNTB neurons, and their ability to be modulated by different environments, are essential for their ability to process auditory information. The system simply would not work the same way if one were to substitute a different set of channels and synaptic components into these neurons.

Spike timing–dependent plasticity. The outputs of each of the networks that we have discussed thus far have been honed through evolution to generate different patterns of motor output or to respond appropriately to sensory inputs. In such cases, it is easy to see that the intrinsic electrical properties, such as the ability to generate patterned bursts of action potentials, can be adapted to shape the eventual behavior of an animal. Having the right types of channels, receptors, and synaptic structures is, however, equally important for stimulation-induced alterations in synaptic transmission.

In Chapter 17 we discussed Hebb's rule and found that, during development, stabilization of synapses occurs when presynaptic action potentials are linked in time to action potentials in the postsynaptic neuron (see Fig. 17–11). As we shall see in Chapter 19, changes in synaptic strength in adults are subject to the same sorts of rules. Simply linking the pre- and postsynaptic action potentials is, however, not sufficient. Their relative timing is critical. For example, at many connections, the presynaptic action potential must be generated a few milliseconds before the postsynaptic action potential for the synapse to become strengthened. If presynaptic action potentials occur immediately after postsynaptic spikes, the connections from those presynaptic neurons may actually become depressed or

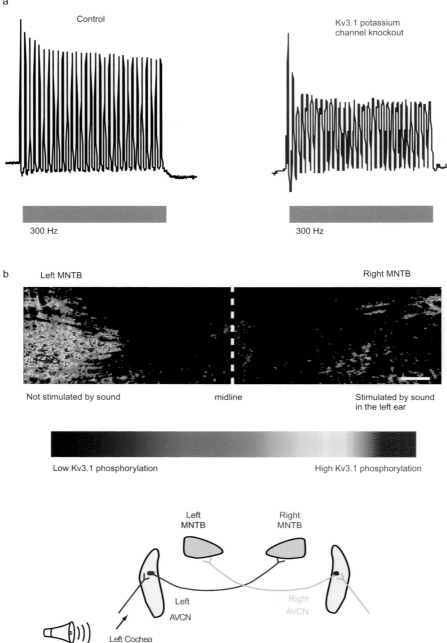

Figure 18–13. Medial nucleus of the trapezoid body (MNTB) neurons. *a*: Firing pattern of a control MNTB neuron and one in which a delayed rectifier potassium channel, termed the *Kv3.1 channel*, has been deleted by genetic manipulation.

eliminated. This makes sense biologically because, in general, only terminals that fire before their postsynaptic cell fires are likely to be part of the neuronal circuit that is successfully activating the postsynaptic target.

The rule that the precise timing of action potentials determines whether a synapse will be stabilized or suppressed is called *spike timing–dependent plasticity*. While many synapses follow rules of spike timing–dependent plasticity, not all synapses follow the simple Hebbian rule. At some well-studied central nervous system synapses, spike timing–dependent plasticity can be *anti-Hebbian*, resulting in the depression of connections from neurons that fire immediately before their targets. This topic will be covered again in the next chapter. The point here, however, is that, whichever rule the synapse obeys, the precise timing of both pre- and postsynaptic action potentials has to be regulated very carefully. The number and the timing of action potentials that are generated by a single stimulus is an intrinsic electrical property of a neuron. Moreover, this intrinsic property is changed through modulation of ion channels.

Dendritic action potentials. How does a synaptic connection actually know that it has been successful in triggering a postsynaptic spike? This is not a trivial question. Most synaptic connections in the brain are made on dendrites that, in electrical terms, are quite far from the cell body and the initial segment where the axonal action potential is generated. Without some means of passing information about how the whole neuron has responded, how does a remote dendrite know that it should stabilize one synaptic input or suppress another?

The answer is that the intrinsic electrical properties of the dendrites are as important as those of the soma or axon in shaping neuronal responses and allowing the neuron to change its synaptic connections. Sodium channels are distributed uniformly throughout the dendritic trees of many neurons (Fig. 18–14a). Calcium channels, nonselective cation channels, and a wide variety of potassium channels also exist in dendrites.

Figure 18–13. (Continued) When stimulated at 300 Hz, the control neuron responds to every stimulus, while in the knockout, full action potentials are generated only at the onset of stimulation (Macica et al., 2003). *b*: When animals are kept in a quiet environment, most Kv3.1 channels in MNTB neurons are phosphorylated. The image shows the effect of 5 minutes of sound stimulation (a series of clicks at 600 Hz) applied at a comfortable sound level (70 dB sound pressure level) to the left ear only (*bottom*). This produces dephosphorylation of Kv3.1 channel in neurons of the stimulated (*right*) MNTB. Phosphorylation was detected with an antibody that detects only phosphorylated channels (Song et al., 2005). AVCN, anterior ventral cochlear nucleus.

Some of these are located in a very specific spatial pattern. For example, the levels of the Kv4.2 channel, which produces an inactivating A-current in the dendrites, increase about seven-fold as one moves from the soma toward the most distant dendrite.

   The existence of dendritic channels has three important consequences (Fig. 18–14a). First, once an action potential has been trigged at the soma, it can propagate back into the dendrites, even into dendrites that were not strongly activated by the stimulus. Such *backpropagating* action

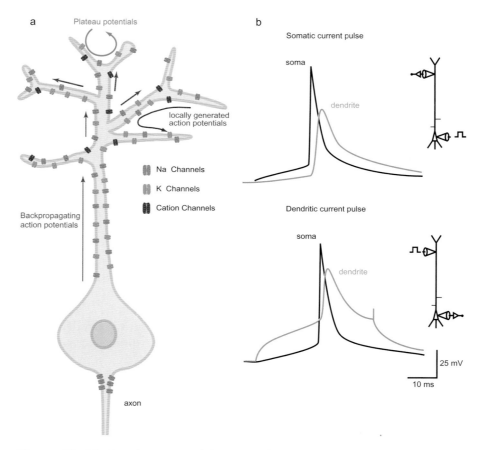

Figure 18–14. Dendritic excitability. *a*: Sodium, potassium, calcium, and non-selective cation channels are widely distributed throughout the dendrites of many neurons. *b*: Recordings of backpropagating action potentials made by placing electrodes on the soma and dendrites of the same neuron. An action potential travels from the soma to the dendrites regardless of whether a depolarization is first applied to either the soma (*top traces*) or the dendrites (*bottom traces*) (Stuart and Sackmann, 1994).

potentials send a signal back into the dendrites to stabilize or inhibit recently activated synapses by producing an elevation of postsynaptic calcium that is linked in time to presynaptic activity. Backpropagating action potentials have been detected both by imaging techniques and by electrical recordings (Fig. 18–14b). Secondly, action potentials can be generated locally within a dendritic tree, without propagating all the way to the cell body. This may allow local stabilization of a set of synaptic inputs on one dendritic branch without influencing other branches. Finally, dendrites can produce their own local plateau potentials, oscillations, or other rhythmic patterns.

Perhaps the most important conclusion from these findings is that regulation of the intrinsic excitability of dendrites by phosphorylation or other mechanisms influences both the timing and the amplitude of dendritic depolarization. For example, levels of dendritic Kv4.2 channels in dendrites and spines of hippocampal neurons are continually regulated by activity in the neurons, both by FMRP-regulated local synthesis (see Chapter 17) and by insertion and removal from the plasma membrane. Modulation of these and other channels will, in turn, determine whether the neuron decides to make a permanent change in the group of fellow neurons it will continue to listen to. ❧

Summary

In previous chapters we discussed the properties of individual neurons, or pairs of neurons connected by a single synapse. More complex interactions among larger numbers of neurons are, however, required to generate most behaviors. Studies in biological model systems, such as the stomatogastric ganglion of lobsters and crabs, circuits for swimming in leeches and lampreys, and neurons controlling reproduction in *Aplysia*, have provided insights into how the intrinsic electrical properties of neurons shape network activity and animal behavior. Some neurons can participate simultaneously in more than a single network, and the properties of a network may be modulated by the actions of neurotransmitters and hormones. Changes in the intrinsic excitability of a single command neuron or command systems of neurons can trigger a complicated and long-lasting behavior. The cellular mechanisms that regulate the accuracy of timing of action potentials, both within a network and in different parts of a dendritic tree, are also important for the interpretation of sensory information and for the ability of a neuron to modify the strength of the connections it makes with other neurons. In the next chapter we shall consider some neuronal properties that may be required for learning and memory.

19

Learning and Memory

S ome of the networks we covered in the previous chapter regulate *fixed-action patterns*, behaviors that always occur in a fixed and stereotyped manner once they are triggered. The trigger for a particular fixed-action pattern may arise from within the animal, as, for example, in the case of a chemical cue that appears at a certain time during development. Alternatively, fixed-action patterns may be triggered by a specific set of environmental conditions. It is thought that the neural circuitry underlying these invariant behaviors is more or less *hard-wired*, that it is specified by the genome and is only to a limited extent subject to modulation. Many, but by no means all, behaviors in invertebrates and lower vertebrates tend to fall into this hard-wired category. We have seen that modulation by neurotransmitters allows a considerable variety of behavioral outputs to be generated by a single hard-wired circuit.

This pattern changes as we move up the phylogenetic tree. Although many behaviors, particularly rhythmic ones such as breathing and locomotion, remain essentially hard-wired in higher vertebrates, including humans, many more examples of *adaptive behavior* begin to appear. In this chapter we will consider two closely related behavioral phenomena that are crucial for animals to survive: learning and memory. We may define *learning* in very broad terms as *a change in behavior as a result of experience*, and *memory* as *the ability to store and recall learned experiences*.

How the brain encodes, stores, and retrieves memories has fascinated not only scientists but the lay public for thousands of years. There is good reason for this broad interest in learning and memory. They are essential ingredients in defining an animal (or human being) as an individual. In addition, the complexity of the nervous system makes the understanding of such higher functions a challenging and exciting intellectual goal.

Accordingly, many scientists from different disciplines have devoted their careers to the study of learning and memory, including the search for the *engram*, the physical memory trace in the brain; and technical and conceptual advances during the last few decades have produced substantial and exciting progress. An in-depth treatment of the various forms of learning and memory that have been defined and characterized by neuropsychologists is well beyond the scope of this book. We will focus here on several simple kinds of behavioral phenomena, the mechanisms of which are beginning to be clarified with the techniques of cell and molecular biology.

Different Kinds of Learning and Memory

Before proceeding further we must define several different classes of behavioral modification exhibited by nervous systems. Historically, it has been useful to divide learning into two categories: *nonassociative* and *associative* learning (Fig. 19–1). Three different types of memory can also be distinguished, *sensory, short-term*, and *long-term* memory.

Nonassociative learning: habituation and sensitization. Habituation and sensitization are two simple forms of learning that involve a change in the intensity of response to a stimulus. *Habituation* can be defined simply as a *decrement* in the behavioral response during repeated presentations of the same stimulus. It is a form of learning that is observed in invertebrates, and in all vertebrate species, including humans. An example might be the diligent student of cellular and molecular neurobiology who is trying to study when he or she is interrupted by some distracting noise, for example, a radio playing in the next room. Although initially the stimulus—the noise—interferes with the ability to concentrate, after many repetitions of the noise, the nervous system for all intents and purposes stops responding, and the contemplation of ion channels can resume.

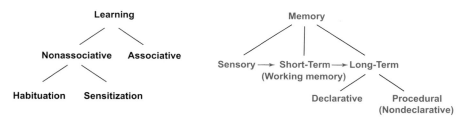

Figure 19–1. Categories of learning and memory.

Sensitization is functionally the opposite of habituation, in that it can be defined as the *enhancement* of a reflex response by the introduction of a strong or noxious stimulus. It also differs from habituation in that the sensitizing stimulus is different from the stimulus that elicits the reflex. Furthermore, there is no specificity—there is a general arousal of the nervous system, and all reflex pathways are strengthened. To carry our analogy further, if the diligent student is then startled by a false-alarm fire bell, he or she may subsequently be more distracted by the radio than when it was first turned on. Sensitization has important adaptive value, in that it allows a novel stimulus to alert an animal to possible predators and other potentially harmful stimuli in its environment.

Associative learning. Associative learning is more complex than habituation or sensitization, in that two stimuli must be closely associated in time for learning to occur. When a normally ineffective or neutral *conditioned* stimulus is paired temporally with a meaningful *unconditioned* stimulus, the animal learns to respond to the former as if it were the latter. The archetypal example of such *classical conditioning* is that of Pavlov's dogs, who learned to associate the ringing of a bell (the conditioned stimulus) with the presentation of food (the unconditioned stimulus), and would salivate in response to the bell alone. The unconditioned stimulus might be *reinforcing*, as in the previous example, or *aversive*, for example, an electric shock, in which case the animal will take the conditioned stimulus as a cue and attempt to escape the aversive unconditioned stimulus associated with it. In contrast with nonassociative learning, which does not involve any temporal relationship between stimuli, associative conditioning allows the animal to draw conclusions about causal relationships in its environment. There is another category of conditioning with different characteristics, called *operant conditioning*, which we will not consider here.

Sensory, short-term, and long-term memories: different mechanisms. Another important concept, which has arisen from studies in invertebrates and a variety of vertebrates including humans, is that there are at least three temporally distinct forms of memory (Fig. 19–1). The most fleeting form, lasting from less than a second to only a few seconds, is *sensory memory*. This is the memory required to interpret sensory information. For example, to understand a spoken word one must remember the first syllable until all the other syllables have been heard. This form of memory is not under conscious control, but is essential for cognition and is required for any subsequent form of learning. Because sensory memory is so rapid and works differently for each sensory modality, little is known in general about the cellular and molecular properties that are required for neurons to encode this form of memory.

Short-term memory is the ability to acquire new information and retain it for periods of time ranging from a few seconds to some minutes. The term *working memory* is very closely related to short-term memory and is generally used when an active attempt is being made to keep information in short-term memory. An example in real life involves looking up a new telephone number and retaining it in short-term memory (often with rehearsal, for example, by repeating it continuously) long enough to dial it. A few minutes after dialing, the number will no longer be available in memory. In contrast, *long-term memory* can involve retention for hours, days, years, or even a lifetime. In the previous example, it is only those numbers that are dialed (and hence rehearsed) often that eventually find their way into long-term memory storage. Both associative and nonassociative learning have short-term and long-term components.

Finally, in our efforts to understand the neural basis of learning, it is important to distinguish among different kinds of long-term memory (Fig. 19–1). *Procedural* or *nondeclarative* knowledge includes memory for skills and procedures—knowing *how*—for example, learning to ride a bicycle, play the piano, knowing the rules of a game. In contrast, *declarative* knowledge involves the memory of specific facts or events—knowing *that*—for example, remembering that you sat in the front row of the opera a week ago. As we shall see later, the memories produced by procedural and declarative learning are laid down and stored by neurons in different parts of the brain.

To understand any type of learning we need to know two things: (*1*) where are the neurons and synapses whose properties are changed by learning and, (*2*) what are the cellular and molecular events that produce the change? It has been evident for a long time that the answers to these questions are quite different for each type of learning. For example, damage to parts of the brain caused by accidents or strokes can selectively disrupt short-term versus long-term memory or procedural versus declarative memory. Electroconvulsive shock selectively prevents the setting down of long-term memory traces. Similarly, inhibition of protein synthesis has no effect on the acquisition and short-term retrieval of a new behavior, but it does inhibit long-term memory. We shall now describe approaches that have been and continue to be used to identify the cells that participate in the memory trace.

Organization of Memory in the Brain:
The Search for the Engram

The difficulty of locating the memory trace has been the greatest barrier to progress in understanding molecular mechanisms of learning and memory. The quest for the physical basis of the memory trace can be pursued at many levels of organization. The first step is localizing the engram to a particular organ. Although we now take it for granted that the brain is the right organ, there is evidence that the early Egyptians thought the heart and liver were the seat of human emotions and behavior. This conclusion was based on simple kinds of experiments—for example, if you remove the heart from an animal, it stops behaving. By the time of Hippocrates, a millennium later, it was recognized by many investigators that organs like the heart are necessary simply to keep the brain alive (kidney chauvinists may object to this simplification), and at the end of the nineteenth century Santiago Ramón y Cajal expressed the central role of the brain, in lyrical and eloquent terms:

> To know the brain is the same thing as knowing the material course of thought and will, the same thing as discovering the intimate history of life in its perpetual duel with eternal forces, a history summarized and literally engraved in the defensive nervous coordination of the reflex, the instinct, and the association of ideas.

But localization of the memory trace to the brain was only the beginning. Narrowing things down further to a particular brain region, and to individual neurons within that region, has proven to be a much more formidable task.

Contributions of human amnesic patients and imaging techniques. Many of the early insights into brain regions required for specific types of learning came from the study of human subjects who are amnesic as a result of accidental or surgical brain damage. Declarative knowledge cannot be acquired by these amnesic patients. In contrast, procedural learning, which depends on many different kinds of information that are processed and localized separately in different brain regions, can often still be acquired even by severely amnesic human subjects. Some of the most striking evidence in support of a different mechanism for these two types of learning comes from studies with amnesic patients who can remember new factual information for only a very short time and cannot

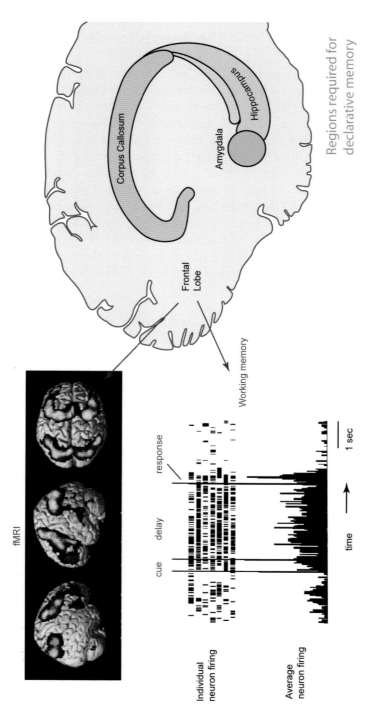

Figure 19–2. Structures in the brain important for working memory and long-term declarative memory. Diagram at *right* after one drawn by Larry Squire, depicting the medial aspect of the human brain to illustrate structures important in the setting down of long-term memories. fMRI images at *top left* indicate regions of cerebral cortex activated during a working memory task (Schöning et al., 2009). *Bottom left* shows recordings made by the laboratory of Pat Goldman-Rakic demonstrating neuronal firing during a working memory task (Funahashi et al., 1989).

store new long-term declarative memories. However, when such patients are taught to read words in a mirror, they learn to carry out this complex task at a normal rate, and they retain the mirror-reading skill when retested months later. Interestingly, this (procedural) skill is retained, although they do not remember the specific words themselves, or even the (declarative) fact that they have ever been trained to perform the task. When asked why they are able to perform so well, they may reply that they are "just good at that sort of thing."

It is now evident from anatomical studies, either at autopsy or using in vivo imaging techniques, that many of these amnesic patients have suffered damage to the *limbic system*, a group of cortical structures that includes the *amygdala*, the *hippocampus*, and anatomically related structures (Fig. 19–2). When similar lesions are produced surgically in nonhuman primates, similar defects in declarative (but not in nondeclarative) memory tasks are observed. A wide variety of other animal studies have also implicated the hippocampus in certain kinds of learning and memory. It is believed that the long-term memory traces themselves are not stored there but rather that the hippocampus participates in memory acquisition and in establishing an enduring and retrievable memory elsewhere. Furthermore, the role of the hippocampus is limited to the acquisition and storage of declarative knowledge.

Another approach to identifying regions of the brain involved in the acquisition of different types of memories is the use of *in vivo* imaging techniques, such as *functional magnetic resonance imaging* (fMRI). This technique follows changes in local blood flow to define brain regions in which neuronal electrical activity changes during learning. It is in widespread use in clinical medicine and can be used to compare changes in blood flow in normal people and those with disorders of learning. Figure 19–2 shows that during a working memory task in humans (remembering if a letter of the alphabet is the same as one presented two trials earlier), a brain region that is strongly activated is the frontal cortex. Recordings of action potential firing in neurons in the frontal cortex of monkeys carrying out a similar task (remembering a point in space where they previously saw a point of light) have shown that many neurons fire repetitively during the entire time that the animal is holding onto the information (Fig. 19–2). Once the information is no longer needed (when the animal receives a reward for remembering the position accurately), the neurons stop firing. The activity of these neurons may be considered similar to that of gating neurons we covered in Chapter 18.

Similar combined lesioning/imaging/electrophysiology experiments can be applied to understand regions involved in other learning tasks. For example, the cerebellum is involved in certain kinds of learned behaviors, particularly *motor learning*. One analysis has been carried out on a learned eyeblink response in the rabbit. Through lesioning studies it has been found that a portion of the cerebellum, on the opposite side of the brain from the trained eye, is essential for the learning and retrieval of the response. Another kind of learning that has been localized to the cerebellum is adaptation of the *vestibulo-ocular reflex*, a compensatory eye motion in response to turning of the head. ❷

Model Systems for the Cellular and Molecular Analysis of Learning and Memory

Since the time of Ramón y Cajal and Sherrington it has been widely accepted that the synapse must be an important site of neuronal plasticity, and that plastic changes in synaptic efficacy might contribute to behavioral plasticity. It has been known for more than 50 years that certain synapses in the peripheral nervous system can undergo changes in synaptic efficacy that can last for seconds or minutes (see Chapter 9). Such changes, in sympathetic ganglia and at the neuromuscular junction, have been thoroughly studied and are fairly well understood. But what does this have to do with learning and memory? One cannot claim that the neuromuscular junction or the sympathetic ganglion has the capacity to "behave." In this case, the synapse that is modulated is accessible for study, but there is no behavioral correlate.

Thus it will be evident that workers interested in mechanisms of learning and memory are faced with a dilemma. The ideal model system would be a two-neuron/one-synapse organism with the behavioral repertoire of humans. Needless to say, this is not available. Accordingly, many investigators have compromised on both sides of the issue and have chosen model systems that exhibit reasonably sophisticated behavioral plasticity, yet have reasonably accessible nervous systems. We shall now discuss some of these models.

Model Systems I: Fruit Flies Can Learn to Avoid Odors: The Genetic Approach

The rationale for investigating animals whose genetic makeup can easily be manipulated is simple and straightforward: If one can generate mutants that exhibit aberrant patterns of behavior, an examination of the mutated gene might provide clues to the molecular mechanisms underlying the behavior. This approach is valid even if the precise location of the engram is not known. Such genetic studies of memory have focused on the fruit fly *Drosophila* and the nematode worm *Caenorhabditis elegans,* because their genetics are better understood than those of any other multicellular organism. We shall focus here on *Drosophila*, which can undergo both associative and nonassociative forms of learning, because a large and growing series of behavioral mutants of different types exists (Table 19–1). As an example, we will discuss the first of these behavioral mutants to be isolated, a stupid fly called *dunce.*

Fruit flies like bananas. If, however, the smell of a banana is followed by a strong electric shock, a fly learns, quite reasonably, to avoid this odor, but not to avoid other odors. A *dunce* fly can learn to associate a particular odorant with an electric shock, but retains this information for only a remarkably short time. *Dunce* was isolated when flies were treated with a chemical mutagen and their offspring screened in this behavioral assay. *Dunce* has normal sensory and motor capacities. Nevertheless, when the time course of memory decay is examined (Fig. 19–3), it is found that its short-term memory decays unusually rapidly, and there is little sign of long-term memory. However, when some low level of residual long-term memory is detected, it appears to decay at the same rate as in normal flies. This suggests that the primary effect of the *dunce* mutation is to severely

Table 19–1 Selected Examples of *Drosophila* Mutants That Exhibit Defects in Learning Memory

Mutant	Behavioral Phenotype	Gene Product or Affected Pathway
Dunce	Rapid decay of short-term memory	cAMP phosphodiesterase
Rutabaga	Rapid decay of short-term memory	Adenylate cyclase
Amnesiac	Defective memory	Neuropeptide modulator of adenylate cyclase
DCO	Defective short- and intermediate-term memory	cAMP-dependent protein kinase
Turnip	Defective acquisition and retention	Modulator of PKC
CREB	Defective long-term memory	cAMP and calcium response element binding protein
Leonardo	Defective Pavlovian learning	14-3-3, a signaling protein scaffold

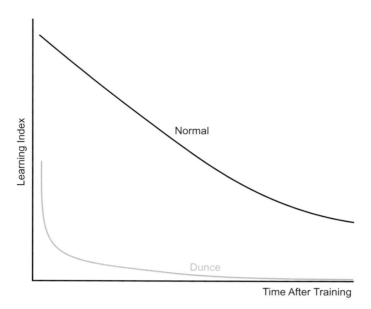

Figure 19–3. Odorant-based shock avoidance in *Drosophila*. Decay of memory as a function of time after training, in normal flies and those carrying the *dunce* mutation. (After a drawing by Tully, 1987.)

attenuate short-term memory, and the loss of long-term memory may be a secondary effect.

What is the protein encoded by the *dunce* genetic locus? The surprise is that it is not a new protein reserved for higher nervous system functions, but part of one of the most common and ubiquitous signaling pathways throughout the entire body. The *dunce* gene codes for a form of the enzyme *cyclic AMP phosphodiesterase* (see Fig. 12–9), which contributes a large percentage of the cyclic AMP hydrolyzing activity in *Drosophila*. The levels of cyclic AMP in dunce are as much as six-fold higher than those in normal flies, which suggests that the ability to signal through cyclic AMP is in some way linked to memory formation. This suggestion is strongly reinforced by the finding that many other memory mutations reside in components of the cyclic AMP signaling pathway (Table 19–1). For example, like *dunce*, the memory mutant *rutabaga* exhibits an abnormally rapid decay of short-term memory in various associative conditioning paradigms. However, in this mutant the phosphodiesterase activity is normal. Instead, *rutabaga* exhibits a defect in a calcium/calmodulin-dependent adenylate cyclase, which is responsible for the synthesis of cyclic AMP (see Chapter 12). Similarly, mutations in the cyclic AMP–dependent protein kinase also disrupt normal memory acquisition.

Other fly learning and memory mutants exist, and not all of them involve defects in the cyclic AMP pathway (Table 19–1). Interruption of other signaling pathways can also disrupt various features of memory acquisition, consolidation, and recall. Nevertheless, the emerging picture is that cyclic AMP plays an important role in both short-term and long-term memory in the fruit fly (and, we shall see, in other organisms). For example, it is now apparent that regulation of gene expression by cyclic AMP is critical for long-term memory. Remember from Chapter 17 that one of the ways that neuronal activity alters gene transcription is by causing the phosphorylation of the cyclic AMP response element binding protein CREB at a specific serine residue (see Fig. 18–24a). This transcription factor can be phosphorylated both by elevations of calcium (acting through Ca^{2+}/CaM kinase IV) and by the cyclic AMP–dependent protein kinase. In flies that have been genetically engineered to express a blocker of CREB function, long-term memory is specifically disrupted, a finding suggesting that CREB-regulated gene expression may act as a molecular switch for certain kinds of long-term memory.

The next stage for understanding the role of cyclic AMP and the products of the other memory mutant genes is to determine which neurons are used in the learning process and what cellular changes these undergo. Imaging techniques are very helpful here. For example, using the Synapto-pHluorin technique, described in Chapter 8 (Fig. 8–12g,h), one can visualize neurons that are actively releasing neurotransmitter, while indicators of intracellular calcium, similar to those presented in Chapter 9, provide pictures of which neurons are active. Both of these methods have been used to establish where changes occur during olfactory learning.

Figure 19–4a shows a map of the olfactory regions in the fly brain. Odors activate olfactory receptor neurons that project into the *antennal lobe* to make connections with projection neurons through large synaptic glomeruli. Just as in the vertebrate nervous system, the axons of receptor neurons containing the same olfactory receptor protein project to the same glomerulus (see Fig. 13–20). Each odor activates only a subset of the ~43 glomeruli in the antennal lobe. An electric shock to the animal, however, activates all of the glomeruli. When a single exposure to an odorant (the conditioned stimulus, or *CS*) is coupled with a small number of shocks to the animal (the unconditioned stimulus, or *US*), the animal learns to avoid this particular odor in subsequent behavioral tests. Imaging with Synapto-pHluorin has revealed that one of the earliest events produced by this "one-trial" learning is that the number of glomeruli

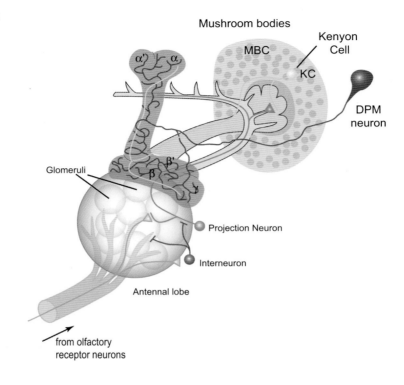

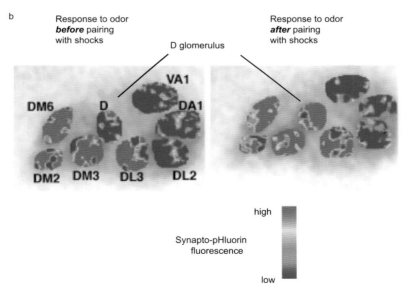

Figure 19–4. Fruit fly olfactory learning. *a*: Diagram of brain regions involved in olfaction (modified from Busto et al., 2010). *b*: Imaging of a sustained increase in neurotransmitter release in a synaptic glomerulus (labeled D) after a single pairing of an odor with electric shocks (Yu et al., 2004).

activated by the CS odor is increased, indicating that the normal pattern of synaptic connections has been altered (Fig. 19–4b).

The increased number of glomeruli activated by the odor after pairing with the shocks represents only a short-term memory trace; the normal pattern of activation by the odor is reinstated within about 7 minutes. The altered behavior lasts longer than this but is gone by 24 hours. Other parts of the brain olfactory system, the *mushroom bodies* and DPM (dorsal paired medial) neurons appear to hold this longer-lasting memory trace. Neurons in the mushroom bodies, termed *Kenyon cells*, receive inputs from the projection neurons in the antennal lobe and also from neurons that use the neurotransmitter dopamine and are activated by the electric shock (Fig. 19–4a). After the pairing, a sustained but rather delayed increase in calcium levels can be detected in a subset of axons of Kenyon cells and in the DPM neurons. This elevation of calcium can last as long as 60 minutes. If the odor is paired with the shocks several times, giving the animal a 15-minute break between pairings, the animal really learns its lesson and avoids the odor for up to a week. Again, it is the mushroom bodies that are required for this long-term olfactory learning, and it is in these neurons that the cyclic AMP machinery, as well as CREB, is needed. The current view is that these neurons are the site at which the coincidence of a specific odor with the electric shocks is converted into the synthesis of a new set of synaptic proteins that alter circuitry in the antennal lobe and elsewhere, leading to a long-lasting aversion to the odor.

Model Systems II: The Gastropod Mollusc *Aplysia*

The virtues of molluscan nervous systems for cellular neurobiology have already been emphasized in different contexts throughout this book. The gastropod molluscs not only provide highly accessible nervous systems with a relatively small number of large, identifiable neurons, they also exhibit a surprisingly varied repertoire of behaviors, including both nonassociative and associative learning. In Chapter 18, we saw how patterns of synaptic connections between identified neurons could explain the control of a variety of behaviors in different creatures. Now we shall discuss the cellular and molecular analysis of short- and long-term memory in the marine snail *Aplysia* (Fig. 19–5). This work was pioneered by Eric Kandel and his colleagues, and was honored by the Nobel Prize in Physiology or Medicine to Kandel in 2000.

Defensive withdrawal reflexes in Aplysia. *Aplysia* can withdraw from strong tactile stimuli, an effect that is analogous to reflex escape

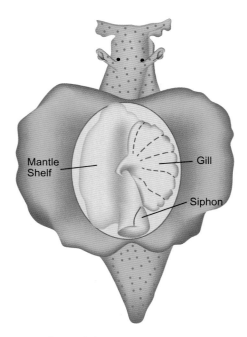

Figure 19–5. Drawing of an adult *Aplysia californica* to show the gill, siphon, and mantle shelf. A tactile stimulus to the mantle or siphon area results in contraction and withdrawal of the gill. (After a drawing by Kandel, 1979.)

and withdrawal observed in vertebrates. The tail withdrawal reflex is a contraction of the tail musculature in response to a stimulus to the skin on the posterior portion of the animal. The gill and siphon withdrawal is a reflex evoked by touching the siphon or mantle shelf (these are external organs of the mantle cavity, a respiratory chamber that contains and protects the gill—Fig. 19–5). An analogous reflex response in humans would be the rapid withdrawal of the arm and hand from a hot stove. Both of the *Aplysia* withdrawal reflexes can be modified by experience, and they appear to be similar with respect to cellular and molecular mechanisms of the behavioral modifications. We focus here on the gill and siphon withdrawal.

The essential neuronal circuitry that underlies the gill and siphon withdrawal reflex has been more or less identified (Fig. 19–6). There are approximately 50 sensory neurons (SN) that have their sensory receptive fields in the skin of the siphon and mantle shelf. These sensory neurons make both monosynaptic and polysynaptic connections (the latter via interneurons, IN) with a group of motor neurons (MN). The latter, in turn, synapse directly onto the gill and siphon musculature, which contracts to produce the reflex withdrawal. The cell bodies of all of these neurons are within the *Aplysia* abdominal ganglion, but these identified central neurons do not tell the entire story. There is also a peripheral circuit that participates in the gill and siphon withdrawal, but the extent of its contribution to the reflex and to the behavioral plasticity is not known.

Nonassociative plasticity in the gill withdrawal reflex: habituation and sensitization. This simple form of behavior can undergo both habituation and sensitization, as well as associative conditioning. When a weak tactile stimulus to the siphon is presented repeatedly, the reflex withdrawal is initially robust but becomes weaker with each subsequent stimulus (Fig. 19–7a). That is, the response habituates. An examination of the wiring diagram for the reflex (Fig. 19–6) reveals several potential loci for cellular plasticity that might account for this behavioral habituation. In principle, the cellular change might lie in (*1*) the sensitivity of the sensory neurons

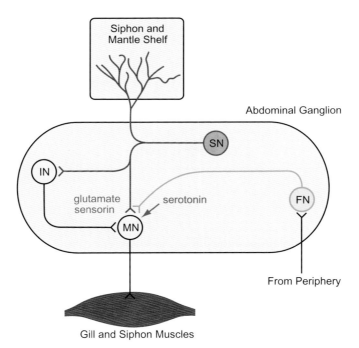

Figure 19–6. Wiring diagram for the gill withdrawal response. Sensory neurons (SN) synapse either directly or indirectly via interneurons (IN) with the motor neurons (MN) that innervate the gill and siphon musculature. Facilitatory neurons (FN) receive synaptic input from the periphery and influence the SN to MN synapse.

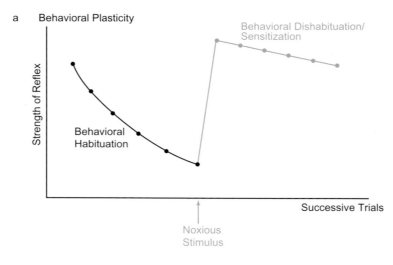

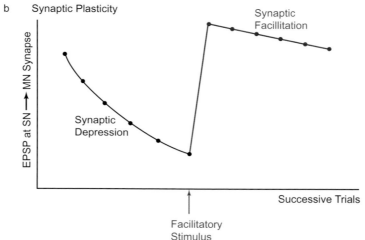

Figure 19–7. Modulation of the gill withdrawal reflex. *a*: The strength of the reflex can be measured by determining the amount of time the gill remains withdrawn following a stimulus. *b*: A change in the efficacy of the sensory neuron (SN)–to motor neuron (MN) synapse accompanies behavioral habituation and dishabituation/sensitization (see Kandel and Schwartz, 1982). EPSP, excitatory postsynaptic potential.

to the stimulus; (2) the sensory or motor neuron firing patterns, spike amplitudes, or durations; or (3) the efficacy of the sensory neuron–to–motor neuron or motor neuron–to–muscle synapses. By surveying the various components of the circuit in the behaving animal, Kandel and his colleagues found that a decrement in synaptic transmission from the sensory to the motor neuron accompanies and, in large part, accounts for the habituation (Fig. 19–7b). This *synaptic depression*, the cellular correlate of behavioral habituation, appears to be *homosynaptic*; that is, it is a property of the sensory/motor pathway itself and can be evoked simply by repeated stimulation of the presynaptic sensory neuron.

When one delivers a strong noxious stimulus such as an electric shock to the animal's head or tail, a large increase in the gill and siphon withdrawal reflex occurs. Depending on how much stimulation the siphon has received previously, this increase is termed either *sensitization* or *dishabituation*. As shown in Figure 19–7a, the reflex response is markedly enhanced, and this is accompanied by an increase in transmission at the sensory-to-motor neuron synapse (Fig. 19–7b). The *synaptic facilitation*, the cellular correlate of the enhanced reflex, is *heterosynaptic*. It results from the activation by the noxious stimulus of facilitatory neurons (FN) that synapse on the sensory neurons (Fig. 19–6), releasing their neurotransmitter to alter the properties of the sensory-to-motor neuron synapse. Both depression and facilitation of the synapse can be observed in the isolated abdominal ganglion, the former by repeated stimulation of the sensory neurons with an intracellular electrode, and the latter by stimulation of a nerve trunk that contains the axons of the facilitatory neurons (Fig. 19–7b). The actions of the facilitatory neurons can also be mimicked exactly by applying the neurotransmitter serotonin directly to the sensory neurons. With appropriately spaced stimuli, both the enhancement of the behavioral reflex and the heterosynaptic facilitation can be made to last 24 hours or longer. In other words, the animal exhibits long-term memory for this nonassociative behavioral modification, and the accompanying synaptic plasticity can also last for a very long time.

Associative plasticity in the gill withdrawal reflex: classical conditioning.
The gill and siphon withdrawal reflex can be enhanced not only by nonassociative sensitization but also by associative conditioning. The conditioning is carried out by pairing a mild tactile stimulus to the siphon, the conditioned stimulus, with a strong electric shock to the tail, the unconditioned stimulus. Prior to conditioning, the conditioned stimulus elicits a weak withdrawal response and the unconditioned stimulus elicits a powerful one. After the stimuli have been paired in time, the conditioned stimulus elicits a powerful response and the conditioning can persist for days (Fig. 19–8a).

What are the cellular correlates of this associative behavioral plasticity? Again, there is an increase in the efficacy of the sensory-to-motor neuron synapse, and again, a similar associative change in synaptic efficacy can be elicited in the isolated nervous system (Fig. 19–8b). When action potentials in one sensory neuron (SN_1—the conditioned stimulus) are paired temporally with stimulation of nerves from the tail (FN—the

a   Behavioral Plasticity

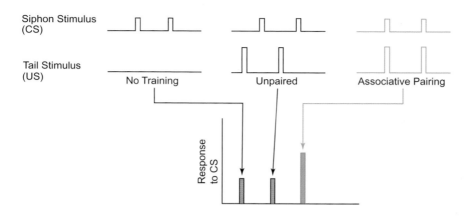

b   Synaptic Plasticity

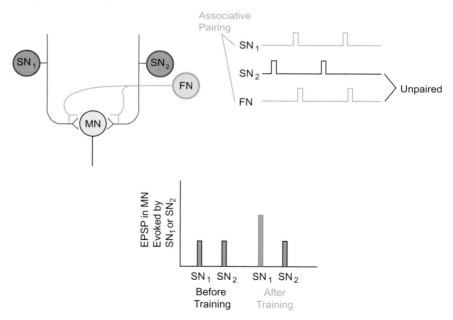

Figure 19–8. Gill and siphon withdrawal can be enhanced by associative conditioning. *a*: When a conditioned stimulus (CS) to the siphon is paired in time with an unconditioned stimulus (US) to the tail, the response to the conditioned stimulus is enhanced. *b*: An associative change in synaptic strength accompanies this behavioral response (see Kandel and Schwartz, 1982). EPSP, excitatory postsynaptic potential; MN, motor neuron; SN, sensory neuron.

unconditioned stimulus), after several trials there is a dramatic increase in the excitatory postsynaptic potential, in the motor neuron, that is evoked by the conditioned stimulus SN_1. However, the response to stimulation of all the *other* sensory neurons, which have *not* been paired with the unconditioned stimulus, remains unchanged (represented by SN_2 in Fig. 19–8b). A similar result is observed in neurons that participate in the tail withdrawal reflex.

Short-term memory is produced by a rapid intrinsic change in excitability of presynaptic cell. All of these plastic changes in synaptic strength involve a change in the amount of excitatory neurotransmitter released from the sensory neuron. Earlier classical work on mammalian and crustacean neuromuscular junctions, by Bernard Katz and Stephen Kuffler, respectively, had pointed to modulation of transmitter release as an important mechanism of synaptic plasticity. The technique of quantal analysis, pioneered by Katz (see Chapter 9), was used by Kandel and his colleagues to demonstrate that the synaptic depression that underlies habituation of the gill withdrawal reflex is accompanied by a decrement in transmitter release.

The rapid synaptic facilitation that is responsible for short-term sensitization results in part from an increase in the width of the action potential of the sensory neuron together with an overall increase in its excitability. This produces an enhancement of transmitter release. One of the advantages of the use of these molluscan neurons is that they can be maintained in culture for many days or weeks, and the cellular changes that occur during short-term and long-term facilitation can be mimicked by single or multiple applications, respectively, of serotonin. In common with olfactory learning in fruit flies, cyclic AMP signaling is a key first step. A single application of serotonin rapidly increases cyclic AMP levels, activating the cyclic AMP–dependent protein kinase. The first effect of activating this enzyme is to produce a decrease in potassium current, resulting in the broadening of the presynaptic action potential. In the sensory neurons, this potassium current was termed the *S-current* (for Serotonin-current). Protein kinase C also plays a role here, particularly at synapses that are substantially depressed before becoming sensitized, by increasing calcium current. In addition to increasing excitability, these signaling pathways may also act directly on the secretory machinery. Remember from Chapter 18 that the regulation of potassium and calcium channels by cyclic AMP and protein kinase C pathways is also used by *Aplysia* bag cell neurons to produce a rapid enhancement of neuropeptide secretion and appears to be widespread mechanism for altering neurotransmitter release.

A small number of repeated stimuli produce intermediate-term memory. A single shock to the animal (mimicked by a single application of serotonin) produces facilitation of transmitter release that lasts for less than 30 minutes. Repeated shocks or applications of serotonin produce much longer-lasting changes that involve several independent mechanisms. It is useful here to make the distinction between *intermediate-term* and *long-term* memory. Intermediate-term learning usually, but not always, requires the synthesis of new proteins and lasts for many hours. It does not, however, require new gene transcription. In contrast, long-term learning lasts for many days or weeks and requires gene transcription. In the *Aplysia* sensory neurons, as in the fly mushroom bodies, activation of the CREB transcription factor is absolutely required for long-term learning.

The *Aplysia* sensory neurons have provided valuable insights into the mechanisms of intermediate-term facilitation of release. It turns out that, before stimulation, the sensory neurons have a large number of empty varicosities that for all the world look like normal presynaptic terminals that synapse onto the motor neurons, except for the fact that they lack synaptic vesicles. A small number of applications of serotonin, which produces intermediate-term facilitation, cause many of these varicosities to be filled with synaptic vesicles and become functional, as judged by synapto-pHluorin measurements (Fig. 19–9a). The unfilled varicosities can be considered preformed "ready-to-go" *silent* synapses that are unsilenced by repeated stimulation.

It appears also that there are different forms of intermediate-term memory that depend on how the learning was initiated. For example, a small number of shocks to the tail produces enhanced transmitter release at most sensory neurons that do not receive input from the tail. For intermediate-term storage, this requires persistent activation of the cyclic AMP–dependent protein kinase and the synthesis of new proteins in these sensory neurons. If, however, one tests the very same sensory neurons that were directly activated by the shocks, these only need a single shock to become sensitized for many hours. These sensory neurons do not need to make new proteins to hang on to the memory. Instead, the proteolytic cleavage of protein kinase C to a persistently activated form termed *PKM* is all that is required to maintain increased transmitter release from these neurons. This latter form of intermediate-term memory can be mimicked in isolated cells by pairing a single application of serotonin with a train of action potentials in the sensory neuron.

Finally, intermediate-term memories require the postsynaptic motor neuron to collaborate with the terminals of the sensory neurons to enhance transmission. It does this by inserting new AMPA receptors into the postsynaptic membrane and enhancing its own synthesis of receptors.

We shall see later that insertion of new AMPA receptors is a common mechanism for enhancing transmission at many other synapses.

Multiple growth factor-related signaling pathways are activated to produce long-term memory. As described earlier, long-term memory, which is produced by a greater number of shocks, is distinguished from the other forms of memory by its persistence for days and weeks, and by the need for transcription of new genes in the nucleus. As in fruit fly learning, phosphorylation of CREB is required for this. The transcription of genes regulated by CREB can be halted by injecting the nucleus of a sensory neuron with an excess of DNA containing a CRE (see Fig. 17–24a). This binds up all the phosphorylated CREB and blocks long-term facilitation.

The signaling pathways that are activated by multiple applications of serotonin resemble those we have covered in previous chapters on neuronal development. In particular, the MAPK pathway used by neurotrophins to induce neurite outgrowth (see Figs. 15–10, 15–11) is required for both intermediate- and long-term facilitation. In part, this is achieved by a neuropeptide termed *sensorin* that is specific to the sensory neurons (Fig. 19–6). In response to repeated serotonin applications, mRNA for sensorin is translated locally in the presynaptic terminals of the sensory neurons. The newly made neuropeptide is then released to act at autoreceptors on sensory neurons, activating the MAPK kinase pathway.

Because growth factor–like pathways are activated during long-term learning, it will come as no surprise that these produce the outgrowth of new neurites and the formation of entirely new synaptic connections. Figure 19–9b shows that, in addition to the filling of previously unfilled varicosities (Fig. 19–9a), treatments that produce long-term facilitation result in the production of brand new release sites. At the same time, the number of axonal branches is increased to accommodate the increase in release sites (Fig. 19–9c, d). These morphological rearrangements require a new set of biochemical reactions, beginning with the remodeling of the actin cytoskeleton. New outgrowth also requires that previously existing cell–cell contacts be dismantled. Accordingly, multiple applications of serotonin cause the endocytosis of a cell adhesion molecule, *apCAM*, which is closely related to the N-CAM molecule discussed in Chapter 16. Interestingly, removal of apCAM appears to liberate a molecule termed *CAMAP* (**CAM-A**ssociated **P**rotein), which is normally bound to apCAM, allowing it to enter the nucleus to promote CREB-dependent gene transcription.

The overall conclusion from these experiments is that there exist parallel pathways leading to short-term, intermediate-term, and long-term learning, and multiple molecular mechanisms can contribute to each of

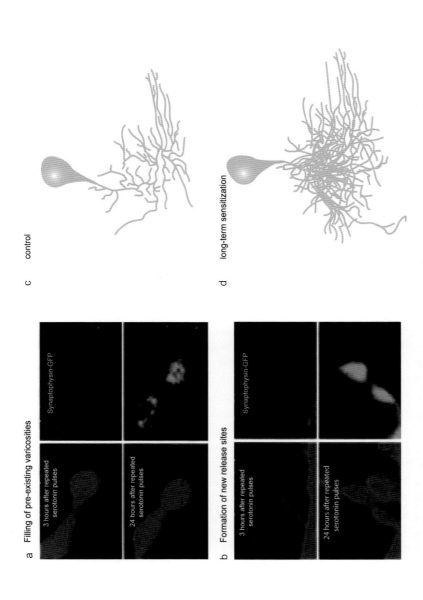

Figure 19–9. Intermediate- and long-term facilitation at sensory-to-motor neuron synapses in *Aplysia*. *a* and *b*: Sensory neurons were labeled with a fluorescent red dye to show the position of neurites and made to express the synaptic vesicle protein synaptophysin linked to the green fluorescent protein (GFP) (Kim et al., 2003). *a*: Two preexisting but empty varicosities become enriched in synaptophysin 24 hours after repeated application of serotonin. *b*: Two new varicosities containing synaptophysin are formed 24 hours after repeated serotonin exposure. *c, d*: Composite drawings of the full extent of the neurites of a sensory neuron taken from a control animal and one subjected to long-term sensitization (Bailey and Chen, 1988).

these forms of plasticity. Each of the signaling cascades is a tried and true biochemical signaling pathway that regulates growth and development not only in neurons but also in most other cell types.

Model Systems III: Long-Term Potentiation and Long-Term Depression

We vertebrates pride ourselves that our brains are more complex and much more subtle in their choice of behavioral actions than those of worms, flies, and molluscs. This has an upside and a downside. It means that some of us vertebrates can begin to figure out how brains work. On the other hand, the complexity of the vertebrate central nervous system and the lack of readily identifiable neurons present a formidable barrier to identifying the molecular or cellular basis of any specific memory trace. Nevertheless, it is evident that stimulation of neurons in vertebrates triggers biochemical events that produce short-term and long-term changes in both intrinsic excitability and in synaptic strength. One example of this is *long-term potentiation* (LTP), a phenomenon that occurs at very many central and peripheral synapses but has been investigated most thoroughly in the hippocampus, a brain region known to be important for learning and memory. In addition it has an associative component and hence provides a model for the kinds of synaptic modulation that might be involved in long-term associative learning. A related phenomenon, *long-term depression* (LTD), has also been widely investigated.

What is LTP? Although in the vertebrate brain one cannot identify single cells as unique individuals, as is the case in gastropod molluscs, particular classes of neurons and the synapses between them can be recognized reliably. As shown in Figure 19–10a, there are three major synapses in the hippocampus at which LTP has been investigated. In 1966 it was shown in the rabbit that the strength of one of these synaptic connections, between the (presynaptic) *perforant fibers* and (postsynaptic) *granule cell neurons*, can be markedly potentiated following a brief high-frequency (tetanic) stimulus to the presynaptic axons (Fig. 19–10b). This potentiation can last as long as several weeks in intact animals.

The study of LTP became much easier (and hence much more popular) with the development of the in vitro hippocampal slice. The slice is a cross section through the hippocampus in which the pathways and synaptic organization, depicted in Figure 19–10a, remain intact. In the slice it is most convenient to study LTP at the connection between the CA3 and

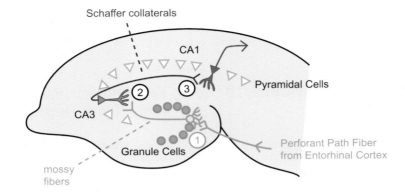

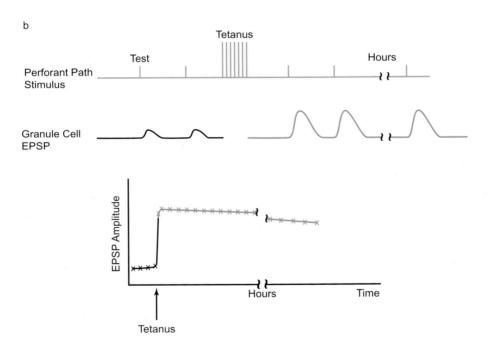

Figure 19–10. Long-term potentiation (LTP) in the hippocampus. *a*: Schematic drawing of a hippocampal slice to illustrate three distinct synapses. Fibers from the entorhinal cortex enter the hippocampus via the perforant path and synapse on dendrites of granule cell neurons (*1*). These in turn synapse on pyramidal cell neurons in the CA3 region of the hippocampus (*2*). The CA3 pyramidal cells synapse on other pyramidal cell neurons in the CA1 region (*3*). *b*: LTP of the perforant path to granule cell neuron synapse. EPSP, excitatory postsynaptic potential. (Modified from Nicoll et al., 1988; see also Bliss and Lomo, 1973.)

the CA1 pyramidal cells by stimulating the Schaffer collaterals (axons of the CA3 neurons). We shall see later that there are many different forms of LTP, produced by very different cellular mechanisms. Nevertheless, LTP at the CA3–CA1 synapse closely resembles that produced in granule cells by perforant path stimulation. Although LTP in slices cannot last for weeks because the slice dies within a matter of hours, it can persist as long as the slice does (Fig. 19–10b). In this preparation it is relatively easy to carry out intracellular recording (and even voltage clamping), and thus LTP can be examined in individual postsynaptic neurons, as well as in populations of postsynaptic neurons, via an extracellular electrode. This has made it possible to investigate the mechanisms of several temporally distinct components of LTP: *initiation, storage*, and *expression.*

Initiation of LTP requires postsynaptic depolarization during neurotransmitter release. From the earliest experiments it was evident that the induction of LTP requires high-frequency stimulation of the presynaptic fibers at a stimulus intensity that is high enough to activate a large number of inputs to the postsynaptic cell. We now know that these conditions reflect the fact that the postsynaptic cells must become sufficiently depolarized for LTP to be initiated. In fact, many experiments today no longer use a single high-frequency stimulus but a repeated train of short bursts termed *theta-burst* stimulation, which more closely resembles patterns of stimulation that occur in the hippocampus in living animals. These repeated short bursts are sufficient to produce the required postsynaptic depolarization. Further evidence that depolarization is required is provided by the finding that LTP induction by a strong tetanic stimulus can be prevented by voltage clamping the postsynaptic cell to prevent the depolarization. Moreover, weak stimuli can produce LTP if the postsynaptic cell is subject to a strong depolarization while the neurotransmitter is being released from the presynaptic terminal (Fig. 19–11).

The requirement for synaptic activation coupled with depolarization is explained by the properties of the NMDA receptor, described in Chapters 11 and 17. The neurotransmitter at these hippocampal synapses is glutamate, which binds to AMPA and NMDA receptors. The excitatory postsynaptic potential evoked by a weak stimulus is normally mediated by glutamate binding to the AMPA receptors. Only the pronounced depolarization produced by strong tetanic or theta burst stimulation can relieve the magnesium block of the NMDA class of receptor. The key consequence is that calcium then flows through the NMDA channel into the postsynaptic dendritic spine (see Fig. 11–10). The NMDA receptor can therefore be thought of as a coincidence detector. *Both* depolarization and synaptic activation are required; neither alone can activate the NMDA

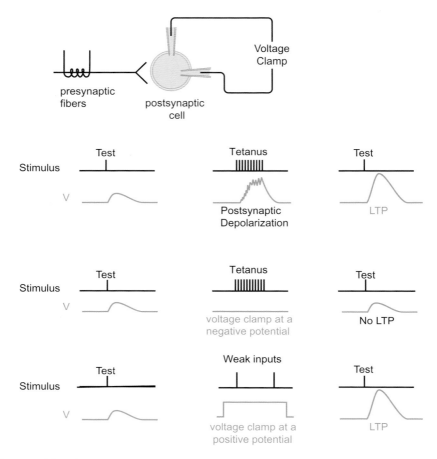

Figure 19–11. Presynaptic transmitter release together with postsynaptic depolarization is required for long-term potentiation. During a tetanus, the membrane potential of the postsynaptic cell (V) becomes depolarized and LTP results (*top traces*). When the postsynaptic cell is voltage clamped to prevent depolarization, even a strong tetanic stimulus does not produce LTP (*center traces*). When the postsynaptic cell is clamped at a positive potential, even weak inputs can become potentiated (*bottom traces*).

receptor channel. It is calcium entry through this channel that initiates LTP.

Associative LTP: Hebb's rule. As discussed in Chapter 17 in the context of synapse stabilization during development, Donald Hebb suggested that synaptic strength might be enhanced by concurrent activity in the pre- and postsynaptic neurons. He postulated further that this might provide a mechanism for associative learning. We can see that homosynaptic LTP fulfills the requirements of Hebb's rule, in that simultaneous presynaptic activity and postsynaptic depolarization (such as postsynaptic action potentials) are necessary for the long-lasting change in synaptic strength. There is also an associative form of LTP, in which activity at one synapse can contribute to the generation of LTP at another. Consider the situation illustrated in Figure 19–12a, in which a single weak input from one presynaptic neuron and a strong input, provided by the simultaneous firing of many other axons from other neurons, impinge on the same target neuron. We know that stimulation of the weak input alone does not produce LTP, because it cannot depolarize the dendrite sufficiently to allow calcium entry through NMDA receptors. In contrast, the simultaneous activation of the many presynaptic axons in the second pathway provides sufficient depolarization to trigger a postsynaptic action potential. If firing in the single weak input occurs at the same time as the second synchronous input, the spreading postsynaptic depolarization will allow NMDA receptors to open at the first input, leading to LTP at this synapse.

The initially weak synapses do not need to be physically close to the synapses made by the synchronously active synapses to undergo LTP. As we saw in Chapter 18, dendrites can generate action potentials, allowing depolarization to spread throughout a dendritic tree (see Fig. 18–14). The time at which the weak input is activated, however, is critical. The hippocampal CA3-to-CA1 synapse can be said to undergo *Hebbian spike timing–dependent plasticity*. In order for the weak synapse to become strengthened, transmitter release at this synapse must occur during the 5 milliseconds before the postsynaptic action potential (Fig 19–12b). If the timing is either slightly too early or too late, the synapse can actually undergo long-term depression, a phenomenon we shall consider shortly. The requirements for temporal pairing of the two stimuli to produce LTP are identical to those required for associative learning.

Calcium, together with protein kinases, produces early LTP. Like memory itself, LTP can be separated into short-term and long-term phases, termed *early LTP* and *late LTP*. As might be expected from all we have learned about short-term learning in invertebrates, the cyclic AMP–dependent protein kinase and protein kinase C play important roles in allowing an enhancement of synaptic transmission immediately after tetanic or theta-burst stimulation. The real star, however, is Ca^{2+}/Cam kinase II, the calcium/calmodulin-dependent protein kinase we first encountered in Chapter 9. Activation of this enzyme by calcium entry through NMDA channels, sometimes aided by calcium from other sources, does two things. First, it directly phosphorylates postsynaptic AMPA receptors to increase

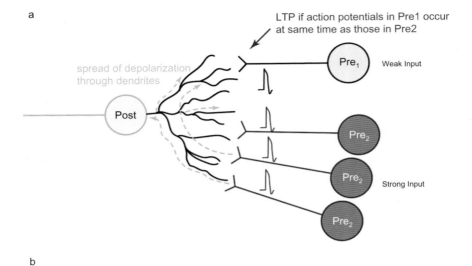

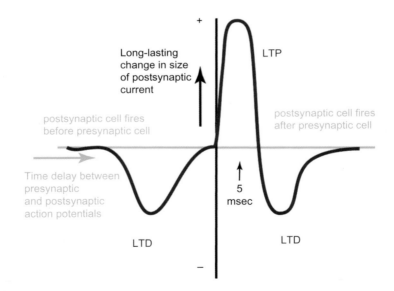

Figure 19–12. Associative long-term potentiation (LTP). *a*: Diagram depicting the spatial relationships between a weak (Pre$_1$) and a strong (Pre$_2$) presynaptic input. When the strong and the weak inputs are paired, the depolarization produced by the strong input spreads to the site of the weak input and contributes to the induction of LTP. *b*: Hebbian spike timing–dependent plasticity. The relative timing of pre- and postsynaptic action potentials determines whether LTP or long-term depression (LTD) of synaptic responses will occur.

their probability of opening. Second, it triggers the insertion of new AMPA receptors into the postsynaptic membrane from a pool of intracellular vesicles (Fig. 19–13a). The insertion of new channels can occur at functional postsynaptic densities that already have some preexisting AMPA receptors, increasing the size of postsynaptic currents evoked by glutamate release, or at postsynaptic spines that initially have only NMDA but no AMPA receptors (Fig. 19–13b). In the latter case this represents the unmasking of previously silent synapses. Functionally, this produces the same effect as the filling of previously/unfilled varicosities in *Aplysia* sensory neurons (Fig. 19–9a), but with hippocampal LTP the action happens on the postsynaptic side of the terminal. The insertion of new AMPA receptors, which often results in the swelling of the postsynaptic spines, has been visualized using a variety of imaging techniques (Fig. 19–13c).

Multiple phases of late LTP. Early LTP does not require any new protein synthesis and lasts for only tens of minutes. To keep things going longer, new mechanisms are brought into play. As with intermediate- and long-term learning described earlier, there are two phases of late LTP, the first of which requires new protein synthesis but not gene transcription. Once again, the MAPK pathway, which is brought into play by the simultaneous activation of the protein kinases responsible for early LTP, is required if LTP is to be maintained for over an hour. An important mediator of the transition between early and late LTP is the neurotrophin BDNF. Genetic manipulations that reduce levels of BDNF suppress the formation of late LTP. As we saw in Chapters 15 and 17, BDNF is secreted during neuronal stimulation, and its actions are linked to the MAPK pathway. The transition from early to late LTP is accompanied by an increase in the synthesis of new BDNF protein. Moreover, external application of BNDF can maintain the increased strength of synaptic transmission into the late phase of LTP, even when the synthesis of other proteins has been blocked. Another protein that begins to be made during the transition from early to late LTP is a constitutively activated protein kinase C. In ways that are still being determined, these biochemical changes act to ensure that the AMPA receptors that were recruited into postsynaptic densities during early LTP stay there.

Synaptic tagging. To keep LTP going longer than an hour or so, gene transcription needs to be engaged. If one recalls our discussion of flies and mollusks, one will not be shocked by the news that the CREB transcription factor is required for long-lasting hippocampal synaptic plasticity. Also, among the target genes whose mRNAs are transcribed following stimulation that results in LTP are the genes for BDNF and for the

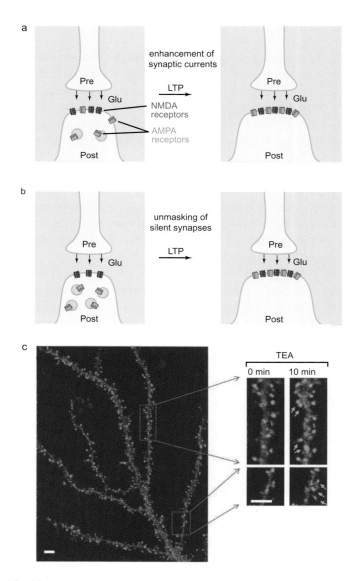

Figure 19–13. Insertion of AMPA receptors during LTP. *a*: Early LTP results from the membrane insertion of AMPA receptors from a readily available store of receptor-containing membrane vesicles. *b*: At silent synapses containing NMDA receptors but no AMPA receptors, insertion of AMPA receptors produces new sites of strong synaptic transmission. *c*: Imaging of the rapid addition of AMPA receptors to the plasma membrane of dendritic spines of an isolated hippocampal neuron. Green surface AMPA receptors were visualized by coupling the GluR1 subunit of AMPA receptors to a fluorescent protein similar to synapto-pHluorin (see Chapter 8). The induction of LTP was mimicked in the isolated cell by application of the drug TEA to block potassium channels for 10 minutes. The increase in surface AMPA receptors can be seen in the insets on the right (Gu et al., 2010).

activity-regulated cytoskeletal protein Arc, the product of an immediate-early gene that we discussed in Chapter 17.

The fact that new genes are transcribed poses a bit of a problem for associative LTP and for learning and memory. Remember that LTP happens locally at the dendrites when there is a coincidence of presynaptic firing and postsynaptic depolarization. Nothing happens at nearby synapses that are not active at this time. This is not a problem for early LTP, because AMPA receptors are inserted locally only at the simultaneously active connections. It may also not be much of a problem for cellular changes that require protein synthesis, because as we saw in Figure 17–25, proteins can be synthesized locally in dendrites. But changes in gene transcription are a different matter. There is only one nucleus in a neuron, and the mRNAs that are made there, as well as any proteins made in the cell body, are likely to be transported equally into all dendrites and reach all synapses, including active ones and inactive ones. How does the neuron know which postsynaptic sites are to be stabilized or enhanced and which are to be eliminated by the incoming RNAs and gene products?

The hypothesis adopted by many researchers to explain the selective actions of new components that come into all dendritic branches is termed *synaptic tagging* (Fig. 19–14). The idea is that when early LTP occurs at a local site in a dendrite, it creates a molecular "tag" that remains at the postsynaptic site for about 3 hours. Proteins or RNAs arriving at these sites during this time recognize the tag and begin the business of building a more permanent postsynaptic site.

The molecular identity of such tags is not yet known. It is clear, however, that a unique set of events dependent on transcription occur specifically at and near recently stimulated postsynaptic sites. This can be seen by examining the synthesis of Arc. Arc mRNA is made in the nucleus under the control of phosphorylated CREB (and other transcription factors) by stimuli that activate the cyclic AMP–dependent protein kinase and the other kinases that initially trigger LTP (Fig. 19–14). Then, bound to proteins, the mRNA is transported to the dendrites and accumulates at the synapses that have been recently activated, where it may be translated into Arc protein. Preventing the synthesis of Arc or deletion of the *Arc* gene prevents late LTP without preventing early LTP, indicating that Arc is absolutely necessary for the transition to the late phase.

What actually happens to stabilize the initial insertion of AMPA receptors so that the short-term increases in excitatory synaptic transmission during early LTP become permanent? The answer to this is not really known, but it may require structural changes produced by the increase in Arc. During the transition to late LTP, there is an increase in the size of spine heads and an elongation of the postsynaptic densities. The Arc

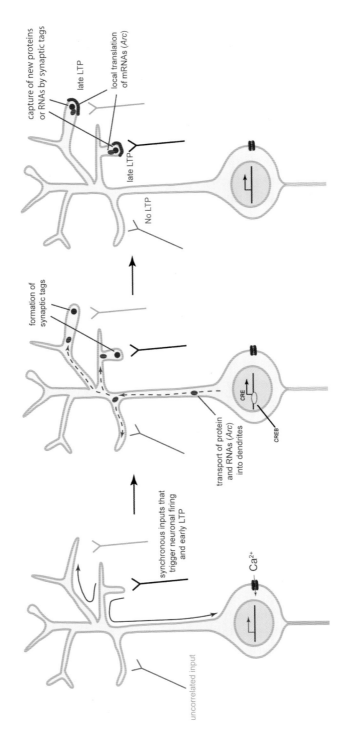

Figure 19–14. Synaptic tagging. The concept that the formation of early LTP at selected synapses onto dendrites results in the formation of tags was proposed by Morris and Frey in 1997. Activation of neuronal firing during the initiation of LTP triggers transcription at the nucleus and the transport of proteins and RNAs into all the dendrites. Only where they encounter a synaptic tag do these newly arrived components aid in producing late LTP.

protein interacts with proteins that regulate endocytosis. It accumulates in postsynaptic densities and on vesicles that contain AMPA receptors. Thus its actions are likely to contribute to these structural changes and to stabilization of AMPA receptors in the postsynaptic density, permitting the progression to late LTP.

Long-term depression. If learning and memory involved simply strengthening selected excitatory synapses, life would involve a continual increase in excitability, which might constitute a problem for those of us who are already rather excitable. As with synapse formation during development, some connections have to be weakened or lost. In fact, we saw in Figure 19–12b that if the relative timing of the presynaptic and postsynaptic action potentials at the CA1-to-CA3 hippocampal synapse is not quite right, the strength of that synaptic connection undergoes a long-lasting decrease, called *long-term depression*, or LTD. The most common way to induce homosynaptic (nonassociative) LTD at this synapse, however, is to stimulate the presynaptic fibers at a low rate of 1–3 Hz for a period of many minutes. In contrast to a rapid tetanus, which gives rise to LTP, the low-frequency train produces a depression of transmission that, in a hippocampal slice, lasts for many hours (Fig. 19–15).

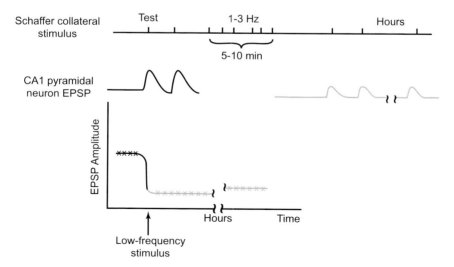

Figure 19–15. Long-term depression (LTD) in the hippocampus. Prolonged low-frequency stimulation of the Schaffer collateral pathway from region CA3 to region CA1 (Fig. 19–10a) produces a long-lasting, NMDA receptor–dependent depression of the synaptic response in the CA1 pyramidal neurons (compare with Figure 19–10b). EPSP, excitatory postsynaptic potential.

Interestingly, LTD requires many of the same molecular players as LTP and is triggered by calcium entry through NMDA receptor/channels. Blocking these receptors prevents LTD. How can the same inducing signal, calcium entry through NMDA receptor/channels, trigger both LTP and LTD? It appears to be the *amount* and the *timing* of the calcium that accumulates in the postsynaptic dendritic spines that is critical. A lower level of calcium in the dendritic spines is produced by stimuli that produce LTD. Instead of activating a collection of protein kinases, this sustained lower level of calcium triggers the calcium-dependent activation of phosphatases, resulting in the dephosphorylation of AMPA receptors and many other proteins. The initial targets of the lower calcium level are the calcium-dependent *phosphatase-2B*, also termed *calcineurin*, and *protein phosphatase-1* (PP1) (Fig. 19–16). The latter is not directly regulated by calcium but is normally inhibited by a phosphoprotein termed *inhibitor 1*, which is dephosphorylated by calcineurin, resulting in activation of PP1. Inhibition of these phosphatases prevents LTD in the CA1 region of the hippocampus.

The dephosphorylation of the AMPA receptors decreases their opening probability and this effect contributes to LTD. As might be guessed, however, the major event that produces the decrease in transmission is the removal of AMPA receptors from the postsynaptic density (Fig. 19–16). The removal of receptors by endocytosis during LTD (as well as their insertion during LTP) does not occur directly in the postsynaptic density but to the side of the density. The dephosphorylation by PP1 of proteins such as *stargazin*, a regulator of AMPA receptors within the postsynaptic density, may allow the receptors to leave the density and move laterally to sites where endocytosis occurs. Once they have been removed, the receptors may reside for a while in internal vesicles. Eventually, these will either be reinserted into the postsynaptic density in response to a new pattern of stimulation or move to late endosomes for degradation. The long-term stabilization of LTD is associated with a decrease in size of the spines, perhaps a direct result of the endocytosis of the receptors, and, like late LTP, appears to require protein synthesis. It should also be pointed out that, although the immediate cause of both LTP and LTD in the CA1 region is a change in the number of postsynaptic AMPA receptors, there are likely to be alterations in the presynaptic terminals in the late phases to accommodate changes in the size of the dendritic spines.

There are many different types of LTP and LTD. Because there has been an immense amount of work carried out on synapses on the CA1 pyramidal neurons, one might be tempted to think that this is how learning and memory work in the mammalian brain. This would be equivalent to

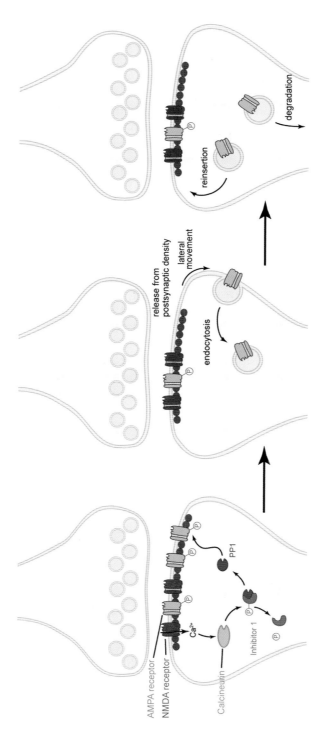

Figure 19–16. Activation of calcineurin and protein phosphatase-1 during stimuli that produce LTD results in removal of AMPA receptors from previously active synapses.

thinking that all synaptic connections work the same way as the neuro-muscular junction. It turns out that there are many different mechanisms for producing long-lasting changes in synaptic strength, and these also occur in many neurons that use transmitters other than glutamate. Moreover, as we pointed out in the last chapter, different patterns of stimulation also produce persistent changes in intrinsic excitability.

As one example, the connection between the granule cells and the CA3 pyramidal neurons in the hippocampus (Fig. 19–10a) also undergoes LTP. Tetanic stimulation of this pathway, however, produces *presynaptic LTP*. Instead of changing the number of postsynaptic AMPA receptors, LTP-inducing stimuli produce a persistent increase in the amount of neurotransmitter release from the terminals at the end of the mossy fibers. Such a long-lasting facilitation of transmitter release is also observed at a wide variety of other synapses in the brain. As with the *Aplysia* sensory neurons, cyclic AMP and activation of the cyclic AMP–dependent protein kinase appear to be important steps in producing the presynaptic change in mossy fibers. Trans-synaptic interactions between ephins and Ephs, similar to those that initially shape synaptic development (see Chapter 17), may also be required for the long-term enhancement of neurotransmitter release.

Just as there are multiple cellular mechanisms that can produce LTP, very different mechanisms exist for generating LTD. In the cerebellum, synapses onto Purkinje cells undergo an associative form of LTD that does not depend on NMDA receptors. Instead, it requires two different synapses to be activated at the same time, one of which activates postsynaptic metabotropic glutamate receptors and another that produces calcium influx. This mechanism has been termed *mGluR LTD*; it requires activation of protein kinase C and other messenger pathways that lead to an increase in protein synthesis controlled by FMRP, the regulator of local protein translation (see Fig. 17–25). mGluR LTD occurs at many other synapses and can be expressed as a decrease in postsynaptic AMPA receptors or a decrease in neurotransmitter release, depending on the location of the synapse and the age of the animals. Indeed, mGluR LTD can be produced at the CA1 synapse discussed earlier, by changing the stimulation to a more "bursty" pattern of action potentials.

Yet other forms of LTD require endocannabinoids (see Chapter 12) to act as retrograde messengers that are released from the postsynaptic cells to act on the presynaptic terminals. At different synapses the requirement for coincidence of presynaptic and postsynaptic activity can be quite different for endocannabinoid-mediated LTD. Indeed, for all of the different types of synaptic plasticity that have been described, the timing requirements can be quite different from the Hebbian spike timing–dependent plasticity we saw in Figure 19–12b. Several different timing relations encountered at different synapses are depicted in Figure 19–17. The precise requirements for Hebbian association can vary (compare Fig. 19–17a with Fig. 19–12b), while other synapses can be anti-Hebbian, such that LTP occurs only if the presynaptic terminal fires after the postsynaptic cell (Fig. 19–17b). Still others undergo only LTD whether the postsynaptic cell fires before or after the presynaptic cell (Fig. 19–17c). In the latter case, the synapse is at its strongest if the postsynaptic cell is actively inhibited by some other input during firing of the presynaptic cell. Moreover, these rules are not steadfast for each synapse but depend on the overall firing rate and pattern of firing, such as bursting or pacing.

Synaptic scaling. In addition to LTP and LTD, both of which produce a very rapid change in numbers of postsynaptic receptors, there is a much slower plastic process that, over periods of many hours or days, regulates the overall number of neurotransmitter receptors on a neuron. This is very closely related to the homeostatic control of excitability that we discussed in Chapter 18. We said that every neuron has a target pattern of firing with which it is comfortable, and that changes in levels of ion channels occur in response to conditions that perturb this pattern. So it is with synaptic connections. If the overall level of firing is reduced in a network, for example, by application of tetrodotoxin to neurons in culture, or by sensory deprivation in intact animals, the numbers of AMPA and NMDA receptors at postsynaptic densities increase in an attempt to increase excitation and to restore the target level of firing. The process works both ways, and overstimulation produces decreases in these receptors. An important aspect of this process is that the relative weights of each of the synaptic inputs onto a neuron are preserved—hence the name *synaptic scaling* for this form of plasticity. Many of the pathways proposed to regulate levels of receptors during LTP have also been implicated in synaptic scaling, but the essential mechanism, and how it is coordinated among all the different types of neurons in a network, is not yet understood.

Relation of LTP and LTD to learned behaviors. In contrast to what we said about some of the invertebrate learning models, we cannot be extremely specific about the behavioral role of a specific synaptic alteration in vertebrates. Disruption of LTP and LTD in the hippocampus by pharmacological or genetic disruption of NMDA receptors prevents rats and mice from learning tasks that require them to remember their physical location in their environment. Moreover, the phosphorylation of AMPA receptors, and even in vivo recordings of LTP, can be correlated

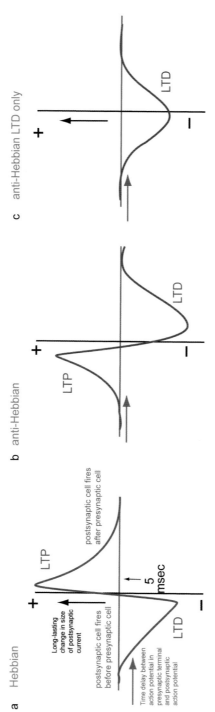

Figure 19–17. Different forms of spike timing–dependent plasticity that produce either long-term potentiation (LTP) or long-term depression (LTD).

with learning tasks that depend on the function of the hippocampus. Evidence also exists showing that LTP in other brain regions, such as the amygdala, is required for classical conditioning in which animals learn to associate a sound with an imminent electric shock. As we come to understand the various circuits involved in complex learning tasks it will be possible to provide more specific tests of how alterations in neurotransmitter receptors or intrinsic excitability contribute to the production of a specific memory.

Summary

The goal of neurobiologists, whether their experimental system is the single ion channel or the behaving human, is to understand how the brain works. One aspect of brain function that has most fascinated scientists and the lay public alike is learning and memory. Psychologists have described different kinds of learning and memory, and there is an ongoing search for the physical basis of these distinctions and for the cellular and molecular mechanisms responsible. Because of the complexity of most nervous systems, the search has focused to a large extent on animals with relatively simple nervous systems and on reduced preparations. Among others, olfactory learning in fruit flies, defensive withdrawal reflexes in *Aplysia*, and the hippocampal slice preparation have contributed to our current understanding of molecular mechanisms in learning and memory. Common themes have emerged, such as the requirement for signaling pathways linked to calcium and cyclic AMP, and the fact that pathways used in normal development continue to be used for plasticity in adults. At the same time, it is clear that there is an enormous diversity of cellular mechanisms that contribute to short-term and long-term phases of memory formation. Each type of synaptic connection has its own personality such that, in response to a particular pattern of stimulation, one synapse increases its postsynaptic receptors while another expands its presynaptic terminals. Much remains to be understood. We can anticipate that the search for the cellular and molecular underpinnings of learning and memory will occupy neurobiologists for a long time to come.

Bibliography

Chapter 1

Recommended Reading

D.A. Fortin, T. Srivastata, and T.R. Soderling. Structural modulation of dendritic spines during synaptic plasticity. *Neuroscientist* 18:326–41 (2012).

A. Peters, S.L. Palay, and H. De F. Webster. *The Fine Structure of the Nervous System: The Neurons and Supporting Cells*. Philadelphia: W.B. Saunders (1976).

S. Ramon y Cajal. Croonian Lecture. La fine structure des centres nerveux. *Proc. R. Soc. Lond.* 55:444–68 (1894).

C. Sherrington. *The Integrative Action of the Nervous System*. Cambridge, UK: Cambridge University Press (1948).

References

M. Fischer, S. Kaech, D. Knutti, and A. Matus. Rapid actin-based plasticity of dendritic spines. *Neuron* 20:847–54 (1998).

O. Loewi. Uber humorale Ubertragbarkeit der Herznervenwirkung. *Pflugers Arch.* 189:239–42 (1921).

R.A. McKinney. Excitatory amino acid involvement in dendritic spine formation, maintenance and remodeling. *J. Physiol.* 588:107–16 (2010).

S.L. Mills and S.C. Massey. Distribution and coverage of A- and B-type horizontal cells stained with neurobiotin in the rabbit retina. *Vis. Neurosci.* 11:549–60 (1994).

Chapter 2

Recommended Reading

N. Hirokawa, S. Niwa, and Y. Tanaka. Molecular motors in neurons: transport mechanisms and roles in brain function, development and disease. *Neuron* 68:610–38 (2010).

A.J. Roberts, T. Kon, P.J. Knight, K. Sutoh, and S.A. Burgess. Functions and dynamics of dynein motor proteins. *Nat. Rev. Mol. Cell Biol.* 14:713–26 (2013).

J.M. Scholey. Kinesin-2: a family of heterotrimeric and homodimeric motors with diverse intracellular transport functions. *Annu. Rev. Cell Dev. Biol.* 29:417–41 (2013).

M. Simons and D.A. Lyons. Axonal selection and myelin sheath generation in the central nervous system. *Curr. Opin. Cell Biol.* 25:512–19 (2013).

References

G. Carmignoto and P.G. Haydon. Astrocyte calcium signaling and epilepsy. *Glia* 60:1227–33 (2012).

P.G. Haydon and G. Carmignoto. Astrocyte control of synaptic transmission and neurovascular coupling. *Physiol. Rev.* 86:1009–31 (2006).

M.R. Metea and E.A. Newman. Calcium signaling in specialized glial cells. *Glia* 47:268–74 (2006).

A. Peters, S.L. Palay, and H. De F. Webster. *The Fine Structure of the Nervous System: The Neurons and Supporting Cells*. Philadelphia: W.B. Saunders (1976).

K.A. Vossel, K. Zhang, J. Brodbeck, A.C. Daub, P. Sharma, S. Finkbeiner, B. Cui, and L. Mucke. Tau reduction prevents Aβ-induced defects in axonal transport. *Science* 330:198 (2010). doi: 10.1126/science.1194653.

Chapter 3

Recommended Reading

B. Hille. *Ion Channels of Excitable Membranes*, 3rd ed. Sunderland, MA: Sinauer Associates (2001).

B. Katz. *Nerve, Muscle and Synapse*. New York: McGraw-Hill (1966).

M.N. Rasband and J.S. Trimmer. Developmental clustering of ion channels at and near the node of Ranvier. *Dev. Biol.* 236:5–16 (2001).

Reference

Y. Ogawa, J. Oses-Prieto, M.Y. Kim, I. Horresh, E. Peles, A.L. Burlingame, J.S. Trimmer, D. Meijer, and M.N. Rasband. ADAM22, a Kv1 channel-interacting protein, recruits membrane-associated guanylate cyclases to juxtaparanodes of myelinated axons. *J. Neurosci.* 30:1038–48 (2010).

Chapter 4

Recommended Reading

K. Deisseroth. Controlling the brain with light. *Sci. Am.* 303:48–55 (2010).
K. Deisseroth. Optogenetics. *Nat. Methods* 8:26–9 (2011).
B. Hille. *Ion Channels of Excitable Membranes*, 3rd ed. Sunderland,
    MA: Sinauer Associates (2001).
F. Qin. Principles of single channel kinetic analysis. *Methods Mol. Biol.*
    *403:*253–86 (2007).
B. Sakmann and E. Neher (Eds.). *Single-Channel Recording*, 2nd ed.
    New York: Plenum (1995).

Reference

O.P. Hamill, A. Marty, E. Neher, B. Sakmann, and F.J. Sigworth. Improved
    patch-clamp techniques for high-resolution current recording from cells
    and cell-free membrane patches. *Pflugers Arch. 391:*85–100 (1981).

Chapter 5

Recommended Reading

C.M. Armstrong. Sodium channels and gating currents. *Physiol. Rev.*
    *61:*644–83 (1981).
W.A. Catterall. Ion channel voltage sensors: structure, function and patho-
    physiology. *Neuron 67:*915–28 (2010).
B. Hille. *Ion Channels of Excitable Membranes*, 3rd ed. Sunderland,
    MA: Sinauer Associates (2001).
T. Hoshi, W.N. Zagotta, and R.W. Aldrich. Biophysical and molecular mech-
    anisms of *Shaker* potassium channel inactivation. *Science 250:*533–8
    (1990).

References

D.A. Doyle, J. Morais Cabral, R.A. Pfuetzner, A. Kuo, J.M. Gulbis, S.L.
    Cohen, B.T. Chait, and R. MacKinnon. The structure of the potas-
    sium channel: molecular basis of K^+ conduction and selectivity. *Science*
    *280:*69–77 (1998).
Y. Jiang, A. Lee, J. Chen, M. Cadene, B.T. Chait, and R. MacKinnon. Crystal
    structure and mechanism of a calcium-gated potassium channel. *Nature*
    *417:*515–22 (2002).

J. Payandeh, T. Scheuer, N. Zheng, and W.A. Catterall. The crystal structure of a voltage-gated sodium channel. *Nature* 475:353–8 (2011).

T.L. Schwarz, B.L. Tempel, D.M. Papazian, Y.N. Jan, and L.Y. Jan. Multiple potassium-channel components are produced by alternative splicing at the Shaker locus in *Drosophila*. *Nature* 331:137–42 (1988).

S. Ye, Y. Li, L. Chen, and Y. Jiang. Crystal structures of a ligand-free MthK gating ring: insights into the ligand gating mechanism of K+ channels. *Cell* 126:1161–73 (2006).

X. Zhang, W. Ren, P. DeCaen, C. Yan, X. Tao, L. Tang, J. Wang, K. Hasegawa, T. Kumasaka, J. He, J. Wang, D.E. Clapham, and N. Yan. Crystal structure of an orthologue of the NaChBac voltage-gated sodium channel. *Nature* 486:130–4 (2012).

Chapter 6

Recommended Reading

B. Hille. *Ion Channels of Excitable Membranes*, 3rd ed. Sunderland, MA: Sinauer Associates (2001).

References

A.L. Hodgkin and A.F. Huxley. Currents carried by sodium and potassium ions through the membrane of the giant axon of *Loligo*. *J. Physiol. (London)* 116:449–72 (1952a).

A.L. Hodgkin and A.F. Huxley. The components of membrane conductance in the giant axon of *Loligo*. *J. Physiol. (London)* 116:473–96 (1952b).

A.L. Hodgkin and A.F. Huxley. The dual effect of membrane potential on sodium conductance in the giant axon of *Loligo*. *J. Physiol. (London)*: 116:497–506 (1952c).

A.L. Hodgkin and A.F. Huxley. A quantitative description of membrane current and its application to conduction and excitation in nerve. *J. Physiol. (London)* 117:500–44 (1952d).

A.L. Hodgkin, A.F. Huxley, and B. Katz. Measurements of current-voltage relations in the membrane of the giant axon of *Loligo*. *J. Physiol. (London)* 116:424–8 (1952).

Chapter 7

Recommended Reading

W.A. Catterall. Voltage-gated calcium channels. *Cold Spring Harb. Perspect. Biol.* 3(8):a003947 (2011). doi: 10.1101/cshperspect.a003947

L.Y. Jan and Y.N. Jan. Voltage-gated potassium channels and the diversity of electrical signaling. *J. Physiol.* *590*:2591–9 (2012).

L.K. Kaczmarek. *Slack*, *Slick* and sodium-activated potassium channels. *ISRN Neurosci.* pii: 354262 (2013). doi: 10.1155/2013/354262

B. Nilius and G. Owsianik. The transient receptor potential family of ion channels. *Genome Biol.* *12*:218 (2011).

References

J.M. Gulbis, M. Zhou, S. Mann, and R. MacKinnon. Structure of the cytoplasmic beta subunit-T1 assembly of voltage-dependent K$^+$ channels. *Science* *289*:123–7 (2000).

R.W. Meech and N.B. Standen. Potassium activation in *Helix aspersa* neurons under voltage clamp: a component mediated by calcium influx. *J. Physiol.* *249*:211–39 (1975).

Multiple authors, 11 articles. IUPHAR compendium of voltage-gated ion channels. *Pharmacol. Rev.* *57*:385–540 (2005).

P.H. Reinhart, S. Chung, and I.B. Levitan. A family of calcium-dependent potassium channels from rat brain. *Neuron* *2*:1031–41 (1989).

Chapter 8

Recommended Reading

N. Brose. For better or for worse: complexins regulate SNARE function and vesicle fusion. *Traffic* *9*:1403–13 (2008).

R.D. Burgoyne, J.W. Barclay, L.F. Ciufo, M.E. Graham, M.T. Handley, and A. Morgan. The functions of Munc18-1 in regulated exocytosis. *Ann. N. Y. Acad. Sci.* *1152*:76–86 (2009).

J.J. Chua, S. Kindler, J. Boyken, and R. Jahn. The architecture of an excitatory synapse. *J. Cell Sci.* *123*:819–23 (2010).

J. Dittman and T.A. Ryan. Molecular circuitry of endocytosis at nerve terminals. *Annu. Rev. Cell Dev. Biol.* *25*:133–60 (2009).

R. Jahn and R.H. Scheller. SNAREs—engines for membrane fusion. *Nat. Rev. Mol. Cell Biol.* *7*:631–43 (2006).

J. Rizo and C. Rosenmund. Synaptic vesicle fusion. *Nat. Struct. Mol. Biol.* *15*:665–74 (2008).

E. Scemes, D.C. Spray, and P. Meda. Connexins, pannexins, innexins: novel roles of "hemi-channels". *Pflugers Arch.* *457*:1207–26 (2009).

S. Schoch and E.D. Gundelfinger. Molecular organization of the presynaptic active zone. *Cell Tissue Res.* *326*:379–91 (2006).

G. Sohl, S. Maxeiner, and K. Willecke. Expression and functions of neuronal gap junctions. *Nat. Rev. Neurosci.* *6*:191–200 (2005).

C. Zhao, J.T. Slevin, and S.W. Whiteheart. Cellular functions of NSF: not just SNAPs and SNAREs. *FEBS Lett.* *581*:2140–9 (2007).

References

L.J. Breckenridge and W. Almers. Current through the fusion pore that forms during exocytosis of a secretory vesicle. *Nature 328*:814–7 (1987).

J.M. Burt and D.C. Spray. Single channel events and gating behavior of the cardiac gap junction channel. *Proc. Natl. Acad. Sci. U.S.A. 85*:3431–4 (1988).

A.J. Cochilla, J.K. Angleson, and W.J. Betz. Monitoring secretory membrane with FM1–43 fluorescence. *Annu. Rev. Neurosci. 22*:1–10 (1999).

P.I. Hanson, R. Roth, H. Morisaki, R. Jahn, and J.E. Heuser. Structural and conformational changes in NSF and its membrane receptor complexes visualized by quick-freeze/deep-etch electron microscopy. *Cell 90*:523–35 (1997).

J.E. Heuser, T.S. Reese, M.J. Dennis, Y. Jan, L. Jan, and L. Evans. Synaptic vesicle exocytosis captured by quick freezing and correlated with quantal transmitter release. *J. Cell Biol. 81*:275–300 (1979).

T.M. Hohl, F. Paralti, C. Wimmer, J.E. Rothman, T.H. Söllner, and H. Engelhardt. Arrangement of subunits in 20S particles consisting of NSF, SNAPs and SNARE complexes. *Mol. Cell 2*:539–48 (1998).

L.K. Kaczmarek, M. Finbow, J.-P. Revel, and F. Strumwasser. The morphology and coupling of *Aplysia* bag cells within the abdominal ganglion and in cell culture. *J. Neurobiol. 10*:525–50 (1979).

L. Makowski, D.L.D. Caspar, W.C. Phillips, and D.A. Goodenough. Gap junction structures. II. Analysis of the X-ray diffraction data. *J. Cell Biol. 74*:629–45 (1977).

A. Pereda, J. O'Brien, J.I. Nagy, F. Bukauskas, K.G. Davidson, N. Kamasawa, T. Yasumura, and J.E. Rash. Connexin35 mediates electrical transmission at mixed synapses on Mauthner cells. *J. Neurosci. 23*:7489–503 (2003).

R.B. Sutton, D. Fasshauer, R. Jahn, and A.T. Brunger. Crystal structure of a SNARE complex involved in synaptic exocytosis at 2.4 Angstrom resolution. *Nature 395*:347–53 (1998).

Y. Wu, W. Wang, A. Diez-Sampedro, and G.B. Richerson. Nonvesicular inhibitory neurotransmission via reversal of the GABA transporter GAT-1. *Neuron 56*:851–65 (2007).

Chapter 9

Recommended Reading

F. Cesca, P. Baldelli, F. Valtorta, and F. Benfenati. The synapsins: key actors of synapse function and plasticity. *Prog. Neurobiol. 91*:313–48 (2010).

E.R. Chapman. How does synaptotagmin trigger neurotransmitter release? *Annu. Rev. Biochem. 77*:615–41 (2008).

E. Eggermann, I. Bucurenciu, S.P. Goswami, and P. Jonas. Nanodomain coupling between Ca^{2+} channels and sensors of exocytosis at fast mammalian synapses. *Nat. Rev. Neurosci.* 13:7–21 (2012).

D. Fioravante and W.G. Regehr. Short-term forms of presynaptic plasticity. *Curr. Opin. Neurobiol.* 21:269–74 (2011).

O. Kochubey, X. Lou, and R. Schneggenburger. Regulation of transmitter release by Ca^{2+} and synaptotagmin: insights from a large CNS synapse. *Trends Neurosci.* 34:237–46 (2011).

R. Schneggenburger and I.D. Forsythe. The calyx of Held. *Cell Tissue Res.* 326:311–37 (2006).

References

G.J. Augustine, M.P. Charlton, and S.J. Smith. Calcium entry and transmitter release at voltage-clamped nerve terminals of squid. *J. Physiol.* 369:163–81 (1985).

I.A. Boyd and A.R. Martin. The end-plate potential in mammalian muscle. *J. Physiol.* 132:74–91 (1956).

M. Geppert, V.Y. Bolshakov, S.A. Siegelbaum, K. Takel, P. DeCamilli, R.E. Hammer, and T.C. Südhof. The role of Rab3A in neurotransmitter release. *Nature* 369:493–7 (1994).

M. Geppert, Y. Goda, R.E. Hammer, C. Li, T.W. Rosahl, C.F. Stevens, and T.C. Sudhof. Synaptotagmin I: a major Ca^{2+} sensor for transmitter release at a central synapse. *Cell* 79:717–27 (1994).

G. Grynkiewicz, M. Poemie, and R.Y. Tsien. A new generation of Ca^{2+} indicators with greatly improved fluorescence properties. *J. Biol. Chem.* 260:3440–50 (1985).

R. Llinas, T.L. McGuinness, C.S. Leonard, M. Sugimori, and P. Greengard. Intraterminal injection of synapsin I or calcium/calmodulin-dependent protein kinase II alters neurotransmitter release at the squid giant synapse. *Proc. Natl. Acad. Sci. U.S.A.* 82:3035–9 (1985).

S. Mochida, A.P. Few, T. Scheuer, and W.A. Catterall. Regulation of presynaptic Ca(V)2.1 channels by Ca^{2+} sensor proteins mediates short-term synaptic plasticity. *Neuron* 57:210–16 (2008).

T. Nishiki and G.J. Augustine. Synaptotagmin I synchronizes transmitter release in mouse hippocampal neurons. *J. Neurosci.* 24:6127–32 (2004).

L. Siksou , P, Rostaing , J.P. Lechaire, T. Boudier, T, Ohtsuka, A. Fejtova, H.T. Kao, P. Greengard, E.D. Gundelfinger, A. Triller, and S. Marty. Three-dimensional architecture of presynaptic terminal cytomatrix. *J. Neurosci.* 27:6868–6877 (2007).

S.J. Smith, J. Buchanan, L.R. Osses, M.P. Charleton, and G. Augustine. The spatial distribution of calcium signals in squid presynaptic terminals. *J. Physiol.* 472:573–93 (1993).

E.F. Stanley. Single calcium channels and acetylcholine release at a presynaptic nerve terminal. *Neuron* 11:1007–11 (1993).

Y.-G. Tang and R.S. Zucker. Mitochondrial involvement in post-tetanic potentiation of synaptic transmission. *Neuron 18*:483–91 (1997).

E.S. Wachman, R.E. Poage, J.R. Stiles, D.L. Farkas, and S.D. Meriney. Spatial distribution of calcium entry evoked by single action potentials within the presynaptic active zone. *J. Neurosci. 24*:2877–85 (2004).

L.Y. Wang and L.K. Kaczmarek. High-frequency firing helps replenish the readily releasable pool of synaptic vesicles. *Nature 394*:384–8 (1998).

Chapter 10

Recommended Reading

H.R. Berthoud and H. Münzberg. The lateral hypothalamus as integrator of metabolic and environmental needs: from electrical self-stimulation to optogenetics. *Physiol. Behav. 104*:29–39 (2011).

P.J. Focke, X. Wang, and H.P. Larsson. Neurotransmitter transporters: structure meets function. *Structure 21*:694–705 (2013).

P. Greengard. The neurobiology of slow synaptic transmission. *Science 294*:1024–30 (2001).

L. Iverson, S. Iverson, F.E. Bloom, and R.H. Roth. *Introduction to Neuropsychopharmacology*. New York: Oxford University Press (2008).

S.W. Kuffler. Slow synaptic responses in autonomic ganglia and the pursuit of a peptidergic transmitter. *J. Exp. Biol. 89*:257–86 (1980).

Chapter 11

Recommended Reading

M. Sheng and E. Kim. The postsynaptic organization of synapses. *Cold Spring Harb. Perspect. Biol. 3*(12) pii:a005678 (2011). doi: 10.1101/cshperspect. a005678

E. Sigel and M.E. Steinmann. Structure, function and modulation of GABA$_A$ receptors. *J. Biol. Chem. 287*:40224–31 (2012).

S.F. Traynelis, L.P. Wollmuth, C.J. McBain, F.S. Menniti, K.M. Vance, K.K. Ogden, K.B. Hansen, H. Yuan, S.J. Myers, and R. Dingledine. Glutamate receptor ion channels: structure, regulation and function. *Pharmacol. Rev. 62*:405–96 (2010).

References

J.E. Baenziger and P.-J. Corringer. 3D structure and allosteric modulation of the transmembrane domain of pentameric ligand-gated ion channels. *Neuropharmacology 60*:116–25 (2011).

E.A. Barnard, R. Miledi, and K. Sumikawa. Translation of exogenous messenger RNA coding for nicotinic acetylcholine receptors produces functional receptors in *Xenopus* oocytes. *Proc. R. Soc. London Ser. B 215*:241–6 (1982).

M.L. Mayer, G.L. Westbrook, and P.B. Guthrie. Voltage-dependent block by Mg^{2+} of NMDA responses in spinal cord neurones. *Nature 309*:261–3 (1984).

L. Nowak, P. Bregestovski, P. Ascher, A. Herbet, and A. Prochiantz. Magnesium gates glutamate-activated channels in mouse central neurones. *Nature 307*:462–5 (1984).

B. Sakmann, C. Methfessel, M. Mishina, T. Takahashi, T. Takai, M. Kurasaki, K. Fukuda, and S. Numa. Role of acetylcholine receptor subunits in gating of the channel. *Nature 318*:538–43 (1985).

A.I. Sobolevsky, M.P. Rosconi, and E. Gouaux. X-ray structure, symmetry and mechanism of an AMPA-subtype glutamate receptor. *Nature 462*:745–56 (2009).

N. Unwin. Refined structure of the nicotinic acetylcholine receptor at 4 Å resolution. *J. Mol. Biol. 346*:967–89 (2005).

L.P. Wollmuth and S.F. Traynelis. Neuroscience: excitatory view of a receptor. *Nature 462*:729–31 (2009).

Chapter 12

Recommended Reading

D.E. Clapham. Calcium signaling. *Cell 131*:1047–58 (2007).

N. Gamper and M.S. Shapiro. Target-specific PIP_2 signalling: how might it work? *J. Physiol. 582*:967–75 (2007).

N.V. Kovalevskaya, M. van de Waterbeemd, F.M. Bokhovchuk, N. Bate, R.J. Bindels, J.G. Hoenderop, and G.W. Vuister. Structural analysis of calmodulin binding to ion channels demonstrates the role of its plasticity in regulation. *Pflugers Arch. 465*:1507–19 (2013).

R. Lujan, E.M. Fernandez de Valasco, C. Aguado, and K. Wickman. New insights into the therapeutic potential of Girk channels. *Trends Neurosci. 37*:20–9 (2014).

K. Mackie. Signaling via CNS cannabinoid receptors. *Mol. Cell. Endocrinol. 286*: S60–5 (2008).

D. Piomelli. More surprises lying ahead. The endocannabinoids keep us guessing. *Neuropharmacol. 76*:228–34 (2014).

References

M.J. Berridge. Inositol trisphosphate and calcium: two interacting second messengers. *Am. J. Nephrol. 17*:1–11 (1997).

G.E. Breitwieser and G. Szabo. Uncoupling of cardiac muscarinic and
β-adrenergic receptors from ion channels by a guanine nucleotide
analogue. *Nature 317*:538–40 (1985).

I.B. Levitan. It is calmodulin after all! Mediator of the calcium modulation of
multiple ion channels. *Neuron 22*:645–8 (1999).

D.E. Logothetis, Y. Kurachi, J. Galper, E.J. Neer, and D.E. Clapham. The βγ
subunits of GTP-binding proteins activate the muscarinic K$^+$ channel in
heart. *Nature 325*:321–6 (1987).

K. Mikoshiba. IP$_3$ receptor/Ca^{2+} channel: from discovery to new signaling
concepts. *J. Neurochem. 102*:1426–46 (2007).

W.M. Oldham and H.E. Hamm. Heterotrimeric G protein activation by
G-protein-coupled receptors. *Nat. Rev. Mol. Cell Biol. 9*:60–71 (2008).

P.J. Pfaffinger, J.M. Martin, D.D. Hunter, N.M. Nathanson, and B. Hille.
GTP-binding proteins couple cardiac muscarinic receptors to a K$^+$ channel.
Nature 317:536–38 (1985).

M.R. Whorton and R. MacKinnon. X-ray structure of the mammalian
GIRK2-βγ G-protein complex. *Nature 498*:190–7 (2013).

Chapter 13

Recommended Reading

M.E. Burns and E.N. Pugh, Jr. Lessons from photoreceptors: turning off
G-protein signaling in living cells. *Physiology 25*:72–84 (2010).

J. Chandrashekar, M.A. Hoon, N.J. Ryba, C.S. Zuker. The receptors and cells
for mammalian taste. *Nature 444*:288–94 (2006).

S. DeMaria and J. Ngai. The cell biology of smell. *J. Cell Biol. 191*:443–52
(2010).

R. Fettiplace R and C.M. Hackney. The sensory and motor roles of auditory
hair cells. *Nat. Rev. Neurosci. 7*:19–29 (2006).

Y. Ishimaru. Molecular mechanisms of taste transduction in vertebrates.
Odontology 97:1–7 (2009).

P. Kazmierczak and U. Müller. Sensing sound: molecules that orchestrate
mechanotransduction by hair cells. *Trends Neurosci. 35*:220–29 (2012).

S.C. Kinnamon. Taste receptor signaling—from tongues to lungs. *Acta
Physiol. 204*:158–68 (2012).

E.R. Liman. Thermal gating of TRP ion channels: food for thought? Science's
STKE: signal transduction knowledge environment. 2006:pe12 (2006).

K. Mori K and H. Sakano. How is the olfactory map formed and interpreted
in the mammalian brain? *Annu. Rev. Neurosci. 34*:467–99 (2011).

B. Nilius and E. Honoré. Sensing pressure with ion channels. *Trends
Neurosci. 35*:477–86 (2012).

B. Nilius B and G. Owsianik. The transient receptor potential family of ion channels. *Genome Biol. 12*:218 (2011).

H. Ripps. Light to sight: milestones in phototransduction. *FASEB J. 24*:970–5 (2010).

M. Spehr and S.D. Munger. Olfactory receptors: G protein–coupled receptors and beyond. *J. Neurochem. 109*:1570–83 (2009).

C.H. Sung and J.Z. Chuang. The cell biology of vision. *J. Cell Biol. 190*:953–63 (2010).

References

J.J. Art and R. Fettiplace. Variation of membrane properties in hair cells isolated from the turtle cochlea. *J. Physiol. 385*:207–242 (1987).

B. Coste, B. Xiao, J.S. Santos, R. Syeda, J. Grandl, K.S. Spencer, S.E. Kim, M. Schmidt, J. Mathur, A.E. Dubin, M. Montal, and A. Patapoutian. Piezo proteins are pore-forming subunits of mechanically activated channels. *Nature 483*:176–81 (2012).

E.F. Fesenko, S.S. Kolesnikov, and A.L. Lyubarsky. Induction by cyclic GMP of cationic conductance in plasma membrane of retinal rod outer segments. *Nature (London) 313*:310–13 (1985).

P.G. Gillespie and R.G. Walker. Molecular basis of mechanosensory transduction. *Nature 413*:194–202 (2001).

L.W. Haynes, A.R. Kay, and K.-W. Yau. Single cyclic GMP-activated channel activity in excised patches of rod outer segment membrane. *Nature (London) 321*:66–70 (1986).

S.E. Kim, B. Coste, A. Chadha, B. Cook, and A. Patapoutian. The role of *Drosophila* Piezo in mechanical nociception. *Nature 483*:209–12 (2012).

S. Kinnamon. Taste transduction: A diversity of mechanisms. *Trends Neurosci. 11*:491–6 (1988).

R.S. Lewis and A.J. Hudspeth. Voltage- and ion-dependent conductances in solitary vertebrate hair cells. *Nature 304*:538–41 (1983).

J.H. Martin. Somatic sensory system I: Receptor physiology and submodality coding. In *Principles of Neural Science*, E. R. Kandel and J. H. Schwartz (Eds.). New York: Elsevier North Holland, pp. 157–69 (1981).

R. O'Hagan, M. Chalfie, and M.B. Goodman. The MEC-4 DEG/ENaC channel of *Caenorhabditis elegans* touch receptor neurons transduces mechanical signals. *Nat. Neurosci. 8*:43–50 (2005).

K. Palczewski, T. Kumasaka, T. Hori, C.A. Behnke, H. Motoshima, B.A. Fox, I. Le Trong, D.C. Teller, T. Okada, R.E. Stenkamp, M. Yamamoto, and M. Miyano. Crystal structure of rhodopsin: A G protein–coupled receptor. *Science 289*:739–45 (2000).

K.J. Ressler, S.L. Sullivan, and L.B. Buck. A molecular dissection of spatial patterning in the olfactory system. *Curr. Opin. Neurobiol. 4*:588–96 (1994).

L. Song L and J. Santos-Sacchi. Disparities in voltage-sensor charge and electromotility imply slow chloride-driven state transitions in the solute carrier SLC26a5. *Proc. Natl. Acad. Sci. U. S. A. 110*:3883–8 (2013).

N. Yu, M.L. Zhu, and H.B. Zhao. Prestin is expressed on the whole outer hair cell basolateral surface. *Brain Res. 1095*:51–8 (2006).

Chapter 14

Recommended Reading

N. Ballas and G. Mandel. The many faces of REST oversee epigenetic programming of neuronal genes. *Curr. Opin. Neurobiol. 15*:500–6 (2005).

A.M. Bond, O.G. Bhalala, and J.A. Kessler. The dynamic role of bone morphogenetic proteins in neural stem cell fate and maturation. *Dev. Neurobiol. 72*:1068–84 (2012).

Bragdon, O. Moseychuk, S. Saldanha, D. King, J. Julian, and A. Nohe. Bone morphogenetic proteins: a critical review. *Cell. Signal. 23*:609–20 (2011).

M.C. Chang and A. Hemmati-Brivanlou. Cell fate determination in embryonic ectoderm. *J. Neurobiol. 36*:128–51 (1998).

Gotz and W.B. Huttner. The cell biology of neurogenesis. *Nat. Rev. Mol. Cell Biol. 6*:777–88 (2005).

L. Grabel. Developmental origin of neural stem cells: the glial cell that could. *Stem Cell Rev. 8*:577–85 (2012).

D.R. Green and J.C. Reed. Mitochondria and apoptosis. *Science 28*:1309–12 (1998).

F. Guillemot. Spatial and temporal specification of neural fates by transcription factor codes. *Development 134*:3771–80 (2007).

A.S. Hazell. Excitotoxic mechanisms in stroke: an update of concepts and treatment strategies. *Neurochem. Int. 50*:941–53 (2007).

R. Kageyama, T. Ohtsuka, J. Hatakeyama, and R. Ohsawa. Roles of bHLH genes in neural stem cell differentiation. *Exp. Cell Res. 306*:343–8 (2005).

A.R. Kriegstein and S.C. Noctor. Patterns of neuronal migration in the embryonic cortex. *Trends Neurosci. 27*:392–9 (2004).

J.-M. Lee, G.J. Zipfel, and D.W. Choi. The changing landscape of ischaemic brain injury mechanisms. *Nature 399*:A7–14 (1999).

M.F. Mehler, P.C. Mabie, D. Zhang, and J.C. Kessler. Bone morphogenetic proteins in the nervous system. *Trends Neurosci. 20*:309–17 (1997).

D. Purves and J. W. Lichtman. *Principles of Neural Development.* Sunderland, MA: Sinauer Associates (1985).

S.D. Skaper. Alzheimer's disease and amyloid: culprit or coincidence? *Int. Rev. Neurobiol. 102*:277–316 (2012).

References

F.H. Gage. Mammalian neural stem cells. *Science 287*:1433–8 (2000).

M. Jacobson. *Developmental Neurobiology*. New York: Plenum (1978).

P. Rakic. Neuron–glia relationship during ganglion cell migration in developing cerebellar cortex. A Golgi and electron microscopic study in *Macacus
rhesus. J. Comp. Neurol. 141*:283–312 (1971).

R. Sattler and M. Tymianski. Molecular mechanisms of calcium-dependent
excitotoxicity. *J. Mol. Med. 78*:3–13 (2000).

D.J. Solecki, N. Trivedi, E.E. Govek, R.A. Kerekes, S.S. Gleason, and M.E.
Hatten. Myosin II motors and F-actin dynamics drive the coordinated
movement of the centrosome and soma during CNS glial-guided neuronal
migration. *Neuron 63*:63–80 (2009).

D. van der Kooy and S. Weiss. Why stem cells? *Science 287*:1439–41 (2000).

Chapter 15

Recommended Reading

J.L. Ables, J.J. Breunig, A.J. Eisch, and P. Rakic. Not(ch) just development: Notch signalling in the adult brain. *Nat. Rev. Neurosci. 12*:269–83
(2011).

Apostolova and G. Dechant. Development of neurotransmitter phenotypes in
sympathetic neurons. *Auton. Neurosci. 151*:30–8 (2009).

M.V. Chao. Neurotrophins and their receptors: a convergence point for many
signalling pathways. *Nat. Rev. Neurosci. 4*:299–309 (2003).

T.W. Gould and H. Enomoto. Neurotrophic modulation of motor neuron
development. *Neuroscientist 15*:105–16 (2009).

E.J. Huang and L.F. Reichardt LF. Trk receptors: roles in neuronal signal
transduction. *Annu. Rev. Biochem. 72*:609–42 (2003).

G. Le Dreau and E. Marti E. Dorsal-ventral patterning of the neural tube: a
tale of three signals. *Dev. Neurobiol. 72*:1471–81 (2012).

S.E. London, L. Remage-Healey, and B.A. Schlinger BA. Neurosteroid
production in the songbird brain: a re-evaluation of core principles.
Front. Neuroendocrinol. 30:302–14 (2009).

P. Micevych P and A. Christensen. Membrane-initiated estradiol actions
mediate structural plasticity and reproduction. *Front. Neuroendocrinol.
33*:331–41 (2012).

K.A. Mulligan KA and B.N. Cheyette. Wnt signaling in vertebrate neural
development and function. *J. Neuroimmune Pharmacol. 7*:774–87 (2012).

H. Park and M.M. Poo. Neurotrophin regulation of neural circuit development and function. *Nat. Rev. Neurosci. 14*:7–23 (2013).

T. Pierfelice, L. Alberi, and N. Gaiano. Notch in the vertebrate nervous
system: an old dog with new tricks. *Neuron 69*:840–55 (2011).

D.S. Reddy. Neurosteroids: endogenous role in the human brain and therapeutic potentials. *Prog. Brain Res.* 186:113–37 (2010).

M.M. Wang. Notch signaling and Notch signaling modifiers. *Int. J. Biochem.
Cell Biol.* 43:1550–62 (2011).

References

S. Artavanis-Tsakonas, K. Matsuno, and M.E. Fortini. Notch signaling.
Science 268:225–32 (1995).

D. Bar-Sagi and J. R. Feramisco. Microinjection of the *ras* oncogene protein
into PC12 cells induces morphological differentiation. *Cell* 42:841–8
(1985).

R.M. Evans. The steroid and thyroid hormone receptor superfamily. *Science*
240:889–95 (1988).

R.A. Gorski. Structural sex differences in the brain: their origin and significance. In *Neural Control of Reproductive Function*, J.M. Lakoski,
J.R. Perez-Polo, and D.K. Rassin (Eds.). New York: Alan R. Liss,
pp. 33–44 (1989).

M.E. Gurney. Hormonal control of cell form and number in the zebra finch
song system. *J. Neurosci.* 1:658–73 (1981).

F. Nottebohm. From bird song to neurogenesis. *Sci. Am.* 260:74–9 (1989).

J. Palka and M. Schubiger. Genes for neural differentiation. *Trends Neurosci.*
11:515–17 (1988).

T. Raabe. The Sevenless signaling pathway: variations of a common theme.
Biochim. Biophys. Acta 1496:151–63 (2000).

J.R. Sanes. Roles of extracellular matrix in neural development. *Annu. Rev.
Physiol.* 45:581–600 (1983).

Chapter 16

Recommended Reading

S.J. Araujo and G. Tear. Axon guidance mechanisms and molecules: lessons
from invertebrates. *Nat. Rev. Neurosci.* 4:910–22 (2003).

G. Bai G and S.L. Pfaff. Protease regulation: the Yin and Yang of neural
development and disease. *Neuron* 72:9–21 (2011).

C.S. Barros, S.J. Franco, and U. Muller. Extracellular matrix: functions in the
nervous system. *Cold Spring Harb. Perspect. Biol.* 3:a005108 (2011).

D.H. Bhatt, S. Zhang, and W.B. Gan. Dendritic spine dynamics. *Annu. Rev.
Physiol.* 71:261–82 (2009).

S.M. Hansen, V. Berezin, and E. Bock. Signaling mechanisms of neurite outgrowth induced by the cell adhesion molecules NCAM and N-cadherin.
Cell. Mol. Life Sci. 65:3809–21 (2008).

A. Holtmaat and K. Svoboda. Experience-dependent structural synaptic plasticity in the mammalian brain. *Nat. Rev. Neurosci. 10*:647–58 (2009).

R. Itofusa and H. Kamiguchi. Polarizing membrane dynamics and adhesion for growth cone navigation. *Mol. Cell. Neurosci. 48*:332–8 (2011).

A.L. Kolodkin and M. Tessier-Lavigne. Mechanisms and molecules of neuronal wiring: a primer. *Cold Spring Harb. Perspect. Biol. 3*:1–14 (2011).

M.I. Mosevitsky. Nerve ending "signal" proteins GAP-43, MARCKS, and BASP1. *Int. Rev. Cytol. 245*:245–325 (2005).

V. Pernet and M.E. Schwab. The role of Nogo-A in axonal plasticity, regrowth and repair. *Cell Tissue Res. 349*:97–104 (2012).

D.M. Suter and K.E. Miller. The emerging role of forces in axonal elongation. *Prog. Neurobiol. 94*:91–101 (2011).

References

L.I. Benowitz and A. Routtenberg. A membrane phosphoprotein associated with neural development, axonal regeneration, phospholipid metabolism and synaptic plasticity. *Trends Neurosci. 10*:527–32 (1987).

G. Chen, J. Sima, M. Jin, K.Y. Wang, X.J. Xue, W. Zheng, Y.Q. Ding, and X.B. Yuan. Semaphorin-3A guides radial migration of cortical neurons during development. *Nat. Neurosci. 11*:36–44 (2008).

Z. Chen, B.B. Gore, H. Long, L. Ma, and M. Tessier-Lavigne M. Alternative splicing of the Robo3 axon guidance receptor governs the midline switch from attraction to repulsion. *Neuron 58*:325–32 (2008).

P. Forscher, L.K. Kaczmarek, J.A. Buchanan, and S.J. Smith. Cyclic AMP induces changes in distribution and transport of organelles within growth cones of *Aplysia* bag cell neurons. *J. Neurosci. 7*:3600–11 (1987).

P. Forscher and S.J. Smith. Actions of cytochalasins on the organization of actin filaments and microtubules in a neuronal growth cone. *J. Cell Biol. 107*:1505–16 (1988).

A.L. Harrelson and C.S. Goodman. Growth cone guidance in insects: Fasciclin II is a member of the immunoglobulin superfamily. *Science 242*:700–8 (1988).

P.G. Haydon, D.P. McCobb, and S.B. Kater. Serotonin selectively inhibits growth cone dynamics and synaptogenesis of specific identified neurons. *Science 226*:561–4 (1984).

A. Jaworski, H. Long, and M. Tessier-Lavigne. Collaborative and specialized functions of Robo1 and Robo2 in spinal commissural axon guidance. *J. Neurosci. 30*:9445–53 (2010).

T.E. Kennedy, H. Wang, W. Marshall, and M. Tessier-Lavigne. Axon guidance by diffusible chemoattractants: a gradient of netrin protein in the developing spinal cord. *J. Neurosci. 26*:8866–74 (2006).

T. Kidd, C. Russell, C.S. Goodman, and G. Tear G. Dosage-sensitive and complementary functions of roundabout and commissureless control axon crossing of the CNS midline. *Neuron 20*:25–33 (1988).

M. Matsunaga, K. Hatta, and M. Takeichi. Role of N-cadherin cell adhesion molecules in the histogenesis of neural retina. *Neuron* 1:289–95 (1988).

F. Polleux, T. Morrow, and A. Ghosh. Semaphorin 3A is a chemoattractant for cortical apical dendrites. *Nature* 404:567–73 (2000).

M. Shelly, L. Cancedda, B.K. Lim, A.T. Popescu, P.L. Cheng, H. Gao, and M.M Poo. Semaphorin3A regulates neuronal polarization by suppressing axon formation and promoting dendrite growth. *Neuron* 71:433–46 (2011).

M. Shelly, B.K. Lim, L. Cancedda, S.C. Heilshorn, H. Gao, and M.M. Poo. Local and long-range reciprocal regulation of cAMP and cGMP in axon/dendrite formation. *Science* 327:547–52 (2010).

J.T. Trachtenberg, B.E. Chen, G.W. Knott, G. Feng, J.R. Sanes, E. Welker, and K. Svoboda. Long-term in vivo imaging of experience-dependent synaptic plasticity in adult cortex. *Nature* 420:788–94 (2002).

M. Westerfield and J.S. Eisen. Neuromuscular specificity: pathfinding by identified motor growth cones in a vertebrate embryo. *Trends Neurosci.* 11:18–22 (1988).

Chapter 17

Recommended Reading

J. Cang and D.A. Feldheim. Developmental mechanisms of topographic map formation and alignment. *Annu. Rev. Neurosci.* 36:51–77 (2013).

S. Cohen-Cory, A.H. Kidane, N.J. Shirkey, and S. Marshak. Brain-derived neurotrophic factor and the development of structural neuronal connectivity. *Dev. Neurobiol.* 70:271–88 (2010).

M. Constantine-Paton, H.T. Cline, and E. Debski. Patterned activity, synaptic convergence, and the NMDA receptor in developing visual pathways. *Annu. Rev. Neurosci.* 13:129–54 (1990).

M.C. Crair. Neuronal activity during development: permissive or instructive. *Curr. Opinion Neurobiol.* 9:88–93 (1999).

T. del Rio T and M.B. Feller. Early retinal activity and visual circuit development. *Neuron* 52:221–5 (2006).

M. Hruska and M.B. Dalva. Ephrin regulation of synapse formation, function and plasticity. *Mol. Cell. Neurosci.* 50:35–44 (2012).

L.C. Katz and J.C. Crowley. Development of cortical circuits: lessons from ocular dominance columns. *Nat. Rev. Neurosci.* 3:34–42 (2002).

T.T. Kummer, T. Misgeld, and J.R. Sanes. Assembly of the postsynaptic membrane at the neuromuscular junction: paradigm lost. *Curr. Opin. Neurobiol.* 16:74–82 (2006).

M.R. Lyons and A.E. West. Mechanisms of specificity in neuronal activity-regulated gene transcription. *Prog. Neurobiol.* 94:259–95 (2011).

H. Nishimune. Active zones of mammalian neuromuscular junctions: formation, density, and aging. *Ann. N. Y. Acad. Sci.* 1274:24–32 (2012).

A.R. Punga and M.A. Ruegg. Signaling and aging at the neuromuscular synapse: lessons learnt from neuromuscular diseases. *Curr. Opin. Pharmacol.* 12:340–6 (2012).

J.R. Sanes and J.W. Lichtman. Development of the vertebrate neuromuscular junction. *Annu. Rev. Neurosci.* 22:389–442 (1999).

N. Singhal and P.T. Martin. Role of extracellular matrix proteins and their receptors in the development of the vertebrate neuromuscular junction. *Dev. Neurobiol.* 71:982–1005 (2011).

L.S. Wijetunge, S. Chattarji, D.J. Wyllie, and P.C. Kind. Fragile X syndrome: from targets to treatments. *Neuropharmacology* 68:83–96 (2013).

A. Yoshii and M. Constantine-Paton. Postsynaptic BDNF-TrkB signaling in synapse maturation, plasticity, and disease. *Dev. Neurobiol.* 70:304–22 (2010).

References

E.D. Apel, S.L. Roberds, K.P. Campbell, and J.P. Merlie. Rapsyn may function as a link between the acetylcholine receptor and the agrin-binding dystrophin-associated glycoprotein complex. *Neuron* 15:115–26 (1995).

S. Bevan and J.H. Steinbach. The distribution of α-bungarotoxin binding sites on mammalian skeletal muscle developing in vivo. *J. Physiol.* 267:195–213 (1977).

M.C. Brown. Sprouting of motor nerves in adult muscles. A recapitulation of ontology. *Trends Neurosci.* 7:10–14 (1984).

A.J. Buller, J.C. Eccles, and R.M. Eccles. Interactions between motoneurons and muscles in respect of the characteristic speed of their responses. *J. Physiol.* 150:417–39 (1960).

M. Constantine-Paton and M.I. Law. Eye-specific termination bands in tecta of three-eyed frogs. *Science* 202:639–41 (1978).

T. Curran and J.I. Morgan. Fos: an immediate-early transcription factor in neurons. *J. Neurobiol.* 26:403–12 (1995).

Y. Dan and M.-M. Poo. Hebbian depression of isolated neuromuscular synapses in vitro. *Science* 256:1570–3 (1992).

U. Drescher. Excitation at the synapse: Eph receptors team up with NMDA receptors. *Cell* 103:1005–8 (2000).

P. Forscher, L.K. Kaczmarek, J. Buchanan, and S.J. Smith. Cyclic AMP induces changes in distribution of organelles within growth cones of *Aplysia* bag cell neurons. *J. Neurosci.* 7:3600–11 (1987).

H. Fujisawa. Persistence of disorganized pathways on tortuous trajectories of regenerating retinal fibers in the adult newt *Cynops pyrrhogaster. Dev. Growth Differ.* 23:215–19 (1981).

M. Hattori, M. Osterfield, and J.G. Flanagan. Regulated cleavage of a contact-mediated axon repellent. *Science* 289:1360–5 (2000).

D.O. Hebb. *The Organization of Behavior*. New York: Wiley (1949).

S.P. Hunt, A. Pini, and G. Evan. Induction of c-fos-like protein in spinal cord neurons following sensory stimulation. *Nature (London) 328*:632–4 (1987).

P.W. Janes, N. Saha, W.A. Barton, M.V. Kolev, S.H. Wimmer-Kleikamp, E. Nievergall, C.P. Blobel, J.P. Himanen, M. Lackmann, and D.B. Nikolov. Adam meets Eph: an ADAM substrate recognition module acts as a molecular switch for ephrin cleavage in trans. *Cell 123*:291–304 (2005).

M.S. Kayser, A.C. McClelland, E.G. Hughes, and M.B. Dalva. Intracellular and trans-synaptic regulation of glutamatergic synaptogenesis by EphB receptors. *J. Neurosci. 26*:12152–64 (2006).

M.S. Kayser, M.J. Nolt, and M.B. Dalva. EphB receptors couple dendritic filopodia motility to synapse formation. *Neuron 59*:56–69 (2008).

B.K. Lim, N. Matsuda, and M.-M. Poo. Ephrin-B reverse signaling promotes structural and functional synaptic maturation in vivo. *Nat. Neurosci. 11*:160–9 (2008).

Y.-J. Lo and M.-M. Poo. Activity-dependent synaptic competition in vitro: heterosynaptic suppression of developing synapses. *Science 254*:1019–22 (1992).

S. Marshak, A.M. Nikolakopoulou, R. Dirks, G.J. Martens, and S. Cohen-Cory. Cell-autonomous TrkB signaling in presynaptic retinal ganglion cells mediates axon arbor growth and synapse maturation during the establishment of retinotectal synaptic connectivity. *J. Neurosci. 27*:2444–56 (2007).

L.M. Marshall, J.R. Sanes, and U.J. McMahan. Reinnervation of original synaptic sites on muscle fiber basement membrane after disruption of the muscle cells. *Proc. Natl. Acad. Sci. U.S.A. 74*:3073–7 (1977).

H. Nishimune, J.R. Sanes, and S.S. Carlson. A synaptic laminin-calcium channel interaction organizes active zones in motor nerve terminals. *Nature 432*:580–7 (2004).

M. Ruggiua, R. Herbst, N. Kimb, M. Jevsek, J.J. Faka, M.A. Mann, G. Fischbach, S.J. Burden, and R.B. Darnell. Rescuing Z^+ agrin splicing in Nova null mice restores synapse formation and unmasks a physiologic defect in motor neuron firing. *Proc. Natl. Acad. Sci. U.S.A. 106*: 3513–18 (2009).

S. Salmons and F.A. Sréter. Significance of impulse activity in the transformation of skeletal muscle type. *Nature 263*:30–4 (1976).

Y. Song and R. Balice-Gordon. New dogs in the dogma: Lrp4 and Tid1 in neuromuscular synapse formation. *Neuron 60*:526–8 (2008).

R.W. Sperry. Chemoaffinity in the orderly growth of nerve fiber patterns and connections. *Proc. Natl. Acad. Sci. U.S.A. 50*:703–10 (1963).

J. Walter, S. Henke-Fahle, and F. Bonhoeffer. Avoidance of posterior tectal membranes by temporal retinal axons. *Development 101*:909–13 (1987).

Chapter 18

Recommended Reading

M.R. Brown and L.K Kaczmarek. Potassium channel modulation and auditory processing. *Hearing Res.* 279:32–42 (2011).

D.E. Feldman. The spike-timing dependence of plasticity. *Neuron* 75:556–71 (2012).

P. A. Getting. Emerging principles governing the operation of neural networks. *Annu. Rev. Neurosci.* 12:185–204 (1989).

D. Granados-Fuentes and E.D. Herzog. The clock shop: coupled circadian oscillators. *Exp. Neurol.* 243:21–7 (2013).

S. Grillner. The motor infrastructure: from ion channels to neuronal networks. *Nat. Rev. Neurosci.* 4:573–86 (2003).

S. Grillner, P. Wallen, R. Hill, L. Cangiano, and A. El Manira. Ion channels of importance for the locomotor pattern generation in the lamprey brainstem-spinal cord. *J. Physiol.* 533:23–30 (2001).

D. Johnston and R. Narayanan. Active dendrites: colorful wings of the mysterious butterflies. *Trends Neurosci.* 31:309–16 (2008).

P.X. Joris, P.H. Smith, and T.C.T. Yin. Coincidence detection in the auditory system: 50 years after Jeffress. *Neuron* 21:1235–8 (1998).

C. Kopp-Scheinpflug, J.R. Steinert, and I.D. Forsythe. Modulation and control of synaptic transmission across the MNTB. *Hearing Res.* 279:22–31 (2011).

E. Marder. Variability, compensation, and modulation in neurons and circuits. *Proc. Natl. Acad. Sci. U. S. A.* 108(Suppl 3):15542–8 (2011).

E. Marder. Neuromodulation of neuronal circuits: back to the future. *Neuron* 76:1–11 (2012).

References

P.D. Brodfuehrer and W.O. Friesen. From stimulation to undulation: A neuronal pathway for swimming in the leech. *Science* 234:1002–4 (1986).

P.J. Conn and L.K. Kaczmarek. The bag cell neurons: a model system for the investigation of prolonged changes in animal behavior. *Mol. Neurobiol.* 3:237–73 (1990).

P.S. Dickinson and E. Marder. Peptidergic modulation of a multioscillator system in the lobster. I. Activation of the cardiac sac motor pattern by the neuropeptides proctolin and red pigment-concentrating hormone. *J. Neurophysiol.* 61:833–44 (1989).

G.V. Di Prisco, E. Pearlstein, D. Le Ray, R. Robitaille, and R. Dubuc. A cellular mechanism for the transformation of a sensory input into a motor command. *J. Neurosci.* 20:8169–76 (2000).

W.O. Friesen. Neuronal control of leech swimming movements. In
   Neuronal and Cellular Oscillators, J.W. Jacklet (Ed.). New York: Dekker,
   pp. 269–316 (1989).

R. Grashow, T. Brookings, and E. Marder. Compensation for variable intrinsic neuronal excitability by circuit-synaptic interactions. *J. Neurosci.*
   30:9145–56 (2010).

S. Grillner and T.M. Jessell. Measured motion: searching for simplicity in
   spinal locomotor networks. *Curr. Opin. Neurobiol. 19*:572–86 (2009).

R.M. Harris-Warrick and R.E. Flamm. Chemical modulation of a small
   central pattern generator. *Trends Neurosci. 9*:432–37 (1986).

J.N. Itri, S. Michel, M.J. Vansteensel, J.H. Meijer, and C.S. Colwell. Fast
   delayed rectifier potassium current is required for circadian neural activity.
   Nat. Neurosci. 8:650–6 (2005).

L.K. Kaczmarek. A model of cell firing patterns during epileptic seizures. *Biol.*
   Cybernetics. 22:229–34 (1976).

J. Kim, S.C. Jung, A.M. Clemens, R.S. Petralia, and D.A. Hoffman DA.
   Regulation of dendritic excitability by activity-dependent trafficking of
   the A-type K^+ channel subunit Kv4.2 in hippocampal neurons. *Neuron*
   54:933–47 (2007).

R.J. Knox, E.A. Quattrocki, J.A. Connor, and L.K. Kaczmarek. Recruitment
   of Ca^{2+} channels by protein kinase C during rapid formation of putative
   neuropeptide release sites in isolated Aplysia neurons. *Neuron 8*:883–9
   (1992).

C.M. Macica, C.A. von Hehn, L.Y. Wang, C.S. Ho, S. Yokoyama, R.H. Joho,
   and L.K. Kaczmarek. Modulation of the Kv3.1b potassium channel
   isoform adjusts the fidelity of the firing pattern of auditory neurons. *J.*
   Neurosci. 23:1133–41 (2003).

J.A. Mohawk, C.B. Green, and J.S. Takahashi. Central and peripheral circadian clocks in mammals. *Annu. Rev. Neurosci. 35*:445–62 (2012).

M. Moulins and F. Nagy. Extrinsic inputs and flexibility in the motor output
   of the lobster pyloric neural network. In *Model Neural Networks and*
   Behavior, A.I. Selverston (Ed.). New York: Plenum, pp. 49–68 (1985).

M.P. Nusbaum and E. Marder. A modulatory proctolin-containing neuron (MPN). II. State-dependent modulation of rhythmic motor activity.
   J. Neurosci. 9:1600–7 (1989).

R.A. Satterlie. Reciprocal inhibition and postinhibitory rebound produce
   reverberation in a locomotor pattern generator. *Science 229*:402–4
   (1985).

R.A. Satterlie, M. LaBarbara, and A.N. Spencer. Swimming in the ptero-
   pod mollusc *Clione limacina*. I. Behavior and morphology. *J. Exp. Biol.*
   116:189–204 (1985).

P. Song, Y. Yang, M. Barnes-Davies, A. Bhattacharjee, M. Hamann, I.D.
   Forsythe, D.L. Oliver, and L.K. Kaczmarek. Acoustic environment determines phosphorylation state of the Kv3.1 potassium channel in auditory
   neurons. *Nat. Neurosci. 8*:1335–42 (2005).

G.J. Stuart and B. Sakmann. Active propagation of somatic action potentials into neocortical pyramidal cell dendrites. *Nature* 367:69–72 (1994).

L.R. Varshney, B.L. Chen, E. Paniagua, D.H. Hall, and D.B. Chklovskii. Structural properties of the *Caenorhabditis elegans* neuronal network. *PLoS Computat. Biol.* 7:e1001066 (2011).

P. Wallen and S. Grillner. N-methyl-D-aspartate receptor-induced, inherent oscillatory activity in neurons active during fictive locomotion in the lamprey. *J. Neurosci.* 7:2745–55 (1987).

Y. Zhang, J.S. Helm, A. Senatore, J.D. Spafford, L.K. Kaczmarek, and E.A. Jonas. PKC-induced intracellular trafficking of Ca(V)2 precedes its rapid recruitment to the plasma membrane. *J. Neurosci.* 28:2601–12 (2008).

Chapter 19

Recommended Reading

V. Anggono and R.L. Huganir. Regulation of AMPA receptor trafficking and synaptic plasticity. *Curr. Opin. Neurobiol.* 22:461–9 (2012).

C.H. Bailey and E.R. Kandel. Synaptic remodeling, synaptic growth and the storage of long-term memory in *Aplysia. Prog. Brain Res.* 169:179–98 (2008).

C.R. Bramham, M.N. Alme, M. Bittins, S.D. Kuipers, R.R. Nair, B. Pai, D. Panja, M. Schubert, J. Soule, A. Tiron, and K. Wibrand. The Arc of synaptic memory. *Exp. Brain Res.* 200:125–40 (2010).

G.U. Busto, I. Cervantes-Sandoval, and R.L. Davis. Olfactory learning in *Drosophila. Physiology* 25:338–46 (2010).

A. Citri and R.C. Malenka. Synaptic plasticity: multiple forms, functions, and mechanisms. *Neuropsychopharmacology* 33:18–41 (2008).

R.L. Davis. Traces of *Drosophila* memory. *Neuron* 70:8–19 (2011).

D.E. Feldman. The spike-timing dependence of plasticity. *Neuron* 75:556–7 (2012).

D.L. Glanzman. New tricks for an old slug: the critical role of postsynaptic mechanisms in learning and memory in *Aplysia. Prog. Brain Res.* 169:277–92 (2008).

P.S. Goldman-Rakic. Cellular basis of working memory. *Neuron* 14:477–85 (1995).

S.J. Kim and D.J. Linden. Ubiquitous plasticity and memory storage. *Neuron* 56:582–92 (2007).

Y. Lu, K. Christian, and B. Lu. BDNF: a key regulator for protein synthesis-dependent LTP and long-term memory? *Neurobiol. Learning Memory* 89:312–23 (2008).

L.R. Squire. *Memory and Brain.* New York: Oxford University Press (1987).

S. Stough, J.L. Shobe, and T.J. Carew. Intermediate-term processes in memory formation. *Curr. Opin. Neurobiol.* 16:672–8 (2006).

G.G. Turrigiano. The self-tuning neuron: synaptic scaling of excitatory synapses. *Cell* 135:422–35 (2008).

References

B.W. Agranoff, R. E. Davis, and J. J. Brink. Memory fixation in the goldfish. *Proc. Natl. Acad. Sci. U.S.A.* 54:788–93 (1965).

C.H. Bailey and M. Chen. Long-term memory in *Aplysia* modulates the total number of varicosities of single identified sensory neurons. *Proc. Natl. Acad. Sci. U. S. A.* 85:2373–7 (1988).

T.V.P. Bliss and T. Lomo. Long lasting potentiation of synaptic transmission in the dentate area of the anaesthetized rabbit following stimulation of the perforant path. *J. Physiol.* 232:331–56 (1973).

J. DeZazzo and T. Tully. Dissection of memory formation: from behavioral pharmacology to molecular genetics. *Trends Neurosci.* 18:212–8 (1995).

U. Frey and R.G. Morris. Synaptic tagging and long-term potentiation. *Nature* 385:533–6 (1997).

S. Funahashi, C.J. Bruce, and P.S. Goldman-Rakic PS. Mnemonic coding of visual space in the monkey's dorsolateral prefrontal cortex. *J. Neurophysiol.* 61:331–49 (1989).

K.P. Giese, J.F. Storm, D. Reuter, N.B. Fedorov, L.R. Shao, T. Leicher, O. Pongs, and A.J. Silva. Reduced K1 channel inactivation, spike broadening, and after-hyperpolarization in Kvb1.1-deficient mice with impaired learning. *Learning and Memory* 5:257–73 (1998).

J. Gu, C.W. Lee, Y. Fan, D. Komlos, X. Tang, C. Sun, K. Yu, H.C. Hartzell, G. Chen, J.R. Bamburg, and J.Q. Zheng. ADF/cofilin-mediated actin dynamics regulate AMPA receptor trafficking during synaptic plasticity. *Nat. Neurosci.* 13:1208–15 (2010).

E.R. Kandel. *Behavioral Biology of Aplysia*. San Francisco: Freeman (1979).

E.R. Kandel and J.H. Schwartz. Molecular biology of learning: modulation of transmitter release. *Science* 218:433–43 (1982).

J.H. Kim, H. Udo, H.L. Li, T.Y. Youn, M. Chen, E.R. Kandel, and C.H. Bailey. Presynaptic activation of silent synapses and growth of new synapses contribute to intermediate and long-term facilitation in *Aplysia*. *Neuron* 40:151–65 (2003).

K.S. Lashley. In search of the engram. *Soc. Exp. Biol. Symp.* 4:454–82 (1950).

C. Luscher, H. Xia, E.C. Beattie, R.C. Carroll, M. von Zastrow, R.C. Malenka, and R.A. Nicoll. Role of AMPA receptor cycling in synaptic transmission and plasticity. *Neuron* 24:649–58 (1999).

L. Minichiello, M. Korte, D. Wolfer, R. Kuhn, K. Unsicker, V. Cestari, C. Rossi-Arnaud, H.-P. Lipp, T. Bonhoeffer, and R. Klein. Essential role for TrkB receptors in hippocampus-mediated learning. *Neuron* 24:401–14 (1999).

R.A. Nicoll, J.A. Kauer, and R.C. Malenka. The current excitement in long-term potentiation. *Neuron 1*:97–103 (1988).

M.E. Raichle. Behind the scenes of functional brain imaging: a historical and physiological perspective. *Proc. Natl. Acad. Sci. U.S.A. 95*:765–72 (1998).

S. Schöning, P. Zwitserlood, A. Engelien, A. Behnken, H. Kugel, H. Schiffbauer, K. Lipina, C. Pachur, A. Kersting, U. Dannlowski, B.T. Baune, P. Zwanzger, T. Reker, W. Heindel, V. Arolt, and C. Konrad. Working-memory fMRI reveals cingulate hyperactivation in euthymic major depression. *Hum. Brain Map. 30*:2746–56 (2009).

C.F. Stevens and I. Verma. A model with good CREdentials. *Curr. Biol. 4*:736–8 (1994).

T. Tully. *Drosophila* learning and memory revisited. *Trends Neurosci. 10*:330–5 (1987).

D. Yu, A. Ponomarev, and R.L. Davis. Altered representation of the spatial code for odors after olfactory classical conditioning; memory trace formation by synaptic recruitment. *Neuron 42*:437–49 (2004).

D. Zamanillo, R. Sprengel, O. Hvalby, V. Jensen, N. Burnashev, A. Rozov, K.M. Kaiser, H.J. Koster, T. Borchardt, P. Worley, J. Lubke, M. Fritscher, P.H. Kelly, B. Sommer, P. Andersen, P.H. Seeburg, and B. Sakmann. Importance of AMPA receptors for hippocampal synaptic plasticity but not for spatial learning. *Science 284*:1805–11 (1999).

Index

Italicized page numbers indicate an illustration on the designated page. Page numbers followed by t indicate a table on the designated page.